Applications of Intelligent Technologies in Renewable Energy

RIVER PUBLISHERS SERIES IN ENERGY SUSTAINABILITY AND EFFICIENCY

Series Editor:

PEDRAM ASEF
Lecturer (Asst. Prof.) in Automotive Engineering,
University of Hertfordshire,
UK

The "River Publishers Series in Sustainability and Efficiency" is a series of comprehensive academic and professional books which focus on theory and applications in sustainable and efficient energy solutions. The books serve as a multi-disciplinary resource linking sustainable energy and society, fulfilling the rapidly growing worldwide interest in energy solutions. All fields of possible sustainable energy solutions and applications are addressed, not only from a technical point of view, but also from economic, social, political, and financial aspects. Books published in the series include research monographs, edited volumes, handbooks and textbooks. They provide professionals, researchers, educators, and advanced students in the field with an invaluable insight into the latest research and developments.

Topics covered in the series include, but are not limited to:

- Sustainable energy development and management;
- Alternate and renewable energies;
- Energy conservation;
- Energy efficiency;
- Carbon reduction;
- Environment.

For a list of other books in this series, visit www.riverpublishers.com

Applications of Intelligent Technologies in Renewable Energy

Editors

Anuradha Tomar

Netaji Subhas University of Technology, India

Yog Raj Sood

JIIT, India

Ritu Kandari

KIET group of Institutions, India

River Publishers

Routledge
Taylor & Francis Group
NEW YORK AND LONDON

Published 2026 by River Publishers
River Publishers
Broagervej 10, 9260 Gistrup, Denmark
www.riverpublishers.com

Distributed exclusively by Routledge
605 Third Avenue, New York, NY 10017, USA
4 Park Square, Milton Park, Abingdon, Oxon OX14 4RN

Applications of Intelligent Technologies in Renewable Energy / by Anuradha Tomar, Yog Raj Sood, Ritu Kandari.

Routledge is an imprint of the Taylor & Francis Group, an informa business

ISBN 978-87-4380-984-5 (hardback)
ISBN 978-87-4380-989-0 (paperback)
ISBN 978-87-4380-986-9 (online)
ISBN 978-87-4380-985-2 (master ebook)

DOI: 10.1201/9788743809852

Contents

Preface

In the face of an escalating climate crisis, depleting fossil fuel reserves, and mounting energy demands, the transition towards renewable and sustainable energy systems has become an urgent global priority. This book, *Sustainable Development Towards Renewable Energy: A Next Generation Energy*, explores the powerful intersection of intelligent technologies and renewable energy systems, offering a comprehensive view of how smart, data-driven innovations can drive a more resilient, efficient, and sustainable energy future.

The chapters collectively examine the role of artificial intelligence, the Internet of Things (IoT), and advanced computing in managing, optimizing, and strengthening renewable energy infrastructures. From the application of intelligent techniques in energy management to state-of-the-art methods for distributed smart grids, the content presented here delves into the most recent advancements in intelligent energy systems.

Special attention is given to critical areas such as solar photovoltaic (PV) recycling, demand-side management, wind energy balancing, and fault ride-through capabilities of hybrid systems. This book also provides in-depth analyses of the circular economy in solar PV use, the resilience of power systems, and the importance of intelligent systems in enabling flexible, reliable, and clean energy solutions.

Designed for researchers, energy professionals, policymakers, and students, this compilation serves both as an academic resource and a practical guide. It equips readers with the knowledge and insights needed to harness emerging technologies in achieving energy sustainability goals.

We hope this book sparks innovation, fosters cross-disciplinary collaboration, and contributes meaningfully to the ongoing global pursuit of a cleaner and more intelligent energy landscape.

List of Figures

List of Tables

List of Contributors

Agarwal, Lalit, *MAIT, GGSIPU, India*

Ansaripour, Shiva, *Energy Modelling and Sustainable Energy System (METSAP) Research Lab, Faculty of New Sciences and Technologies, University of Tehran, Iran*

C. R., Raghavendran, *Assistant Professor, Easwari Engineering College, India*

D., Devaraj, *Director-Academic and Dean-SEET, Kalasalingam Academy of Research and Education, India*

Emblemsvåg, Jan, *NTNU, Ålesund, Norway*

Gupta, Vikalp, *Coventry University, United Kingdom*

Kaldate, Avinash P., *Department of Mechanical Engineering, Sinhgad College of Engineering, Savitribai Phule Pune University, India*

Kanase-Patil, Amarsingh B., *Department of Mechanical Engineering, Sinhgad College of Engineering, Savitribai Phule Pune University, India*

Kumar, Amit, *Electrical Engineering Department, NIT Hamirpur, India*

Lokhande, Shashikant D., *Department of Electronics and Telecommunication Engineering, Sinhgad College of Engineering, Savitribai Phule Pune University, India*

Malik, Rohit, *School of Computer Applications, Lovely Professional University, Phagwara, Punjab-144411, India*

Mohapatra, Purujit, *University of Glasgow, Scotland*

Nain, Ajay, *School of Computer Applications, Lovely Professional University, Phagwara, Punjab-144411, India*

Noorollahi, Younes, *Energy Modelling and Sustainable Energy System (METSAP) Research Lab, Faculty of New Sciences and Technologies, University of Tehran, Iran*

Pattnaik, Amruta, *NIST University, Institute Park, Pallur Hills, Berhampur, Odisha 761008, India*

Roselyn J., Preetha, *Associate Professor, SRM Institute of Science and Technology, India*

Sheikh, Sophiya, *School of Computer Applications, Lovely Professional University, Phagwara, Punjab-144411, India*

Shoran, Preety, *School of Sciences, CHRIST (Deemed to be University), Ghaziabad, Uttar Pradesh-201003, India*

Som, Trina, *Researcher on renewable energy planning & power system, The Celestial Earth Development Council, India*

Sood, Yog Raj, *Electrical Engineering Department, NIT Hamirpur, India*; *National Institute of Technology, India*

Thakur, Shikha, *Electrical Engineering Department, NIT Hamirpur, India*

Tomar, Anuradha, *Netaji Subhash University of Technology (NSUT), Delhi, India*

Verma, Rohit, *National Institute of Technology, India*

List of Abbreviations

AAL	Ambient assistive living
ABC	Artificial bee colony
AGA	Adaptive genetic algorithm
AI	Artificial intelligence
AMI	Advanced metering infrastructure
ANN	Artificial neural network
AR	Augmented reality
BAS	Building automation system
BESS	Battery energy storage system
BKMOFS	Binary kangaroo mob optimization features
BoL	Beginning-of-life
BOS	Balance of system
BPSOFS	Binary particle swarm optimization features
BTI	Bridge to India
CB	Capacitor bank
CCA	Climate change adaptation
CCM	Climate change mitigation
CE	Consumer empowerment
CFD	Computational fluid dynamics
CHP	Combined heat and power
CNN	Convolutional neural network
CPS	Cyber-physical system
CSA	Cuckoo search algorithm
CV	Cross-validation
DAP	Day-ahead pricing
DBN	Deep belief network
DDSG	Direct drive synchronous generator
DE	Differential evolution
DER	Distributed energy resource
DFIG	Doubly fed induction generator

DG	Distributed generation
DGM	Distributed generation management
DL	Deep learning
DNN	Deep neural network
DNO	Distributed network operator
DO	Dissolved oxygen
DOD	Depth of discharge
DQN	Deep Q-network
DR	Demand response
DRP	Demand response program
DSM	Demand side management
DSS	Decision support system
DT	Decision tree
DVR	Dynamic voltage restorer
ECS	Energy capacitor systems
EDA	Exploratory data analysis
EEE	Electrical and electronic equipment
EMS	Energy management system
EoL	End-of-life
EPR	Extended producer responsibility
ESI	Energy services interface
ESS	Energy storage systems
EU	European Union
EV	Electronic vehicle
EVA	Ethylene vinyl acetate
FEA	Finite element analysis
FLC	Fuzzy logic control
FTM	Fault tolerance mechanism
GA	Genetic algorithm
GBT	Gradient-boosted tree
GDP	Gross domestic production
GHG	Greenhouse gas
GPR	Gaussian process regression
GRA	Grey relational analysis
GRU	Gated recurrent unit
GSC	Grid side converter
GWP	Global warming potential
HIL	Hardware-in-the-loop
HPC	High-performance computing

HTGR	High-temperature gas-cooled reactor
HVDC	High-voltage direct current
IEA	International Energy Agency
IoT	Internet-of-Things
IRENA	International Renewable Energy Agency
JNNSM	Jawaharlal Nehru National Solar Mission
KNN	K-nearest neighbor
LCA	Life-cycle assessment
LCOE	Levelized cost of energy
LOLE	Loss of load expectation
LOLP	Loss of load probability
LQR	Linear quadratic regulator
LR	Logistic regression
LSI	Load-shedding incidents
LSTM	Long short-term memory
LSTM-RNN	Long short-term memory recurrent neural network
LVRT	Low voltage ride through
MADM	Multiattribute decision-making
MARS	Multivariate adaptive regression splines
MAS	Multiagent system
MDP	Markov decision process
ML	Machine learning
MLP	Multilayer perceptron
MLR	Multiple linear regression
MLSTM	Multidirectional long short-term memory
MPC	Model predictive control
MPPT	Maximum power point tracking
MW	Megawatts
NCPRE	National centre for photovoltaic research and education
NISE	National institute of solar energy
OECD	Organisation for economic co-operation and development
OTEC	Ocean thermal energy conversion
OWC	Oscillating water column
PCF	Power capacity factor
PGSA	Plant growth simulation algorithm
PIP	Power interface point

PMU	Phasor measurement unit
PRO	Producers responsibility organizations
PSO	Particle swarm optimization
PSO-WNN	Particle swarm optimization-wavelet neural network
PV	Photovoltaic
QOCSOS	Quasi-oppositional chaotic symbiotic organisms search
R & D	Research and development
REE	Rare earth elements
RER	Renewable energy resources
RES	Renewable energy sources
RF	Random forest
RL	Reinforcement learning
RNN	Recurrent neural network
RPS	Renewable portfolio standards
RTP	Real-time pricing
RTU	Remote terminal unit
SAARC	South Asian Association for Regional Cooperation
SC	Supercapacitor
SCADA	Supervisory control and data acquisition
SDG	Sustainable development goal
SFCL	Series fault current limiter
SFLA	Shuffled frog leaping algorithm
SG	Smart grid
SPM	Solar photovoltaic module
SPV	Solar photovoltaic
STHS	Short-term hydropower scheduling
SVM	Support vector machine
SVR	Support vector regression
TERI	The Energy Resource Institute
THD	Total harmonic distortion
TOU	Time of use
UAV	Unmanned aerial vehicle
VPP	Virtual power plant
VR	Virtual reality
VRE	Variable renewable energy
WEC	Wave energy converter

WECS	Wind energy conversion systems
WEEE	Waste electrical and electronic equipment
WtE	Waste-to-energy
XAI	Explainable artificial intelligence

1

Sustainable Development towards Renewable Energy: A Next Generation Energy

Amruta Pattnaik[1] and Anuradha Tomar[2]

[1]NIST University, Institute Park, Pallur Hills, Berhampur, Odisha 761008, India
[2]Netaji Subhash University of Technology (NSUT), Delhi, India
E-mail: amruta.pattnaik@nist.edu; anuradha.tomar@nsut.ac.in

Abstract

Despite the inherent constraints of the Earth's traditional electricity supplies, the need for energy is always increasing as a result of population growth and industrialisation. Currently, the use of fossil energy resources has increasingly negative consequences on the global ecosystem [1, 2]. In this context, there must be an immediate necessity to adopt environmentally secure and environmentally friendly energy sources. Sustainable energy is the huge, underutilized solution to the two main challenges of the twenty-first century, poverty, and global warming. The energy produced from renewable sources not only provides a clean, flexible power source for homes, schools, and hospitals, but it also could provide important and valuable jobs at the micro-to-medium scale. In this paper, different types of renewable based sustainable energy sources are discussed.

Keywords: Renewable energy, Biofuels, Geothermal energy, Hydropower, solar, wind, tidal, wave, ocean, Gasification and Biorefinery.

1.1 Introduction

To ensure continued human progress and growth, sustainable energy sources must be prioritised. Energy is crucial to human well-being and economic

DOI: 10.1201/9788743809852-1

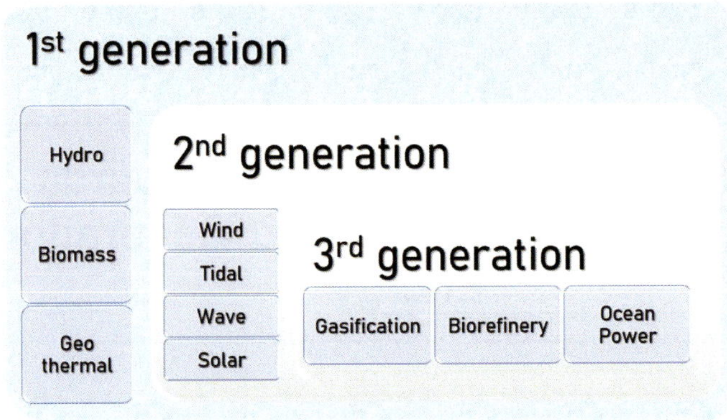

Figure 1.1 Generation of Renewable Energy.

growth, and its consumption is increasing rapidly [3]. To mitigate environmental issues, long term potential actions are needed. Renewable energy is one of the solutions of the long term issues. Greenhouse gas emissions can be reduced in part because of sustainable energy's role in lowering reliance on fossil fuels. Three generations of renewable technologies are possible as shown in Figure 1.1. In the nineteenth century, humans first experimented with hydropower, biomass, and geothermal energy. The 1980s marked the beginning of the second phase, which includes the use of renewable energy sources like wind, tidal, and wave power, and solar. The second-generation energy are tidal energy and wave energy. All the tidal, wave and ocean energy are sound similar but there is a margin line in each word. Tidal energy generates energy by utilising the gravitational pull of the Earth, moon, and sun, whereas wave energy generates electricity by utilising the kinetic forces of waves. The tides at ocean shorelines rise and fall around twice a day. Whereas ocean energy also known as ocean thermal energy and Ocean thermal energy conversions use the advantage of the warmth variation between the ocean's surface water and its depth. Gasification, biorefinery, and ocean power make up the third stage, which is now in the works [4]. In this chapter, some of the renewable energy and its co relation with the sustainable development are discussed.

1.2 Sustainable Development towards Renewable Energy

Energy supply and consumption problems are all related to environmental problems including contamination of the air, ozone loss from the atmosphere, destruction of forests, and nuclear waste releases. All these aspects need to be considered at the same time if humanity is to have a bright energy future with little environmental impact. There is a wealth of evidence indicating that human environmental degradation will have harmful future consequences. The energy sector and the general public have begun to prioritise environmental factors beyond just carbon emissions. There has been widespread acceptance of the idea that individuals bear some of the financial burden associated with pollution. Many energy supplies have seen price increases in various regions over the past decade to two, in part to cover the costs associated with protecting the environment [5].

By the middle of the 21st century, global population growth [6] is forecast to have doubled, and economic progress will almost probably continue to expand. By 2050, primary-energy consumption are projected to rise by 1.5–3 times, while the demand for energy services could increase by as much as an order of magnitude [6]. Acid rain, ozone depletion in the stratosphere, and climate change are just a few of the energy-related environmental issues that people are starting to get worked up about. Use of renewable energy sources and technology can help alleviate the approaching energy deficit. Sometimes people who are passionate about this topic make claims that are both outlandish and unattainable. It is important to take into account the technical feasibility, dependability, applicability, economics, scarcity, and public acceptability of a solution. Of course, all energy on Earth comes from the sun, which warms us, grows crops through photosynthesis, heats the land and sea differentially and causes winds, waves, and rain, which generates hydropower. Geothermal heat comes from radioactive decay deep in the Earth and tidal rise and fall from the moon and sun. All are conceivable energy sources, but even while the physics is recognised, it does not mean that if enough research money is invested, an engineering solution will be developed [7]. Science is easy; engineering is hard.

Development that satisfies the requirements of the present without jeopardising the ability of future generations to do the same is considered sustainable. Environmental, social, and economic variables are sometimes at odds with one another, yet this idea considers equity across countries and generations. Sustainable development depends on several things. Energy sustainability is crucial [8–11]. Energy security is important but not sufficient for society's progress. Sustainable development requires a sustainable supply of energy resources at a reasonable cost that can be used for all required tasks without negative social impacts. Renewable energy and sustainable development are closely linked. Several other definitions of sustainable

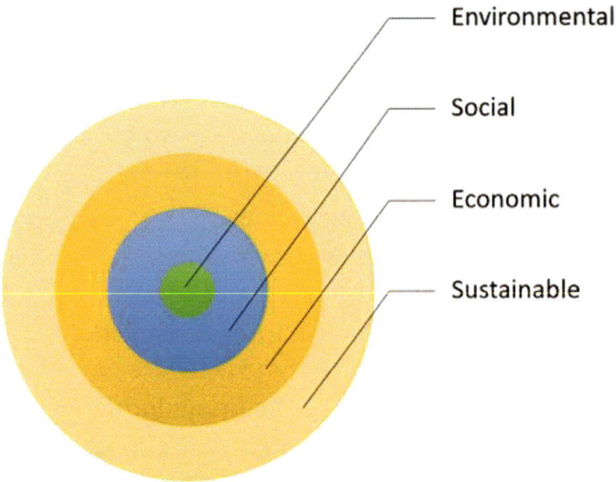

Figure 1.2 Illustration of sustainability.

development have been proposed, but the most frequently cited one comes from the Brundtland Report1 of 1987:For future generations to have the same opportunities we do today, development must be conducted in a way that does not compromise the demands of the present. The key to a sustainable future is through sustainable development. Sustainability consists of environmental, economic, and social as shown in Figure 1.2—must be met simultaneously for an activity, product, or entity to be considered truly sustainable [12].

A long-term supply of energy resources that are easily and sustainably accessible at a fair cost and can be used for all critical tasks without negatively influencing society is required for sustainable development. Energy supplies such as fossil fuels and uranium are universally acknowledged to be finite; yet, other energy sources such as sunlight, wind, and falling water are largely acknowledged to be renewable and hence sustainable over the relatively long term. Waste and biomass fuels are often viewed as environmentally friendly energy sources [11]. There are numerous critical elements that can aid in establishing good long-term development in a society. The following are some examples of such parameters:

I. Social understanding:
II. Knowledge of energy use is required.
III. Training and instruction on environmental issues:
IV. Energy-saving strategies:
V. Renewable resources for energy should be promoted:
VI. Evaluation and surveillance techniques for financing

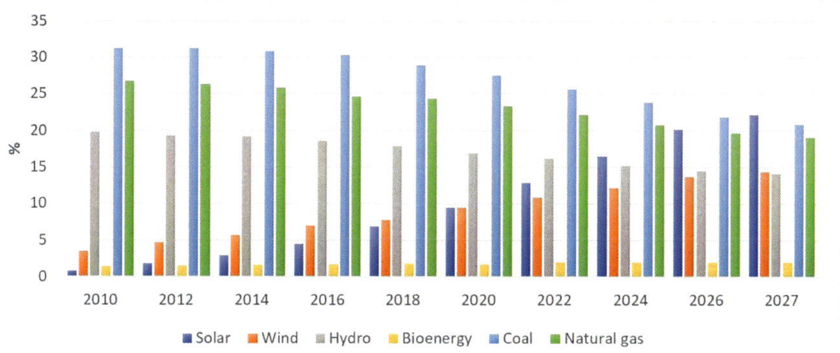

Figure 1.3 IEA, Share of cumulative power capacity by technology, 2010-2027, IEA, Paris adapted from https://www.iea.org/data-and-statistics/charts/share-of-cumulative-power-capacity-by-technology-2010-2027, IEA. Licence: CC BY 4.0.

The use of renewable energy resources and technology is a critical component of long-term development [13]. There are three significant reasons as follows: (a) It is required to feasible if conventional energy regulations were tightened, (b) Resources for fossil fuels and uranium are limited and can be depleted through extraction and consumption. So, it should be used wisely and resourcefully and (c) Improving system flexibility and inexpensive power delivery to small isolated settlements [14].

Figure 1.3 represents the share of cumulative power capacity by technology 2010–2027 where solar PV power generation capacity is expected to exceed coal by 2027, making it the world's largest. According to IEA forecasts, cumulative solar PV capacity nearly triples, increasing by nearly 1500 GW over the timeframe, surpassing natural gas by 2026 and coal by 2027.

1.3 Sustainable Energy Sources

Some of the sustainable energy sources are explained below sections:

1.3.1 Biofuels for sustainable development

Two biomass-based liquefied transportation fuels which could eliminate petrol and diesel fuel are currently being explored globally. Two such instances are bioethanol and biodiesel. Biofuels have been shown to offer several benefits that are including long-term viability, less emission of greenhouse gases, and energy security [15]. The resources of biomass may be implemented to generate a variety of fuels, including fluids like ethanol, methanol, biodiesel, and Fischer-Tropsch diesel, as well as gaseous substances like hydrogen and methane. The oxygen content of biofuels and petroleum feedstocks differs the greatest. Biofuels contain 10–45% oxygen, whereas petroleum has nearly little oxygen, resulting in biofuels' chemical properties being markedly different from those found in petroleum [16].

Unlike petroleum diesel, biodiesel is recyclable and safe, and when used as a fuel, it significantly reduces toxic and other carbon emissions. Biodiesel as diesel fuel has several benefits including versatility, rapid supply, sustainability, greater combustibility, lower sulfur and chemical content, and higher biodegradable properties .The primary advantages of biodiesel are its local origin, which reduces a nation's dependence on petroleum imports, The primary disadvantages of biodiesel include its greater viscosity and lesser energy content. Fuel freezing in cold temperatures, reduced energy density, and fuel degradation, when stored for lengthy periods of time, constitute a few of the technical drawbacks of biodiesel/fossil diesel blends [17].

Biofuels offer numerous advantages over petroleum fuels: (a) readily made from biomass, (b) sustainable, (c) their combustion is based on the CO_2 cycle, and (d) environmentally sound. Biofuel's market share will grow significantly over the next decade due to its environmental benefits. This will surely enhance agricultural productivity and byproducts [18, 19].

By offering a low-carbon alternative for challenging to reduce industries like trucking, shipping, and aviation, biofuels play a crucial role in decarbonizing transportation. They frequently require little to no modification when utilised in already installed engines. The usage of biofuels for transportation increases dramatically under the Net Zero Scenario by 2030, with a significantly greater portion coming from waste, residues, and non-food crops. In order to comply with the Net Zero Scenario by 2030, aviation biofuels, also known as biojet kerosene, would need to make the most significant advancements.

In 2022, biofuel consumption surpassed levels observed in 2019 before the Covid-19 epidemic, rising to a record high of 4.3 EJ (170 000 million litres). According to the NZE Scenario, biofuel production must expand by an average of 11% year to reach over 10 EJ by 2030. The use of advanced feedstocks must increase as well; by 2030, non-food energy crops and biofuels made from waste and residues will account for almost 40% of all biofuel demand, up from about 9% in 2021 as shown in Figure 1.4

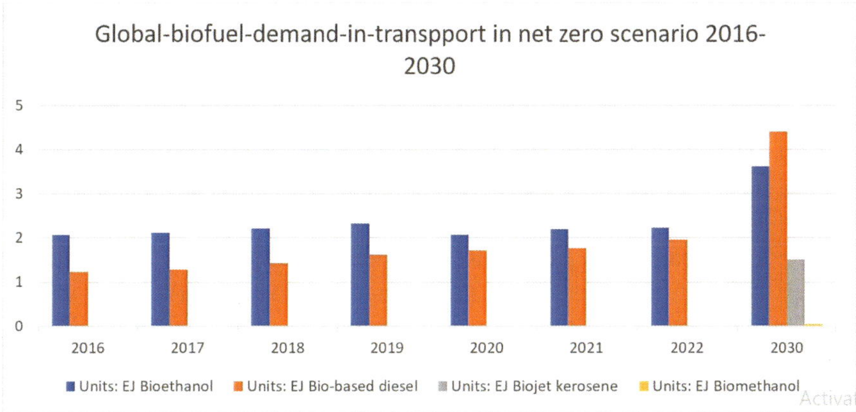

Figure 1.4 IEA, Global biofuel demand in transport in the Net Zero Scenario, 2016-2030, IEA, Paris https://www.iea.org/data-and-statistics/charts/global-biofuel-demand-in-transport-in-the-net-zero-scenario-2016-2030-2, IEA. Licence: CC BY 4.0.

In 2022, biofuels supplied more than 3.5% of the world's transport energy needs, primarily for road transport. Apart from 2020, when it decreased because of the Covid-19 outbreak, the consumption of biofuels has increased at a rate of around 6% year during the past five years. To reach the Net Zero Scenario, aviation biofuels, commonly known as biojet kerosene, would need to make the most significant development between now and 2030. India exceeded its 20% ethanol blending target for 2025 by attaining 10% ethanol blending in 2022. Most biofuels are currently made from so-called conventional feedstocks including sugar cane, maize, and soybeans. Increasing the generation of biofuel using novel raw materials, on the other hand, is critical to limiting the impact on land usage, nutrition and livestock costs, and other environmental factors while increasing biofuel output in line with the NZE Scenario. Expenditure in fluid biofuels, notably renewable diesel, grew dramatically by 2022. Worldwide transit biofuel capacity will grow by 7% in 2022, the biggest annual rise in the last ten years. Because of advantageous regulations in the United States and Europe, biorefineries focusing on renewable diesel accounted for most of the expansion, while ethanol capacity rose rapidly in Brazil, Indonesia, India, and China.

1.3.2 Geothermal energy for sustainable development

Geothermal energy, one of the forthcoming sources of green energy, is clean, and safe, thus its usage is rising for power production, heating, and cooling.

Geothermal energy is a clean energy source that provides power with minimum environmental effect [20, 21]. Since the 1940s, geothermal energy has been generated commercially on a scale of hundreds of megawatts (MW) for both direct usage and the production of electricity. Following the 1973 oil crisis, most the world's geothermal power facilities were constructed in the 1970s and 1980s. The geothermal potential for both power production and direct applications has been estimated by many writers in a variety of ways [22]. Between 35 to 72 GW should be the global geothermal potential for power production. This range might increase to 138 GW [23] with the advancement of technologies like as permeability upgrades and drilling advancements.

Geothermal energy represents one of the renewable energy sources acquired by utilizing the existing heat within the earth, and it can be detected deep within the earth. It employs a hydrothermal convection technique to capture geothermal energy. This process entails digging a hole deep beneath the soil and then inserting a pipe. The trapped steam in the rocks travels through this conduit to the earth's surface. The steam is then used to turn the rotating parts of a turbine into an electric generator. Another approach entails using steam to heat water from an external source, which is then used to rotate the turbine. The three basic types of geothermal electricity production facilities are binary, dry steam, and flash steam. Reservoir pressure and temperature are what control each type's output of electricity [22]. According to forecasts, geothermal energy might supply 3% of the world's electricity needs and 5% of its need for heating by 2050 [23]. In 2008, geothermal energy made up around 0.1% of the world's primary energy supply. Geothermal energy is often seen as a renewable energy source, but its growth and use can have significant, multifaceted sustainability implications [24]. Geothermal resources must be used sustainably, especially for power generation projects, given the likelihood that geothermal energy demand would increase significantly [24].

Geothermal energy has numerous applications, ranging from electricity generation to warmth. To begin, geothermal plants must be erected within a two-mile radius of the geothermal reserve. By using steam from these reserves to turn the turbines of an electrical generator or heat water to make steam. Many frigid countries use geothermal energy to heat greenhouses or water for irrigation. Following then, it has numerous applications in industries such as food dehydration, milk pasteurization, gold mining, and others. It is also used to heat buildings using district heating systems, which transfer hot water straight from springs to buildings via pipelines. There are many advantages and disadvantages of geo thermal energy. Those are as follows: Geo thermal energy is free and abundant, and it is endless and limitless due to the ongoing flow of heat from the Earth. It is environmentally benign and non-polluting too. Transportation and transmission are key drawbacks. Similarly, poisonous chemicals may be released into the environment as a result of transmission. Installation is extremely expensive. Another disadvantage is that geothermal energy is only available specific areas.

1.3.3 Hydropower for sustainable development

Even well into the 2030s, hydropower is expected to remain the world's top source of renewable electricity generation. It already produces more energy than all other renewable technologies put together. After then, it will keep contributing significantly to improving system flexibility and lowering carbon emissions from the electricity system. Although hydro is expected to eventually be replaced by wind and solar, it will still be essential as a readily available power source to support variable renewables. More than 70 TWh (almost 2%) more hydroelectricity was produced in 2022, totalling 4300 TWh as shown in Figure 1.5.

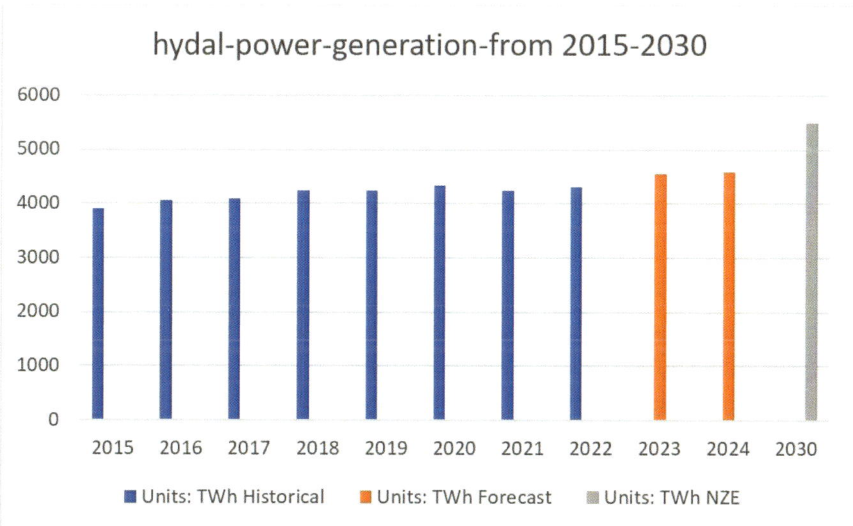

Figure 1.5 IEA, Hydro-power generation in the N, 2015-2030, IEA, Paris https://www.iea.org/data-and-statistics/charts/hydropower-generation-in-the-net-zero-scenario-2015-2030, IEA. Licence: CC BY 4.0.

In the Net Zero Emissions by 2050 Scenario, hydropower continues to expand its output on average each year at a pace of close to 4%, producing around 5500 TWh of energy annually. Due to persistent droughts in hydropower-rich countries like Canada, China, Turkey, and the United States as well as Western Europe, even if generation increased in 2021–2022 as a result of a robust capacity deployment, the global capacity utilisation factor is still below historical levels. The average growth rate over the preceding five years has been less than one-third of what is required, showing the need for significantly expanded efforts, notably to speed up permits and ensure project sustainability. As a stable foundation for upcoming clean electric systems, hydropower facilities should be recognised and developed. According to Karady et al. (2005), [25], hydraulic turbines are used in hydropower projects to transform the potential energy of water head into mechanical energy.

The generator, which is coupled to the hydro-turbines in turn, transforms the mechanical energy into electrical energy [26]. The generator, the turbine, the penstock, and the wicket gates are the main components of the hydroelectric plant. Most of the time, a vertical shaft directly connects the generator and turbine. Water flows quickly through the penstock and into the turbine when there is a high head present. The wicket gates regulate how much water enters the turbine. The benefits of Hydropower Facilities are:

I. A constant supply of water is accessible.
II. Less expensive than comparable power plants in terms of operations and maintenance
III. The price of fuel is zero.
IV. Hydro Plants are created for a variety of functions.
V. There is a lower need for working personnel. Compared to other plants, the cost of expenditures is cheaper.

However, the Drawbacks of Hydropower Facilities are listed below [27]:

i. It costs money to construct an embankment.
ii. Considerable acreage is required for setup.
III. Water must be plentiful for the process to proceed.
IV. Water-life is impacted
V. Areas near embankments must be evacuated in order to avoid flooding.

With over 150 years of history, hydropower is a well-established renewable energy technique. However, innovation in this industry has never stopped and is now largely focused on boosting plant flexibility through changes in turbine design and operational patterns, and through digitisation. The major goal is to allow hydropower facilities to better meet the demands of modern power networks that have greater variable demand and a higher mix of intermittent renewables. Hydropower facilities, particularly reservoir-type hydropower plants, are best adapted to supply the power system with much-needed emissions-free flexibility. Although hydropower is included in worldwide renewable energy development collaborations, it is frequently overlooked. The goal of the Hydropower Technology Collaboration Programme of the International Energy Agency is to increase knowledge of, support for, and awareness of the sustainable use of water resources in hydropower development and management.

1.3.4 Solar energy for sustainable development

Compared to fossil fuel energy sources, solar energy is inherently more ecologically friendly. Utilising the biggest and most sustainable resource on the earth, the sun, it converts solar energy into electrical energy. Sustainability in regard to the environment simply implies that it does not harm it. It only implies that you must start with a positive energy balance [28]. So, "development that meets the needs of the present without compromising the ability of future generations to meet their own" is what is meant by "sustainably." Because the sun's energy may be utilised indefinitely without affecting its future supply, solar energy personifies the concept of sustainability [29]. The benefits of solar energy are as follows [30]:

i. Ensures permanent energy supply
ii. The panel's photovoltaic (PV) cells collect sunlight and convert it into energy. There will always be access to solar energy, which is an infinite supply of power.
iii. Sustainability
iv. The use of solar energy will reduce humankind's impact on climate change because it is the most pure energy source available.
v. Prevents damage to the environment and habitat
vi. Solar energy use can also lessen damage to ecosystems and to land. Contrarily, there is no need to excavate or construct above the land for solar farms because they are constructed directly on top of it.
vii. A power source that is both ecologically responsible and secure.
viii. While fossil and nuclear energy are two of the most frequent power sources, they also have a number of environmental consequences. Solar energy is the cleanest, safest, and best energy source since it emits no pollutants.

Apart from advantages, there are few challenges to encounter the development of solar energy. Some of them are explained here. Those obstacles must be solved before solar energy can become a big role in the global energy industry. Solar panels are currently relatively expensive to manufacture, hence the cost of solar energy stays high [31–32]. Furthermore, solar panels require more area to create a lot of power, which might be difficult in densely populated places. Finally, because solar power is weather dependent, it cannot supply a regular and stable source of energy. Despite these obstacles, solar energy remains a possible future alternative, and more research and development are required to make solar power a more realistic choice for addressing our energy needs. The world's poorest and most environmentally conscious people have the greatest potential to change their lives thanks to solar electricity, but it must first become more affordable. Despite these restrictions, the worldwide momentum is still strong; more than half of all new jobs have been generated since 2010, proving that energy efficiency provides financial sense even in the absence of market incentives.

Solar PV power generation climbed by a record 270 TWh in 2022, up 26% from 2021. Solar PV accounted for 4.5% of total worldwide electricity generation in 2016, and it remains the third largest renewable energy technology after hydropower and wind as shown in Figure 1.6. The IEA Photovoltaic

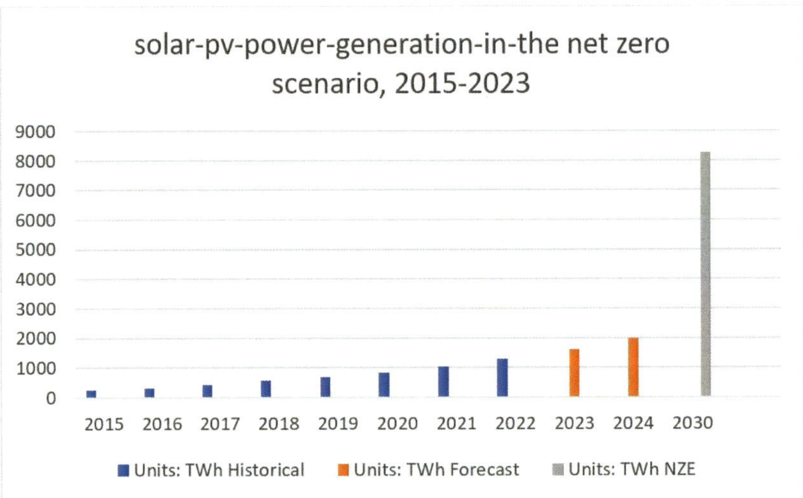

Figure 1.6 IEA, Solar PV power generation in the Net Zero Scenario, 2015-2030, IEA, Paris https://www.iea.org/data-and-statistics/charts/solar-pv-power-generation-in-the-net-zero-scenario-2015-2030, IEA. Licence: CC BY 4.0.

Power Systems Technology performs several joint projects related to solar PV technologies and systems in order to lower prices, analyse hurdles, and raise awareness of the potential of PV electricity. The International Solar Alliance provides a platform for 86 member countries to promote solar energy in a safe, economical, sustainable, and fair manner.

1.3.5 Wind energy for sustainable development

Wind energy is a significant source of renewable energy. Wind energy consumption is crucial to investigate because of the tremendous potential that wind has to offer. Wind exists in every country, and wind turbines can be installed offshore or on land. Wind energy consumption is the greatest component of the renewable energy category aside from hydroelectric power [33]. Wind energy is an important component of renewable energy; it also has various advantages, such being cost-effective, dependable, and ecologically beneficial. It's currently utilised to generate power . Because of the great potential of wind, wind energy consumption is critical to explore. Recent years have seen a significant advancement in wind turbine technology as well. Turbine effectiveness and dependability have increased, helping them to generate more energy. Masdar City was classified a dangerous wind zone with high turbulence intensity, with a wind speed of 19.82% at a height of 10 meters [34].

Utilising renewable energy sources, such as wind energy, that can be used without endangering future generations, is essential for achieving sustainability. Since wind power does not pollute or harm the environment while producing electricity, it is a sustainable and ecologically friendly energy generation technology. In order to attain sustainability, it is essential to use renewable energy sources, like wind energy, which can be used without harming subsequent generations.

A wind farm needs more acreage to run, and it can be challenging to find a place with sufficient wind to achieve perfect efficiency and power, among other challenges presented by wind energy. In addition to providing us with clean air and water, wind energy shields us from natural calamities like storms and floods (which are linked to global warming). There are various problems connected with harnessing wind energy, such as the requirement for more land to operate a wind farm and the difficulty in choosing a location with enough wind to give ideal efficiency and electricity. Most aground wind farms are often situated in rural regions, far from towns that require power, necessitating the installation of transmission lines to carry energy from the wind farm to the city, raising the cost over time. Other downsides include the impact on adjacent wildlife; on many instances, birds and bats have perished as a result of collisions with spinning turbine blades and turbine obstacles in their flight paths [35].

Reaching the annual wind electricity generation levels predicted in the Net Zero Scenario will necessitate additional funding for both onshore and offshore wind turbines. Efforts should be directed toward expediting permitting, obtaining public support, assisting in the identification of viable sites, lowering costs, and shortening project development schedules [36].

In 2022, the production of wind energy increased by a record 265 TWh (up 14% from the previous year), surpassing 2100 TWh. Among all renewable energy technologies, this has the second-fastest growth rate, just behind solar PV. The average annual generation growth rate must reach more than 17% in order to satisfy the Net Zero Emissions by 2050 Scenario, which asks for nearly 7400 TWh of wind power output in 2030. This would need increasing annual capacity expansions from around 75 GW in 2022 to 350 GW in 2030, as indicated in Figure 1.7. To accomplish this level of capacity development, far stronger regulatory and private-sector initiatives are required, with the most crucial areas for improvement being permits for onshore wind

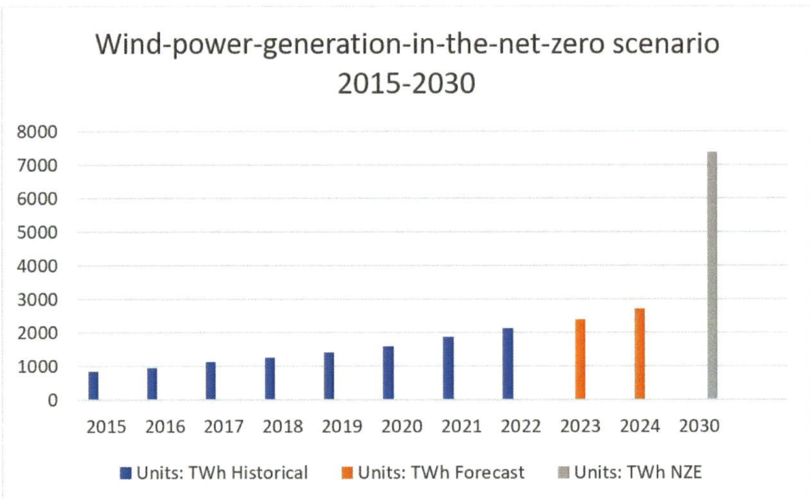

Figure 1.7 IEA, Wind power generation in the Net Zero Scenario, 2015-2030, IEA, Paris https://www.iea.org/data-and-statistics/charts/wind-power-generation-in-the-net-zero-scenario-2015-2030, IEA. Licence: CC BY 4.0.

and cost reductions for offshore wind. Countries and regions making significant advances in advancing wind energy include China, **European Union, US, UK.**

The second-highest growth among all power production technologies in 2022 was seen by wind, which produced 265 TWh more energy (up 14%). In 2022, wind will produce more energy than all other non-hydro renewables combined, making it the industry leader. In 2022, China will contribute around 40% of the increase in wind generation, with the United States contributing 22%. The output in the European Union recovered in 2022, increasing 14%, following unusually long stretches of low wind conditions in 2021.

The goals of improving productivity and lowering costs are still being prioritised in the development of wind power technologies. The IEA Wind Energy Systems Technology Collaboration Programme offers a forum for participating governments and industry leaders to exchange information on cooperative R&D initiatives to reduce the cost of wind energy technologies and improve T&D adaptability..

1.3.6 Tidal energy for sustainable development

The rising and falling waters of the ocean define the amount of tidal power potential. The power production potential of neap and spring tides with a range of 4-12 m along the beach is 1-10 MW/km [37]. An estuary near the seaside can be utilised to calculate tidal power. For a tidal height of R metres above the sea datum line, the energy potential is specified as follows:

$$E = \rho g \int_{h=0}^{h=R} hAdh$$

Where ρ is the density per square metre of seawater (typically around 1025 kg/m^3), and g is the constant of gravity (9.81 m/s^2). The tidal range equals the difference between the highest and minimum basin levels, an average water discharge of (Q=Ah/t) flowing through the turbine will do work (falling from height h).

$$P = \rho Qh$$

In Europe, the first commercial tidal power plant was built. In 1920 [38], the United Kingdom was the first to suggest harnessing the force of the tides to create energy. However, the first commercial tidal power plant was established in France, namely in Brittany's Rance Estuary in 1967 [38, 39]. South Korea started building a second tidal power plant (Uldolmok Tidal

Power Station) in 2009, with a 1.5 MW installed capacity and the potential to produce 2.4 GWh per year (project).

The spring tidal increase in the Gulf of Khambat is more than 10 metres. There are hardly many places in the world with such significant tidal ranges. Tidal energy is a fairly dependable source of electricity. In contrast to the wind and waves, which exhibit seasonal and nonseasonal changes, tides are relatively predictable years in advance. This will make the output predictable and allow for the planning of consumption cycles. There is a temporal lag every day and it is difficult to align the cycles of power generation and human consumption since tides mostly follow the moon. Energy storage becomes important to level out the power supply and demand curves.The danger of interaction with migratory and mobile marine creatures, radio frequency noise, destruction of habitat, visual amenity degradation, and changes in sediment distribution are some of the negative consequences of tidal energy. Technological challenges develop during the creation of tidal power. These include affordability, reliability, durability, operability, unpredictability, and production capacity. Tidal energy is a clean, natural, and environmentally friendly energy source. To optimise the utilisation of materials, science, and cutting-edge manufacturing processes, a more integrated engineering design methodology is required [40].

1.3.7 Wave energy

In different parts of the world, air is heated by the sun, which causes wind to blow across water surfaces. Similar to waves on a beach, surface waves are produced by the wind. Without really losing much energy, the waves can travel for thousands of kilometres before coming to rest. In essence a power plant, a wave energy converter (WEC) transforms wave energy striking the shore into electricity. The operating principle of this power plant is both conventional and novel. It is a sealed chamber with a seafloor opening that allows powerful sea waves to enter and exit the chamber. Full-scale WEC systems are often moored miles offshore in the deep sea, where wave energy is greatest. Wave energy converters gather energy from waves of various sizes.

The following is a list of the various types of wave energy converters.

- ▷ Attenuators
- ▷ Point Absorbers
- ▷ Surge converter for oscillating waves
- ▷ The water column oscillates.

▷ Device for overtopping/termination
▷ Submerged pressure difference
▷ The tidal wave
▷ The rotating mass

Wave energy is very predictable and can be created near load centers to reduce transmission needs and ease grid integration. Furthermore, wave energy could power distributed applications soon, such as desalination plants that extract salt from saltwater to help water-stressed cities and military locations. Although wave energy offers numerous advantages, it also has certain disadvantages. One of the most significant disadvantages of wave energy is the high cost of building wave energy plants. Only a few of the factors pushing up the cost of wave energy are energy plant servicing, T&D system, wave resources, expected fall in energy prices once the infrastructure is operating, and technical shelf life. Because wave energy is still in its early phases, knowing the exact cost is very difficult [https://www.linquip.com/]. There are several methods for converting the energy on the surface waves into useable forms. This can be accomplished by turning a generator's shaft or by pumping air or hydraulics. The phrase "wave energy converters" refers to various plant designs that may capture or convert the energy on surface waves to useful forms such as electricity or useable potential energy. This can be turned to power locally or at the moment of landfall [43]. There are several methods for converting surface wave energy into useable forms. A generator's shaft can be turned in this manner, or air or hydraulic pressure can be pumped. The term "wave energy converters" is a catch-all phrase used to refer to a variety of plant designs that may capture or transform the energy contained in surface waves into forms that can be used, such as electricity or useful potential energy. This might be transformed into power nearby or at the moment of landfall [43].

Most countries with a coast have a large fetch and waves bring a lot of energy to the coast. It is not the breaking waves on the beach, but the unbroken waves about a hundred odd meters away that represents this energy. The following are the benefits of using wave energy as a source of energy [43]:

1. Environmentally friendly energy source
2. Wave power is a renewable energy source, much like all other alternative energy sources [44].
3. Waves are a dependable energy source since they are rarely stopped and are almost always in motion.
4. Massive energy possibilities

5. Less reliance on the use of fossil fuels
6. No land damage

Aside from the benefits, wave energy has a few drawbacks, which are stated below [44, 45]:

1. Ecological consequences like destruction of marine life , the adjacent ecosystems, Local fishery areas and additional coastal corrosion due to wave energy power plants.
2. Wave power installation cost are high [46].
3. High maintenance costs.
4. Unable to generate a significant amount of electricity [47–49].

It's still early days for tidal energy production. Consequently, only a little amount of power has thus far been produced. There are just a handful commercial-scale tidal generating stations in operation across the world. The first occurrence happened at La Rance, France. The largest facility is South Korea's Sihwa Lake Tidal Power Station. There are not any tidal energy facilities in the United States, and there aren't many places where it can be produced affordably. This type of energy has significantly more promise in China, France, England, Canada, and Russia.

1.3.8 Ocean Energy or Ocean Thermal Energy for sustainable development

Because of the usually dramatic destructive affects of waves, ocean wave energy is the most apparent form of ocean energy. Many innovators have been motivated by the concept of converting wave energy into usable energy: by 1980 [46, 47]. Girard, a father and son who submitted the first patent on wave energy in France in 1799 [47–49], is largely regarded as the inventor of modern wave energy technology, with experiments extending back to the 1940s in Japan. He developed a navigation buoy powered by wave energy and equipped with an air turbine, which he named a (floating) oscillating water column (OWC). Since 1965, these buoys have been commercialised in Japan (and later in the United States) [48]. Later, Masuda promoted the creation of a much larger device, the Kaimei (80 m 12 m), in Japan as a floating testing platform housing several OWCs outfitted with different kinds of air turbines [49]. Because the Kaimei testing programme was conducted early, when theoretical knowledge of wave energy absorption was still in its infancy, the power output levels achieved there were not particularly successful.

The high reliance on technology and high standard of living, particularly in most developed countries, has resulted in a sudden increase in energy demand. To meet the world's high energy demand, the consumption of fossil fuels is increasing dramatically. In effect, this leads to ozone layer depletion, abrupt changes in climatic conditions, and environmental pollution, all of which cause health problems for all living species on the planet. Ocean energy is a type of renewable energy. Current renewable energy research is focusing on the utilisation of ocean energy as well, as oceans cover nearly 3/4 of the Earth's surface. Today, specific methods like wave power and tidal currents are being used to investigate this kind of energy. Compared to other renewable energy sources, ocean energy is more suitable for energy generation due to its consistency and predictability [50]. The energy that can be produced from marine energy sources can meet all the world's energy needs, but now, only a small portion of that demand is met by ocean energy. Additionally, extensive research has been done to determine the best locations for capturing ocean energy, the effectiveness of the ocean energy technologies currently in use, the quantity of power produced, and the environmental effects of these technologies.

All renewable energy sources that come from the sea are referred to as ocean energy. Ocean technology is primarily divided into three categories: wave, tidal, ocean thermal, and Salinity gradient. The commercialization of ocean energy in all its forms is still in its early stages. The cost of wave energy continues to exceed that of the other ocean technologies. In areas with significant tidal resources, such as La Rance in France and Sihwa in South Korea, tidal range has been implemented, while tidal stream has only been tested on a small scale. Global marine energy capacity was 524 megawatts as of 2022. South Korea received 256 megawatts, while France received 211 megawatts. In comparison, the Netherlands and Russia each have two megawatts of maritime energy capacity.

By converting the energy contained in ocean waves (swells) into electricity, wave energy is produced. To convert wave energy into electricity, numerous different wave energy technologies are being developed and tested. There are different ways that tidal energy can be used, both of which produce electricity [51]:

i. The potential energy produced by the height difference between high and low tides is harvested by tidal range technologies. Tidal energy is harvested by barricades (dams) at various ranges.

ii. The kinetic energy of currents flowing into and out of tidal areas, such as seashores, is captured by tidal stream (or current) technologies. Like wind turbines, tidal stream equipment runs in arrays.
iii. Ocean thermal energy is created through the conversion of the temperature differential between the surface and deeper waters into energy.
iv. The energy generated by the difference in salt concentration between two liquids is referred to as salinity gradient power. There are two stages to the production of this energy. They are classified as standalone, which involves a plant placed between the sea and a body of water, and hybrid. Desalination or waste water treatment are part of the hybrid process. The main concept is the application of osmotic power [43, 44].

The tropical waters store a massive amount of solar energy, which is then utilised to generate power. Ocean thermal energy can be captured in tropical oceans where the temperature gradient is greatest, roughly 28°C on top and about 5°C at 1000 meters depth. The temperature differential between surface and deep water in the oceans is greatest in the tropics, ranging from 20 to 25°C. Some of the best OTEC locations are along the coasts of Tamil Nadu, Lakshadweep, and Andaman and Nicobar.

The operation of an OTEC plant is based on the thermodynamic principle, which states that if an elevated-temperature heat source and a lower temperature heat sink are available, the variation in temperature is able to be used to power a prime mover (e.g., a turbine), which is capable of turning some of the heat taken from the source into mechanical energy. The remaining heat is dissipated to a lower temperature sink. OTEC has the advantage of providing continuous, renewable, and pollution-free electricity. A portable OTEC generator can generate electricity even in the middle of the ocean and can power businesses like offshore mining and manganese nodule processing.

1.3.9 Gasification and Biorefinery Energy

Energy generation by biomass gasification may appear to be a new technique, although it has been around for over a century. Our predecessors were certainly adept at using fire from existing biomass, as history demonstrates. People used such elements in new occupations needing high heat, such as pottery and metallurgy, due to time and cultural progress. Syngas, a fuel gas abundant in CO and H_2, is created during the gasification process by converting carbonaceous material thermochemically. The composition of syngas can be influenced by gasification parameters as temperature, equivalent ratio, and pressure [52]. It is challenging to create a sound theory that would adequately explain the complete gasification process in general due to the wide range of raw materials that are accessible [53, 54].

Gasification is a particularly active area of effort in utilizing biomass as a fuel right now. There are current projects looking for sustainable gasification systems that generate various types of energy. Electricity generation from gasified biomass has long been used around the world. Biomass gasification is one of the best methods for Combined Heat and Power (CHP), as it offers a direct path to efficiently extracting energy from renewable resources. Producing ethanol from cellulose has the potential to significantly increase the amount of gasoline ethanol that can be produced globally. Hydrogen production from biomass gasification is gaining popularity as a renewable energy source. Because the techniques that use fossil fuels are not very practical, there is increased interest in manufacturing hydrogen from other source materials.

Decarbonization of the globe is one of the targets that governments throughout the world have set for 2050. One way to do this is to decarbonize the production of an element like hydrogen, which would produce green hydrogen. At the moment, hydrogen emissions account for more than 2% of all CO_2 emissions worldwide [55]. This technique is based on electrolysis, a chemical reaction that yields hydrogen, a fuel that is versatile, light, and highly reactive. In this process, an electrical current is used to separate hydrogen from oxygen in water. This energy source has benefits and drawbacks, such as being 100% sustainable, storable, and versatile. However, there are certain drawbacks to green hydrogen, including high costs, high energy consumption, and safety concerns [56].

The word "biorefinery" comes from unprocessed fuel, which is renewable biomass, as well as the conversion processes that are frequently used in the treatment and processing of raw materials. The biorefinery strategy incorporates multistep processes, with the first step usually involving the treatment of biomass for subsequent processing (pretreatment). Following that, the biomass is treated biologically and/or chemically [57]. Feedstock variety, long-term viability preservation of resources, environmental friendliness, catalytic transformation, reliability of energy, financial viability, disposal of waste, and prospects are all advantages of biorefineries. Intricacy, disturbances, substantial recovery cost, sluggish transformation, technological limits, low acceptance, difficulty to handle lignin, possible usage of fossil fuels, and the destruction of forests are all disadvantages of biorefineries.

1.4 Conclusion

Sustainable energy sources and their usage are inextricably linked to long-term development. For civilizations to achieve or attempt to achieve sustainable development, substantial effort should be given to the discovery of renewable energy resources. Concerns about the environment should also be addressed. In order to achieve environmental goals, the economic development of sustainable renewable-energy options must be sustained and orderly. The governing body must support the development of renewable energy technologies as well as their commercialization in the nation. Similar to financing renewable energy research and development, making investments in the results of research programmes is a proper use of public resources. Many of the environmental and societal problems resulting from the use of fossil and nuclear fuels may be resolved using renewable energy. While it looks technically feasible to replace all fossil and nuclear fuels with renewable energy sources, the energy produced would be far more expensive than it is currently, based on current and forecast costs.

References

[1] Sun, Y. Z., et al. "Organic pollution of street dust in the Handan City, China." *Bulletin of environmental contamination and toxicology* 70.3 (2003): 422-429.

[2] Sun, Y. Z., et al. "Pollution extents of organic substances from a coal gangue dump of Jiulong Coal Mine, China." *Environmental Geochemistry and Health* 31 (2009): 81-89.

[3] Rosen, Marc A., and Aida Farsi. *Sustainable Energy Technologies for Seawater Desalination*. Academic Press, 2022.

[4] Bai, Jiping, ed. *Advanced fibre-reinforced polymer (FRP) composites for structural applications*. Woodhead Publishing, 2022.

[5] Kaygusuz, Kamil. "Energy for sustainable development: key issues and challenges." *Energy Sources, Part B: Economics, Planning, and Policy* 2.1 (2007): 73-83.

[6] Nakićenović, Nebojša, Arnulf Grübler, and Alan McDonald, eds. *Global energy perspectives*. Cambridge University Press, 1998.

[7] Anastas, Paul, and Nicolas Eghbali. "Green chemistry: principles and practice." *Chemical Society Reviews* 39.1 (2010): 301-312.

[8] Willers, Bill. "Sustainable development: a new world deception." *Conservation Biology* 8.4 (1994): 1146-1148.

[9] Norton, R. "An overview of a sustainable city strategy." *Report Prepared for the Global Energy Assessment Planning for Cities and Municipalities, Montreal, Quebec* (1991).

[10] MacRae, Morgan. "Realizing the benefits of community integrated energy systems. Study No. 45." (1992).

[11] Dincer, Ibrahim, and Marc A. Rosen. "A worldwide perspective on energy, environment and sustainable development." *International journal of energy research* 22.15 (1998): 1305-1321.

[12] Purvis, Ben, Yong Mao, and Darren Robinson. "Three pillars of sustainability: in search of conceptual origins." *Sustainability science* 14 (2019): 681-695.

[13] Dincer, Ibrahim. "Renewable energy and sustainable development: a crucial review." *Renewable and sustainable energy reviews* 4.2 (2000): 157-175.

[14] Dincer, Ibrahim, and Marc A. Rosen. *Exergy analysis of heating, refrigerating and air conditioning: methods and applications.* Academic Press, 2015.

[15] Reijnders, Lucas. "Conditions for the sustainability of biomass based fuel use." *Energy policy* 34.7 (2006): 863-876.

[16] Kralova, Iva, and Johan Sjöblom. "Biofuels–renewable energy sources: a review." *Journal of Dispersion Science and Technology* 31.3 (2010): 409-425.

[17] Gaurav, N., et al. "Utilization of bioresources for sustainable biofuels: a review." *Renewable and sustainable energy reviews* 73 (2017): 205-214.

[18] Kim, Seungdo, and Bruce E. Dale. "Life cycle assessment of various cropping systems utilized for producing biofuels: bioethanol and biodiesel." *Biomass and Bioenergy* 29.6 (2005): 426-439.

[19] Demirbas, Ayhan. "Biofuels sources, biofuel policy, biofuel economy and global biofuel projections." *Energy conversion and management* 49.8 (2008): 2106-2116.

[20] Fridleifsson, Ingvar B. "Geothermal energy for the benefit of the people." *Renewable and sustainable energy reviews* 5.3 (2001): 299-312.

[21] Barbier, Enrico. "Nature and technology of geothermal energy: a review." *Renewable and sustainable energy reviews* 1.1-2 (1997): 1-69.

[22] Dursun, Bahtiyar, and Cihan Gokcol. "The role of geothermal energy in sustainable development of Turkey." *Energy exploration & exploitation* 30.2 (2012): 207-222.

[23] Bertani, Ruggero. "World geothermal power generation in the period 2001–2005." *Geothermics* 34.6 (2005): 651-690.

[24] Mekhilef, Saad, Rahman Saidur, and Masoud Kamalisarvestani. "Effect of dust, humidity and air velocity on efficiency of photovoltaic cells." *Renewable and sustainable energy reviews* 16.5 (2012): 2920-2925.

[25] Karady, George G., and Keith E. Holbert. *Electrical energy conversion and transport: an interactive computer-based approach*. John Wiley & Sons, 2013.

[26] Naghizadeh, Ramezan Ali, Saeed Jazebi, and Behrooz Vahidi. "Modeling hydro power plants and tuning hydro governors as an educational guideline." *International Review on modelling and simulations* 5.4 (2012): 1780-1790.

[27] Acakpovi, Amevi, Essel Ben Hagan, and Francois Xavier Fifatin. "Review of hydropower plant models." *International Journal of Computer Applications* 108.18 (2014).

[28] Maka, Ali OM, and Jamal M. Alabid. "Solar energy technology and its roles in sustainable development." *Clean Energy* 6.3 (2022): 476-483.

[29] Kannan, Nadarajah, and Divagar Vakeesan. "Solar energy for future world:-A review." *Renewable and sustainable energy reviews* 62 (2016): 1092-1105.

[30] Mekhilef, Saidur, Rahman Saidur, and Azadeh Safari. "A review on solar energy use in industries." *Renewable and sustainable energy reviews* 15.4 (2011): 1777-1790.

[31] Thakur, Parag P., Shriram S. Sonawane, and Hussein A. Mohammed. "Recent Trends in Applications of Nanofluids for Effective Utilization of Solar Energy." *Current Nanoscience* 19.2 (2023): 170-185.

[32] Kumar, Mukesh, Lokesh Rana, and Amruta Pattnaik. "Innovative Design for Efficient Solar Cell." *Advanced Functional Materials and Devices: Select Proceedings of AFMD 2021*. Springer Singapore, 2022.

[33] Sadorsky, Perry. "Wind energy for sustainable development: Driving factors and future outlook." *Journal of Cleaner Production* 289 (2021): 125779.

[34] Olabi, A. G., et al. "Wind Energy Contribution to the Sustainable Development Goals: Case Study on London Array." *Sustainability* 15.5 (2023): 4641.

[35] Kumar, Nitin, and Om Prakash. "Analysis of wind energy resources from high rise building for micro wind turbine: A review." *Wind Engineering* 47.1 (2023): 190-219.

[36] Yasmeen, Rizwana, et al. "The role of wind energy towards sustainable development in top-16 wind energy consumer countries: Evidence from STIRPAT model." *Gondwana Research* 121 (2023): 56-71.
[37] Khan, N. D., et al. "Review of ocean tidal, wave and thermal energy technologies." *Renewable and Sustainable Energy Reviews* 72 (2017): 590-604.
[38] Kirby, Robert, and C. Retière. "Comparing environmental effects of Rance and Severn barrages." *Proceedings of the Institution of Civil Engineers-Maritime Engineering*. Vol. 162. No. 1. Thomas Telford Ltd, 2009.
[39] Kang, Nam Seon, et al. "Red tides in Shiwha Bay, western Korea: a huge dike and tidal power plant established in a semi-enclosed embayment system." *Harmful Algae* 30 (2013): S114-S130.
[40] Chowdhury, M. S., et al. "Current trends and prospects of tidal energy technology." *Environment, development and sustainability* 23 (2021): 8179-8194.
[41] Jollands, Nigel. "International Energy Agency." *Competitive Cities and Climate Change* 136 (2008).
[42] Cruz, Joao. *Ocean wave energy: current status and future prespectives*. Springer Science & Business Media, 2007.
[43] Hussain, Akhtar, Syed Muhammad Arif, and Muhammad Aslam. "Emerging renewable and sustainable energy technologies: State of the art." *Renewable and sustainable energy reviews* 71 (2017): 12-28.
[44] Wilberforce, Tabbi, et al. "Overview of ocean power technology." *Energy* 175 (2019): 165-181.
[45] Falcao, Antonio F. de O. "Wave energy utilization: A review of the technologies." *Renewable and sustainable energy reviews* 14.3 (2010): 899-918.
[46] McCormick, M. E. "Ocean wave energy conversion. Courier Corporation." (2013).
[47] Ross, David. "Power from the Waves." *(No Title)* (1995).
[48] Masuda, Y. "Wave-Activated Generator, Int." *Colloq Exposition Oceans, France, Bordeaux*. 1971.
[49] Masuda, Y. "Experimental full-scale results of wave power machine Kaimei in 1978." *Proc First Symp Wave Energy Utilization, Gothenburg, Sweden*. 1979.
[50] Budar, K., and Johannes Falnes. "A resonant point absorber of ocean-wave power." *Nature* 256.5517 (1975): 478-479.

[51] Alcorn, R., et al. "Results of the testing of the Energetech wave energy plant at Port Kembla." *Energeth Report* 5 (2005): 44.
[52] El-Emam, Rami S., Ibrahim Dincer, and Greg F. Naterer. "Energy and exergy analyses of an integrated SOFC and coal gasification system." *International Journal of Hydrogen Energy* 37.2 (2012): 1689-1697.
[53] Arafat, Hassan A., and Kenan Jijakli. "Modeling and comparative assessment of municipal solid waste gasification for energy production." *Waste management* 33.8 (2013): 1704-1713.
[54] Reed, Thomas B., and Agua Das. *Handbook of biomass downdraft gasifier engine systems*. Biomass Energy Foundation, 1988.
[55] Teske, Sven. *Achieving the Paris Climate Agreement Goals: Part 2: Science-based Target Setting for the Finance industry—Net-Zero Sectoral 1.5°C Pathways for Real Economy Sectors*. Springer Nature, 2022.
[56] Hydrogen insights: https://hydrogencouncil.com/en/hydrogen-insights-2021/
[57] Evans, Annette, Vladimir Strezov, and Tim J. Evans. "Bio-oil applications and processing." *Biomass processing technologies* (2014): 357-378.

2

Application of Intelligent Techniques for Renewable Energy Management

Rohit Verma and Yog Raj Sood

National Institute of Technology, India
E-mail: verma.rohitverma5@gmail.com; yrsood@nith.ac.in

Abstract

In this chapter, an approach for distributed generation (DG) placement in distribution systems is presented. It utilizes a hybrid algorithmic framework that combines k-nearest neighbors (KNN) and PSO. By using similarity and proximity criteria, the KNN algorithm finds possible places for DG and then reduces the size of the search field. Next, taking into account a variety of limitations and goals, the PSO algorithm optimizes the capacities and types of DG units. Enhancing the effectiveness of DG deployment while balancing system reliability, financial considerations, and environmental impact is the goal of this synergistic strategy.

The suggested methodology is validated by simulations conducted on benchmark distribution systems, which show gains in voltage profiles, line losses, and overall system efficiency. KNN and PSO integration provides a strong and practical approach to DG placement optimization, offering insightful information to practitioners, academics, and stakeholders in the power distribution and renewable energy industries. The importance of this strategy in promoting resilient and sustainable distribution networks is shown by the chapter's discussions on prospective uses, benefits, and directions for further study.

Keywords: Artificial intelligence, distributed generation, machine learning, Python language, renewable energy, solar energy.

DOI: 10.1201/9788743809852-2

2.1 Introduction

The integration of distributed generation (DG) into distribution systems represents a pivotal step toward achieving a sustainable and resilient energy infrastructure. Decentralized generating capabilities are provided by DG technologies, which lessen dependency on centralized power plants and increase system flexibility. Examples of these technologies include solar photovoltaics, wind turbines, and microturbines. To optimize their benefits while minimizing operational issues, distributed generation units must be positioned within distribution networks in the best possible way.

Conventional techniques for distributed generation placement frequently depend on heuristic methods or oversimplified models, which may fail to account for the complexity of actual distribution networks. Consequently, there is an increasing interest in creating sophisticated optimization methods to determine the best places and arrangements for distributed generation installations. Of these methods, the combination of the k-nearest neighbors (KNN) algorithm and the PSO offers a viable path for DG placement that is both successful and efficient.

This chapter introduces an approach for DG placement optimization in distribution systems: the integrated KNN-PSO algorithm. This method leverages the synergistic framework of PSO and KNN to identify optimal DG configurations, while also addressing the uncertainties and complexities of distribution systems. Specifically, the method aims to optimize DG placement to improve system sustainability, efficiency, and reliability by utilizing proximity and similarity metrics to identify potential DG locations.

This chapter's remaining sections are organized as follows: In Section 2.2, we provide a comprehensive review of existing methods for DG placement and discuss the motivation for adopting advanced optimization techniques. The suggested KNNPSO algorithm's methodology, including its phases and formulation, is presented in Section 2.3. Section 2.4 shows the algorithm's process overview. Section 2.5 concludes by outlining potential future study possibilities, and Section 2.6 provides concluding remarks.

By means of this chapter, we hope to further the development of ideal distribution grid placement techniques for distributed generation (DG), making it easier to include renewable energy sources in distribution networks and promoting a more sustainable energy future.

2.2 Background and Literature Review

2.2.1 Distributed generation in distribution systems

Small-scale power generation units that are linked to the distribution grid and usually situated near the point of electricity consumption are referred to as distributed generation (DG) units. Combining traditional fossil fuel-based generators like diesel generators and combined heat and power (CHP) systems with renewable energy sources like solar photovoltaic (PV) panels, wind turbines, micro-hydro turbines, and biomass generators is what is meant by distributed generation (DG) systems.

The integration of DG into distribution systems offers several advantages, including

- Decreased transmission losses: Distributed generation (DG) systems minimize energy losses related to transmission lines by producing electricity closer to the point of consumption, which eliminates the need for long-distance power transmission.
- Enhanced system resilience: By supplying backup power during grid disruptions or outages, distributed generation (DG) systems can bolster the resilience of distribution networks.
- Improved integration of renewable energy: By lowering greenhouse gas emissions and reliance on fossil fuels, DG systems based on renewable energy sources help to shift toward a more sustainable energy balance.

For example, rooftop solar PV systems can be installed on residential and commercial buildings in urban regions with high electricity demand to balance peak demand and lessen the burden on the grid during times of high energy usage.

2.2.2 Traditional methods for DG placement

Conventional methods for placing distributed generation (DG) in distribution networks frequently depend on deterministic optimization techniques or heuristic approaches. Usually, these techniques entail

- Rule-based algorithms: Based on variables including load density, voltage limitations, and grid capacity, DG unit location and sizing are decided upon using straightforward rules or guidelines.

- Simplified models: To optimize DG placement and sizing based on predetermined objectives like cost minimization or reliability improvement, linear programming or analytical methods may be utilized.
- Heuristic techniques: To explore the solution space and find close to optimal solutions, metaheuristic algorithms like ant colony optimization, simulated annealing, and genetic algorithms are employed.

However, traditional methods have several limitations, including

- Lack of scalability: Large-scale distribution systems with lots of constraints and decision factors may be too much for traditional approaches to manage.
- Unable to manage uncertainty: Conventional approaches frequently make the assumption that all system parameters and operational conditions are well known, excluding uncertainties related to load fluctuations, the production of renewable energy, and equipment failures.
- Simplistic presumptions: Conventional approaches could rely on simplistic models of the behaviors and components of the system, which could result in unrealistic or subpar solutions.

For example, standard approaches may not adequately account for the intermittent nature of renewable power and its impact on system operation and stability in a distribution network with high levels of renewable energy penetration.

2.2.3 Intelligent techniques for optimization

Utilizing intelligent methods for distribution system management optimization challenges, such as machine learning (ML) and artificial intelligence (AI) [1, 2, 3, 4], has gained popularity in recent years. Comparing intelligent procedures to traditional ways yields many benefits, such as

- Flexibility and adaptability: AI and ML algorithms are well-suited for dynamic optimization challenges because they have the ability to learn from data and adjust to changing system conditions [5].
- Complex constraint management: Nonlinearities, uncertainties, and multi objective optimization criteria can all be accommodated by intelligent algorithms, leading to more reliable and realistic solutions.
- Scalability: The optimization of complicated distribution systems is made possible by the ability of AI and ML algorithms to scale to tackle large-scale optimization issues with a multitude of decision factors and constraints.

Examples of intelligent techniques used in optimization include

- Neural networks: Artificial neural networks (ANNs) [6] are a useful tool for modeling intricate interactions between input and output variables. They may be taught to predict system behavior and optimize the placement and size of distributed generators (DGs).
- Evolutionary algorithms: To find the best solutions in vast solution spaces, evolutionary algorithms [7] like genetic algorithms (GAs), particle swarm optimization (PSO), and differential evolution (DE) imitate natural evolutionary processes.
- Reinforcement learning: Reinforcement learning algorithms [8] are appropriate for adaptive control and optimization tasks in distribution systems because they acquire the best decision-making methods through trial-and-error interactions with the environment.

A reinforcement learning agent, for example, can be trained to optimize the overall efficiency and reliability of a distribution network by dynamically adjusting the operating parameters of each DG unit based on real-time measurements of system conditions and performance goals, thus enabling autonomous decision-making and adaptive control strategies in dynamic operational environments.

2.2.4 Application of KNN and PSO in DG placement

The k-nearest neighbors (KNN) algorithm [9] and particle swarm optimization (PSO) are two intelligent methods that have been used to optimize the placement of distributed generation (DG) in distribution networks.

KNN is a machine learning technique for classification and regression that is straightforward but effective. KNN can be trained on distribution system data in the context of DG installation to forecast how installing DG units at various locations will affect system performance measures including voltage profiles, line losses, and reliability indices. KNN can determine the best places to deploy DG in order to optimize system benefits and minimize negative effects on system functioning by examining the features of nearby data points.

For example, KNN can assist in identifying areas with high potential for renewable energy and low grid congestion in distribution networks with limited capacity for DG integration. This will enable the best possible placement of DG units to optimize the penetration of renewable energy sources and reduce system losses.

PSO is a population-based metaheuristic optimization algorithm that draws social behavior cues from fish schools and flocks of birds. A population of possible solutions, or particles, is iteratively updated by PSO based on nearby particles' best solutions and fitness values. PSO converges to optimal solutions in vast solution spaces through social interaction and exploration.

PSO can be used to find the best mix of DG unit sizes and placements to minimize system losses while meeting requirements like voltage limitations, line capacities, and environmental considerations in the context of DG placement optimization. PSO is an effective technique for finding near-optimal solutions to complicated optimization problems by modeling the collective behavior of particles exploring the solution space.

For example, PSO can adaptively alter the placement and sizing of DG units in response to changing system conditions, optimizing the overall system performance in real time, in a distribution network with fluctuating load needs and uncertain renewable energy supply.

2.2.5 State-of-the-art in optimal DG placement

In order to improve solution quality and scalability, recent developments in the field of optimum DG placement have concentrated on combining sophisticated approaches like KNN and PSO with conventional optimization techniques. Research has indicated that hybrid strategies, which integrate the advantages of many optimization methods, are efficacious in tackling the intricacies of distribution system administration.

Additionally, research has looked into how to best position distributed generation (DG) in distribution systems by applying machine learning techniques like reinforcement learning and deep learning. With the use of enormous data sets, deep learning algorithms like recurrent neural networks (RNNs) and convolutional neural networks (CNNs) can identify intricate patterns and relationships, making it possible to make more precise predictions and optimization choices for distribution system management tasks.

Similar to this, through trial-and-error interactions with the environment, reinforcement learning algorithms like Q-learning and deep Q-networks (DQNs) may develop optimal control policies for DG units, resulting in resilient and adaptive optimization techniques in dynamic and uncertain operating situations.

2.2.6 Challenges and opportunities

Even while the creation and implementation of intelligent approaches for the best placement of distributed generation have advanced significantly, there are still a number of obstacles and opportunities [10].

Challenges:

- Data availability and quality: High-quality data from distribution systems is necessary for machine learning algorithms like KNN and PSO to function effectively. However, gathering and classifying data can be difficult, particularly in operational settings that are real world and may include noisy or sparse data.
- Computational complexity: Distribution system management optimization issues can be quite computationally demanding, especially when dealing with large-scale distribution networks that have a lot of restrictions and decision factors. The computational complexity of real-world applications requires the use of scalable optimization approaches and efficient algorithms.
- Interpretability and trustworthiness of the model: Machine learning models, like KNN and PSO, are sometimes regarded as "black-box" models, which makes it difficult to decipher their judgments and comprehend the underlying processes guiding optimization outcomes. In order to improve the interpretability and reliability of intelligent procedures in distribution system management, explainable artificial intelligence (XAI) techniques are required.

Opportunities:

- Interdisciplinary cooperation: Interdisciplinary research and innovation in distribution system management can be facilitated by cooperation between scholars, practitioners, and stakeholders from a variety of fields, including electrical engineering, computer science, and economics. It is possible to address complicated challenges in optimal DG placement by developing new viewpoints and ways through the integration of expertise from many disciplines.
- Making decisions based on data: Large-scale data from sensors, IoT devices, and smart meters is now available, opening up possibilities for distribution system management decision-making based on data. DG placement optimization tactics can be informed by machine learning

techniques like KNN and PSO, which take advantage of big data analytics to extract insightful information from massive data sets.
- Policy and regulatory support: The adoption of intelligent solutions for optimal deployment of distributed generation (DG) is greatly aided by policy frameworks and regulatory incentives. Policymakers may expedite the shift toward a more resilient and sustainable energy infrastructure by promoting cooperation between industry, government, and academia, financially supporting research and development, and offering incentives for innovation.

2.2.7 Practical use of intelligent techniques

In the context of distribution system management, this section examines real-world uses for intelligent approaches, including the k-nearest neighbors (KNN) algorithm and particle swarm optimization (PSO). We demonstrate the adaptability and efficacy of these strategies in maximizing distributed generation (DG) deployment and boosting distribution networks' operational efficiency by looking at case studies and real-world examples.

2.2.7.1 Demand forecasting and load management

In order to help utilities predict future patterns of energy demand and enhance load management tactics, intelligent techniques are essential to demand forecasting and load management. For example, PSO approaches can improve demand response programs to lower overall energy usage during peak hours, while KNN algorithms can examine past consumption data to find patterns and trends. Case studies have demonstrated how the use of intelligent demand forecasting and load control strategies has significantly reduced energy expenditures and improved system reliability.

2.2.7.2 Regulating voltage and improving power quality

Distribution networks' system stability and dependability depend on effective voltage management and power quality enhancement. To reduce voltage fluctuations and improve power quality, KNN and PSO algorithms can be used to optimize the size and positioning of voltage control devices, such as voltage regulators and capacitor banks. Intelligent approaches have been crucial in lowering voltage violations, increasing system efficiency, and raising customer satisfaction by analyzing system data and optimizing control tactics. Furthermore, by leveraging intelligent algorithms for voltage management and power quality enhancement, distribution utilities can proactively address system vulnerabilities and adapt to dynamic operating conditions, ultimately enhancing the resilience and robustness of the distribution network.

2.2.7.3 Asset management and condition monitoring

Maintenance and dependability of distribution systems depend heavily on asset management and condition monitoring. KNN and PSO algorithms are capable of predicting equipment failures, optimizing maintenance schedules, and extending the life of distribution assets by analyzing sensor data and past maintenance records. Utilities may get more resilience from their distribution infrastructure, save maintenance costs, and cut downtime by implementing proactive asset management and condition monitoring.

2.2.7.4 Grid integration of renewable energy sources

The integration of renewable energy sources, such as wind and solar electricity, into distribution networks has specific problems concerning grid management and stability. For the purpose of improving energy yield, eliminating grid integration problems and optimizing the integration of renewable energy resources, intelligent procedures provide answers. Utilities may decrease their dependency on fossil fuels, minimize carbon emissions, and make the shift to a more sustainable energy future by using KNN and PSO algorithms to optimize the location and functioning of renewable energy systems.

2.2.7.5 Resilience and risk management

In the event of cyberattacks, natural disasters, or other interruptions, distribution networks must be reliable and secure. To this end, resilience and risk management are critical. By minimizing vulnerabilities, strengthening response capabilities, and optimizing system settings, intelligent approaches can help to improve resilience and risk management. Utilities can identify important infrastructure, prioritize investments, and create backup plans by applying KNN and PSO algorithms to reduce the impact of unfavorable occurrences on distribution system functioning.

This section illustrates the broad applicability and promise of intelligent strategies for solving various operational difficulties and optimization goals by looking at these real-world distribution system management applications.

2.2.8 Literature review report of DG placement

The literature on the best locations for distributed generation (DG) and its optimization methods provides insightful information about the difficulties, approaches, and developments in distribution system management.

This section offers a thorough analysis of pertinent research, emphasizing the field's most recent advancements, useful applications, and theoretical underpinnings.

In order to determine the ideal location and size of the DG units so that distribution companies can profit from integrating the DG units into the distribution system, numerous research findings have suggested various optimization techniques and goal functions. In [15], it is explained how to use a genetic algorithm to deploy DG units optimally for IEEE 33 and PG and E 69 bus distribution test systems, with the goal of reducing loss. The Quasi-Oppositional Chaotic Symbiotic Organisms Search (QOCSOS) method is used to reduce loss and determine the best location and size of DG units in IEEE 33 and PG and E 69 bus test systems. This topic is covered in [16]. In IEEE 33 and 69 bus distribution test systems, the best arrangement of DG units, including their sizing, is addressed in [22] and [21] utilizing the firefly algorithm for loss reduction and the crop pollination method for loss reduction.

[17] addressed how to use Improved Raven Roosting Optimization to arrange and size DG units in IEEE 33 and 69 bus distribution test systems optimally based on active power loss, reactive power loss, voltage variation, and line loading. [18] discusses the best location and size for DG units in

Table 2.1 Literature report of DG placement.

Sr. no.	Algorithm	Objective	Reference no.
1	Genetic algorithm	Optimize power losses	[11]
2	Genetic algorithm	Optimize power losses	[12]
3	Ant colony optimization	Multiple objectives	[13]
4	Harmony search	Optimize DG Capacity	[14]
5	Genetic algorithm	Loss reduction and voltage regulation	[15]
6	QOCSOS	Loss reduction	[16]
7	Improved raven roosting	Multiple objectives	[17]
8	Enhanced genetic algorithm	Losses and voltage deviation	[18]
9	Fuzzy logic controller + ALO + PSO	Losses and voltage deviation	[19]
10	Whale optimization	Voltage stability index	[20]
11	Crop pollination algorithm	Loss reduction and line loading	[21]
12	Firefly algorithm	Loss reduction	[22]
13	Genetic algorithm	Loss reduction	[23]
14	Spring search	Multiple objectives	[24]
15	Kalman filter	Loss reduction	[25]

IEEE 33 and PG and E 69 bus test systems based on voltage deviation and loss reduction using enhanced genetic algorithm; [19] discusses particle swarm optimization, fuzzy logic controller, and ant lion optimization algorithm. In IEEE 33 and 69 bus distribution test systems, the ideal placement and dimensions of DG units are determined by the voltage stability index by the use of whale optimization, as described in [20]. [23] addresses how to use a genetic algorithm to identify the best location and size for DG units in IEEE 33 and PG and E 69 bus test systems based on voltage deviation and loss reduction. In [24], it is detailed how to use the spring search algorithm to determine the best location for DG units in an IEEE 33 bus test system based on emissions, voltage variation, and active power loss using the Kalman filter in [25]. Which are in Table 2.1 [11–25].

2.3 Methodology

This section describes the methodology used to use the KNN-PSO strategy for optimal placement of the DG.

2.3.1 Load flow analysis and KNN regressor model

Using the supplied line and bus data, a modified forward-backward sweep method was employed in the first stage of load flow analysis. Important characteristics including voltage profiles, branch currents, power flow, and total active and reactive losses in the system could all be determined thanks to this research. The gathered information was then used to train a k-nearest neighbor (KNN) regressor model.

> Predicting modifications to the IEEE 33-bus distributed system in particular required the use of the KNN regressor model. The trained KNN model made it easier to identify buses that would be good candidates for installing distributed generation (DG) within the distribution system by choosing a value for *k* that was neither too high nor too low.

2.3.2 Particle swarm optimization (PSO) for DG sizing

After the KNN regressor model identified possible locations for DG installation, particle swarm optimization (PSO) [26] was used to calculate the ideal size of DG units to be installed at these chosen buses. PSO, which is renowned for its efficiency in identifying global optima, used the data from the KNN regressor model to initialize optimization particles for the chosen buses.

The optimization process sought to identify the ideal DG unit size within a predefined range (200 to 4600 kW) while minimizing system losses. Strict limitations, such as power balance, voltage in each unit, and DG size were placed in place to guarantee the viability and dependability of the solutions (2)(3)(4).

Our goal with this integrated strategy was to maximize the placement of distributed generation (DG) in distribution systems while taking reliability and system efficiency into account.

Additionally, the integration of KNN-based predictive modeling with PSO-based optimization provided a holistic approach to DG placement optimization, enhancing the accuracy and efficiency of the process. This synergy between predictive modeling and optimization techniques resulted in optimized DG placement solutions that effectively balanced system performance and operational requirements.

2.3.3 Problem formulation

The optimal location and appropriate size for DG to reduce power loss (P_{loss}) of the system is to be found. P_{loss} in the system can be written as shown in eqn (2.1) [27]

$$P_{\text{loss}} = \sum\nolimits_{i}^{n} \left[I_{\text{branch(i)}}\right]^2 \cdot Ri, \tag{2.1}$$

where

$$I_{\text{branch(i)}} = \text{branch current}$$
$$R_{\text{i}} = \text{branch resistance.}$$

The objective function is to

$$\text{Minimize } P_{\text{loss}}$$

Subjected to constraints,

- *Power balance*:

$$P_{\text{sub}} + P_{\text{DG}} = P_{\text{Load}} + P_{\text{Loss}} \tag{2.2}$$

- *Voltage*:

$$V_{\text{pu}}^{\text{max}} \geq V_{\text{pu}} \geq V_{\text{pu}}^{\text{min}} \tag{2.3}$$

- *DG size*:

$$P_{\text{DG}}^{\text{max}} \geq P_{\text{DG}} \geq P_{\text{DG}}^{\text{min}} \tag{2.4}$$

2.4 Process Overview

An overview of the main procedures for utilizing intelligent strategies to install distributed generation (DG) in the best possible location is given in this section.

2.4.1 Data collection and preprocessing

- Gather relevant information from sources, such as load profiles, bus and line data, renewable energy possibilities, and system limitations.
- To guarantee consistency, correctness, and suitability for analysis and modeling, preprocess the data.

To learn more about the distribution and properties of the data that have been gathered, conduct exploratory data analysis (EDA). This will help you find patterns, trends, and possible correlations that could help you with future modeling and optimization techniques.

2.4.2 Load flow analysis

- Perform load flow analysis to find the voltage profile, power flow, and system losses using well-known procedures like the forward-backward sweep method.
- Examine the load flow data to find instances of voltage violations, line overloads, and DG integration prospects.

2.4.3 Intelligent modeling and prediction

- Utilize past data to train machine learning models, like k-nearest neighbor (KNN) regressors, in order to forecast how the location of distributed generators would affect system performance.
- Make use of intelligent techniques to determine where distributed generation (DG) should be installed, taking into account system limitations and anticipated results.

2.4.4 Optimization using metaheuristic algorithms

- To find the best configurations for placing DGs, use metaheuristic optimization methods like particle swarm optimization (PSO).
- To direct the optimization process toward solutions that reduce system losses, optimize the use of renewable energy sources, and meet operational needs, define objective functions and restrictions.

2.4.5 Validation and sensitivity analysis

- Verify the viability and dependability of the optimal DG placement solutions using simulation or validation data sets.
- Analyze the solutions' robustness and performance under various system conditions and uncertainties by doing sensitivity analysis.

2.4.6 Implementation and deployment

- In order to ensure seamless integration with the current infrastructure and compliance with regulatory requirements, apply the best-practiced distributed generation placement solutions to actual distribution networks while keeping operational and practical limitations in mind. This will help to facilitate a smooth implementation and long-term viability.
- In order to maximize system performance and efficiency, track and assess the effectiveness of the implemented solutions over time and make necessary improvements.

2.5 Future Directions and Research Opportunities

This section examines possible directions for further study and research in the field of optimal DG placement, utilizing clever methods like the KNN-PSO approach to take advantage of new opportunities and solve existing problems like voltage regulation challenges, protection system adjustments, grid stability concerns, grid congestion and line losses, power quality issues, compliance with standards, anti-islanding protection modeling and forecasting requirements, and economic viability factors.

2.5.1 Advanced machine learning techniques

- Investigation of more complex machine learning approaches, such as ensemble methods, deep learning, and reinforcement learning, in addition to KNN and PSO, in order to improve the precision and effectiveness of DG placement optimization.
- Examining hybrid or metaheuristic strategies that combine several clever methods in order to make use of their complimentary advantages and enhance optimization results.

2.5.2 Integration of renewable energy sources

- Incorporating additional renewable energy sources, like geothermal, hydro, and biomass, into frameworks for DG placement optimization can lead to increased resilience and diversity in distributed generation portfolios.
- To improve the dependability and effectiveness of renewable energy integration, factors such as weather prediction data, predictive analytics, and the spatial and temporal variability of renewable resources should be taken into account.

Additionally, considering the spatial and temporal variability of renewable resources enables better planning and allocation of DG units to maximize energy yield and minimize operational uncertainties. This proactive approach enhances the reliability and efficiency of distributed generation systems, ensuring optimal utilization of renewable energy sources while effectively managing system dynamics and variability.

2.5.3 Resilience and adaptability

- Development of optimization frameworks with are silience focus to strengthen distribution networks' resistance to extreme weather, natural disasters, and cyber-physical threats.
- Adaptable algorithms and real-time decision-making skills to dynamically modify DG installation plans in response to shifting operating needs and system conditions.
- Also for reducing the cost of DG integration by taking cost into account as a constraint [30].

2.5.4 Demand-side management and flexibility

- In order to maximize system flexibility and resource usage, demand-side management tactics, demand response programs, and flexible load management approaches are integrated into DG placement optimization frameworks.
- Investigating cutting-edge methods for integrating the implementation of DG with demand-side interventions in order to maximize system reliability and efficiency.

- Incorporate difficulties in precisely forecasting demand, restricted involvement from consumers, legal obstacles, and the requirement for advanced technological infrastructure for execution.

2.5.5 Multiobjective optimization and decision support systems

- Development of multiobjective optimization algorithms for DG location optimization that balance competing goals including cost reduction, environmental sustainability, and system dependability.
- The creation of visualization tools and decision support systems to help stakeholders weigh trade-offs, assess potential outcomes, and decide on DG deployment plans.

2.5.6 Policy and regulatory considerations

- Analyzing market processes, regulatory incentives, and policy frameworks to support grid modernization initiatives, encourage the integration of renewable energy sources, and encourage the best possible placement of distributed generation.
- Working together, politicians, regulators, and industry players may remove regulatory obstacles, expedite approval procedures, and promote innovation in the deployment of distributed generation.

Stakeholders may support the transition toward a more sustainable, resilient, and decentralized energy infrastructure and promote optimal DG installation procedures by investigating the future directions and research opportunities indicated in this area. By fostering collaboration between industry, academia, and government entities, stakeholders can drive innovation, shape policy frameworks, and accelerate the adoption of intelligent techniques for optimal DG placement, thereby advancing the goals of energy sustainability and security.

2.6 Conclusion

We have looked at efficient methods in this chapter for distributed generation (DG) placement in distribution systems, using sophisticated approaches like k-nearest neighbors (KNN) and particle swarm optimization (PSO). By using a methodical process that includes load flow analysis, optimization, and data preparation, we have shown how the combined KNN-PSO technology may improve the efficiency, sustainability, and dependability of distribution systems.

Our findings highlight how crucial it is to use state-of-the-art optimization techniques to handle the complexity and unpredictability that come with managing distribution systems. We have effectively found good locations for DG installation, optimized DG unit sizes, and reduced system losses by using KNN and PSO. These successes have reduced operating difficulties while also optimizing the use of renewable energy sources.

The theoretical method described here offers important insights into the possible applications of intelligent procedures in distribution system design, even if this chapter did not feature a concrete case study. Our approach's effectiveness is further validated by simulation results from earlier research that show how clever tactics can result in notable improvements in voltage profiles, line losses, and overall system efficiency.

In the long run, there is a lot of research and development potential on the subject of optimal DG placement. Subsequent endeavors can concentrate on propelling machine learning techniques, incorporating supplementary sustainable energy resources, augmenting system robustness, and tackling legislative and regulatory aspects.

Through interdisciplinary collaboration and pursuit of these objectives, we may sustainably propel innovation, advance sustainability, and influence the trajectory of distribution system management.

To sum up, the data in this chapter adds to the expanding amount of knowledge about the best location for distributed generation (DG) and offers useful perspectives for those involved in the energy sector.

As we work toward a more reliable and sustainable energy future, the integration of intelligent techniques holds great promise for opening up new doors and maximizing the advantages of distributed generation.

References

[1] JUMAA, Fadhel A.; NEDA, Omar Muhammed; MHAWESH, Mustafa A.. Optimal distributed generation placement using artificial intelligence for improving active radial distribution system. Bulletin of Electrical Engineering and Informatics, [S.l.], v. 10, n. 5, p. 2345–2354, oct. 2021.

[2] S. M. Hossain, M. S. Hassan Onim, S. Biswas and A. H. Chowdhury, "Application of Machine Learning for Optimal Placement of Distributed Generation," 2021 International Conference on Information and Communication Technology for Sustainable Development (ICICT4SD), Dhaka, Bangladesh, 2021, pp. 231–234.

[3] Ankur Kumar, Ritika Verma, Niraj Kumar Choudhary and Nitin Singh (2023) "Optimal placement and sizing of distributed generation in power distribution system: a comprehensive review", Energy Sources, Part A: Recovery, Utilization, and Environmental Effects, 45:3, 7160–7185.

[4] A. Maheshwari, Y. R. Sood, S. Jaiswal, S. Sharma and J. Kaur, "Ant Lion Optimization Based OPF Solution Incorporating Wind Turbines and Carbon Emissions," 2021 Innovations in Power and Advanced Computing Technologies (i-PACT), Kuala Lumpur, Malaysia, 2021, pp. 1–6, doi:10.1109/i-PACT52855.2021.9696806.

[5] Azlina Abdullah, Ismail Musirin, Muhammad Murtadha Othman, Siti Rafidah Abdul Rahim, A.V. Sentilkumar. (2023). Multi-DGPV Planning Using Artificial Intelligence. International Journal of Intelligent Systems and Applications in Engineering, 11(4s), 377–391. Retrieved from https://ijisae.org/index.php/IJISAE/article/view/2677

[6] Swati Gupta, Akash Saxena, Bhanu Pratap Soni, Optimal Placement Strategy of Distributed Generators based on Radial Basis Function Neural Network in Distribution Networks, Procedia Computer Science, Volume 57, 2015, Pages 249–257, ISSN 1877-0509, https://doi.org/10.1016/j.procs.2015.07.478.

[7] Pujari, Harish Kumar. "Optimization Techniques and Algorithms for DG placement in Distribution System: A Review." (2020).

[8] Maya K N and E. A. Jasmin, "Multi-stage formulation of optimal Distributed Generation placement using Reinforcement Learning," 2016 IEEE International Conference on Power Electronics, Drives and Energy Systems (PEDES), Trivandrum, India, 2016, pp. 14, doi:10.1109/PEDES.2016.7914234. keywords: Power systems;Distributed power generation;Learning (artificial intelligence);Uncertainty;Optimization;Minimization;Government,

[9] T. Cover, "Estimation by the nearest neighbor rule," in IEEE Transactions on Information Theory, vol. 14, no. 1, pp. 50–55, January 1968, doi:10.1109/TIT.1968.1054098.

[10] Azawii, Ali and Al-Janabi, Sameer and Abdulteef, Omar. (2022). Artificial Intelligence Techniques Applied on Renewable Energy Systems: A Review. 10.1007/978-981-19-0604-625.

[11] Gandomkar M, Vakilian M and Ehsan M. 2005 A genetic–based tabu search algorithm for optimal dg allocation in distribution networks Electric Power Components and Systems 33 1351–1362.

[12] Singh R and Goswami S 2009 Optimum siting and sizing of distributed generations in radial and networked systems Electric Power Components and Systems 37 127–145.
[13] Wang L and Singh C 2008 Reliability-constrained optimum placement of reclosers and distributed generators in distribution networks using an ant colony system algorithm IEEE Transactions on Systems 38 757–764.
[14] Rao R S, Ravindra K, Satish K and Narasimham S 2012 Power loss minimization in distribution system using network reconfiguration in the presence of distributed generation IEEE transactions on power systems 28 317–325.
[15] A. Alhamali, M. E. Farrag, G. Bevan and D. M. Hepburn, "Determination of optimal site and capacity of DG systems in distribution network based on genetic algorithm," 2017 52nd International Universities Power Engineering Conference (UPEC), Heraklion, Greece, 2017, pp. 1–6, doi:10.1109/UPEC.2017.8231996.
[16] Truong KH, Nallagownden P, Elamvazuthi I, Vo DN (2020) An improved meta-heuristic method to maximize the penetration of distributed generation in radial distribution networks. Neural Comput Appl 32(14):10159.
[17] Nagaballi S, Kale VS (2020) Pareto optimality and game theory approach for optimal deployment of dg in radial distribution system to improve techno-economic benefits. Appl Soft Comput 92:106234.
[18] Almabsout EA, El-Sehiemy RA, An ONU, Bayat O (2020) A hybrid local search-genetic algorithm for simultaneous placement of dg units and shunt capacitors in radial distribution systems. IEEE Access 8:54465.
[19] Samala RK, Kotapuri MR (2020) Optimal allocation of distributed generations using hybrid technique with fuzzy logic controller radial distribution system. SN Appl Sci 2(2):1.
[20] Selim A, Kamel S, Nasrat LS, Jurado F (2020) Voltage stability assessment of radial distribution systems including optimal allocation of distributed generators. Int J Interact Multimedia Artif Intell. https://doi.org/10.9781/ijimai.2020.02.004.
[21] Jayasree M, Sreejaya P, Bindu G (2019) Multi-objective metaheuristic algorithm for optimal distributed generator placement and profit analysis. Technol Econ Smart Grids Sustain Energy 4(1):1.

[22] Jalili A, Taheri B (2020) Optimal sizing and sitting of distributed generations in power distribution networks using firefly algorithm. Technol Econ Smart Grids Sustain Energy 5(1):1.
[23] Aravinth A, Vatul VA, Narayanan K, Muthukumar K, Senjyu T (2019) A multi-objective framework to improve voltage stability in a distribution network. Int J Emerg Electr Power Syst. https://doi.org/10.1515/ijeeps -2018-0239.
[24] Dehghani M, Montazeri Z, Malik O (2020) Optimal sizing and placement of capacitor banks and distributed generation in distribution systems using spring search algorithm. Int J Emerg Electr Power Syst. https://doi.org/10.1515/ijeeps-2019-0217.
[25] Lee, Soo-Hyoung and Park, Jung-Wook. (2009). Selection of Optimal Location and Size of Multiple Distributed Generations by Using Kalman Filter Algorithm. Power Systms, IEEE Transactions on. 24. 1393 – 1400. 10.1109/TPWRS.2009.2016540.
[26] Krueasuk, Wichit and Ongsakul, Weerakorn. (2006). Optimal Placement of Distributed Generation Using Particle Swarm Optimization. Australasian Universities Power Engineering Conference - AUPEC.
[27] SALKUTI, Surender Reddy; BATTU, Neelakanteshwar Rao. An effective network reconfiguration approach of radial distribution system for loss minimization and voltage profile improvement. Bulletin of Electrical Engineering and Informatics, [S.l.], v. 10, n. 4, p. 18191827, aug. 2021.
[28] Gadaram Meerimatha, Gattu.Kesavarao, Nagalamadaka.Sreenivasulu, "A Novel Distribution System Power Flow Algorithm using Forward Backward Matrix Method," IOSR Journal of Electrical and Electronics Engineering (IOSR-JEEE), Vol. 10, Issue. 6, pp 46–51, 2015.
[29] M. A. Mhawesh and O. M. Neda, " A Review of the Photovoltaic System Converters and Algorithms," Journal of Global Scientific Research, vol. 6, no.4, pp. 1297–1308, 2021.
[30] Ghanbari, Niloofar and Mokhtari, Hossein and Bhattacharya, Subhashish. (2018). Optimal Distributed Generation Allocation and Sizing for Minimizing losses and Cost Function. 10.1109/IAS.2018.8544588.

3

State-of-the-Art Artificial Intelligent Techniques for Distributed Smart Grids

Trina Som

Researcher on renewable energy planning & power system, The Celestial Earth Development Council, India
E-mail: trinasom@gmail.com

Abstract

This chapter explores the role of smart grid systems in enhancing the performance, reliability, and sustainability of modern electrical distribution networks. It investigates the integration of Distributed Energy Resources (DERs), including renewable energy and storage systems, into the existing grid architecture. The study emphasizes the bidirectional flow of electricity and information enabled by smart grids, which facilitates real-time monitoring, demand response, and decentralized control. The chapter also discusses optimization strategies for distributed generation placement, energy pricing mechanisms, and load management through intelligent decision systems. By incorporating advanced technologies such as neural networks and Internet of Things (IoT) frameworks, smart grids are positioned as key enablers of both climate change mitigation and adaptation. Overall, the chapter outlines how intelligent grid solutions can contribute to energy efficiency, cost reduction, and enhanced grid resilience.

Keywords: Smart grid, distributed energy resources (DERs), renewable energy integration, load management, demand response, distributed generation (DG), energy efficiency, real-time monitoring, bidirectional power flow, intelligent decision-making, artificial neural networks (ANN), internet of things (IoT), voltage profile optimization, power loss minimization, energy

DOI: 10.1201/9788743809852-3

storage systems, climate change mitigation, climate change adaptation, energy pricing strategies, microgrid, grid resilience.

3.1 Introduction

Globally, power systems are undergoing a revolution as a result of the integration of various distributed components, such as distributed energy resources, electric vehicles, communication infrastructures, and advanced metering infrastructures, which would potentially improve the future power system's energy efficiency, reliability, management, and security. Further, these components are getting increasingly entwined with the Internet of Things.

These developments are anticipated to create a large "treasure trove" of data to enable smart grid applications which include, generation forecasting, fault detection distributed energy managing systems, grid health monitoring, and energy management in the home, among others. Techniques based on artificial intelligence may be used to automate and optimize the smart grid's performance using the new hardware components and data. Artificial intelligence solutions for supporting diverse applications in a distributed smart grid have received a lot of interest lately. Artificial-intelligence-based approaches are also used to promote the integration of customers' responses, energy storage devices, renewable resources, grid and management of household energy, and security. Because the smart grid incorporates a variety of players, including energy producers, markets, and consumers, artificial intelligence, with support from market liberalization, may be able to catapult the grid into an agent for overall societal welfare. However, the incorporation and arrangement of automated distributed devices on a large scale to create a genuinely smart grid is still plagued by a number of scientific hurdles. The future outlook in our quest to overcome these hurdles will be an important point of discussion in what follows.

3.2 Distributed Power Generations

As late as 10 years back, almost all electricity was generated in massive central power plants [1]; in another 20 years, a significant portion of this electricity is predicted to be generated by distributed power units with 10- to 100-fold smaller capacities. Due to their proximity to the consumer, such units are projected to better suit their demands [2]. Parallel to the adoption of distributed generation, the legal framework for the electricity sector is undergoing significant changes in a number of developed countries [3].

These reforms are motivated by a desire for greater liberalization in order to foster competition in the market. Decentralized energy generation must compete with centralized energy generation in this competitive setting. However, it is frequently argued that competitive electricity market regulations do not treat centralized and distributed generation equally [4].

These reforms are driven by a desire for further market liberalization and competitiveness. In this competitive environment, decentralized energy generation must compete with centralized energy. However, competitive electricity market regulations are not equitable to distributed generation vis-à-vis centralized generation is a reasonable and oft-repeated argument [4].

3.2.1 Why is distributed generation important?

Distributed generation encompasses a range of technologies that generate power close to or at the point of use.

Historically, electricity-generating systems have been constructed to yield energy from high-voltage levels to a lower-voltage level of consumer networks. Large generating units are interconnected with transmission networks. However, the power generation and distribution system is evolving toward a widespread network of small generators connected to distribution systems [5]. Network advancements are required to integrate this scattered generation efficiently. Active distribution network management must evolve from centralized to more dispersed system management. Infrastructures for information, communication, and control will be required as system administration becomes more complicated. Distributed generation management (DGM) is becoming mandatory in order to enable the distributed network operator (DNO) to address a variety of grid-related issues involving distributed generation units. Additionally, supervisory units for the DGM are advantageous for dealing with a variety of conditions associated with inverter operation within the grid [6].

Electric distribution systems worldwide are seeing an increase in the deployment of intelligent devices and systems owned by utilities and nonutility customers [7]. In response to public demand for more environmentally friendly alternatives, governments have established legal mechanisms such as renewable portfolio standards (RPS) and carbon trading markets that encourage electric utility companies to diversify their generation portfolios with renewable energy resources (RER). However, renewable energy resource (RER) technologies such as photovoltaic (PV) and wind present a number of difficulties. Energy resources, in particular, are stochastic and largely nondispatchable by nature; weather probabilistically dictates availability, and these resources cannot be reliably accessed. This loss of dispatchability impairs a utility's ability to meet demand and ensure system reliability. Hundreds of thousands of distributed energy resources (DER) can be aggregated by a single authority, such as a utility, system operator, or third-party aggregator, and are now widely adopted by the majority of consumers. Self-healing systems, microgrids, and distribution automation are all examples of new utility-owned asset deployments. Solar photovoltaic generation is a nonutility-owned asset [8], energy storage technologies installed behind the meter, and electric automobiles. While these deployments provide data and control points, existing centralized control architectures lack the flexibility and scalability necessary to integrate an increasing number or variety of devices [9]. The communication bandwidth, latency, and scalability requirements of a centralized control architecture preclude these new devices and systems from being used as active resources. Thus, distributed generations are a practical and superior choice for providing a far more automated service.

Large-scale distributed generating deployments may have an effect on grid-wide services like as frequency regulation and reserve allocation [10]. As a result, the grid gains smart grid capabilities, virtual power plants, and grid energy storage, such as electricity to gas stations. It may result in economic savings for utilities and customers alike, as well as environmental benefits. It will facilitate the restoration of electricity following outages and will facilitate the integration of dispersed generating resources, such as renewable energy, into the larger grid [11]. Existing cost-effective distributed generating technologies may be utilized to power homes and businesses using solar and wind as renewable energy resources. Additionally, distributed generating can capture energy that would have been squandered otherwise.

3.2.2 Distributed generation empowerment

Worldwide climate changes propensity to increase energy consumption in emerging countries and inexorably, though slowly, depleting fossil fuel supplies have elevated sustainable energy supply to a global issue that must be handled by virtually every aspect of human activity. Simultaneously, buildings continue to contribute significantly to the global energy balance. They typically account for between 20% and 30% of industrialized countries' overall primary energy demand. With growing awareness of the ecological consequences of energy consumption, the need for energy and environmentally conscious building design has become increasingly urgent [12]. In a retail electricity market that is becoming increasingly complicated, customers who effectively coordinate both their market activity and their usage of distributed energy resources, such as renewable energy and energy storage systems, may benefit significantly. As the marketable rooftop photovoltaic industry expands, the large, communally owned roofs of residential buildings enable economies of scale. The aggregated loads may result in flatter profiles and increased self-consumption due to their diversity and physical proximity. Additionally, this enables access to more advantageous retail provisions and a greater degree of independence from electricity retailers and networks. Numerous studies have examined how photovoltaic systems, energy storage along with load profiles might be aggregated. Roberts et al. noted that by aggregating the residential loads of three houses, self-sufficiency could be increased by 21% and consumption could be decreased by 15% [13]. Freitas et al. examined the aids of aggregating PV power production in multiple rooftops [14] considering a variable range of orientations and tilts, which may lead to a reduction in energy storage requirements.

There are several dispersed energy sources available, including solar energy, hydrogen fuel cells, micro turbines, and wind energy. The notion of distributed energy generation has been applied in a variety of locations with varying degrees of complexity. Plenty of solar energy resources with increasing retail prices and reduced PV system costs have made photovoltaic an attractive investment for many families, resulting in a continuous increase in residential PV adoption. It may come as a surprise that a recent techno-economic study on distributed photovoltaic systems made by Sommerfeldt and Madani depicted that two-thirds of research made no mention of building load [15]. However, a growing number of articles scrutinize the use of energy storage systems to increase SC for detached residences or commercial establishments.

Regulations and economic changes in the European Union (EU) are influencing the future of customer-focused energy policy. Regulations incentivize utilities to adopt innovative energy technology and improve their customers' energy efficiency. Simultaneously, consumers are becoming more aware of the different options available in energy markets, and new technologies are enabling the introduction of new products and services to the market [16]. Consumers will place a higher premium on supply security and the option of producing more cash through dispersed generation in the future. They also want more information to help them manage their energy consumption more effectively [17]. The recent decade has seen a resurgence of interest in distributed generation as a result of technical advancements and a shifting economic and regulatory environment. This is confirmed by the IEA, which identifies five major drivers of this evolution: advances in distributed generation technologies, constraints on the construction of new transmission lines, increased customer demand for highly reliable electricity, electricity market liberalization, and concerns about climate change [18]. Various DG technology choices can be designed for a given region or locale based on the quality and dependability of the power supply. Market participants are free to adapt their choice of DG technology to changing market conditions. In liberalized markets, the most significant worry about the flexibility of DG alternatives is the shifting economic situation. It is commonly agreed that DG is a feasible alternative for resolving overcapacity issues in utility distribution systems [19] in Sydney, Australia [20]. With numerous uncertainties and dangers in a deregulated environment, DG is advantageous owing to its low capital risk and cheap investment expenses.

A prerequisite for the new digital energy world dubbed Energy 2.0 by its creators is for customers to take a more active part [21]. Empowering consumers to be more active participants in the energy market is a challenge for utilities that should be aided by suitable regulatory frameworks. Customer empowerment is also critical for reaching the EU's ambitious energy security aims and advancing the world's sustainable development. The vast potential of renewable energy is constrained by the intermittent nature, unreliability, scalability, and high cost of solar-based storage devices. Storage is now required as part of the photovoltaic mix under new requirements. This is why we anticipate that the future period will be dominated by power electronics and energy integration [22]. Integrated nonconventional generating paired with storage, Internet-of-Things (IoT) devices, and enhanced monitoring and control will form the backbone of the contemporary grid [23]. The subject of digitalization through the use of power electronics will see an increase in novel applications. New trends in research and development are expected to expand to support future designs of modular and intelligent power-processing circuits [24]. This will pave the way for new entrepreneurial ventures and market prospects.

Despite these advantages, distributed generation has been unable to compete with the relatively affordable traditional power grid. To make technology cost-efficient, it is critical to incorporate a smart mode of operation.

3.2.3 Need for smart grid

With today's customer expectations and environmental concerns, utilities must work to minimize consumers' power costs and integrate green energy generation with conventional energy generation on a wide scale [25]. Additionally, it is widely established that daily growing demand overloads existing electrical grids, while typical solution strategies add complexity to existing networks. These crucial difficulties prompted the researchers to consider the smart grid as a more comprehensive and superior solution. Smart grid technology allows electricity providers to reduce (shed) some of the demand on the power grid during times when electricity use is very high and prices increase [26]. Additionally, it can educate people about the need to conserve energy. Additionally, the benefits of a smart grid include increased electricity supply efficiency and dependability, as well as the integration of more renewable energy into the existing network [27].

Additionally, it encourages the commercialization of electric vehicles and provides innovative solutions for clients to maximize their power use and carbon reduction.

The integration of distributed generation and smart grid-enabling technologies and concepts into power systems has been widely accepted by industry and academia as the key to achieving a more reliable, efficient, and secure grid that is also environmentally sustainable [28]. Smart grids are more self-sufficient and enhance the efficacy and efficiency of the transmission of electricity. Thus utility companies may be able to leverage existing infrastructure and avoid the need to build new power plants. Smart grids enable the safe amalgamation of renewable power sources into the grid in order to increase the power supply with energy generated and stored by consumers [29].

Thus, smart grids did not appear out of nowhere; rather, they emerged as a response to a need to upgrade the energy grid, make it more environmentally friendly, and enhance power distribution.

3.3 Smart Grid – Integration of Distributed Generation

While the conventional grid is limited to transmitting and distributing electricity, the contemporary grid is capable of storing, communicating, and making choices. The smart grid reimagines the modern grid as one that operates cooperatively, responsively, and organically [30]. Additionally, it may be characterized as a sophisticated self-healing two-way digital power flow system, adaptive, robust, and enduring, with foresight for prediction in the face of a variety of uncertainties.

The fast growth of DG in a variety of ways and capabilities is reshaping traditional distribution network planning. Despite the benefits of distributed renewable energy technologies, the incompatibility of distributed generation with existing distribution networks can create a number of economic and technical issues. Thus, effective distributed generation planning is critical to ensuring that the distribution network's performance meets the desired levels of voltage stability, power quality, dependability, profitability, and power loss reduction. Distributed energy management is enabled by a smart grid. The smart grid is envisaged as a planned countrywide network that utilizes information technology to effectively, reliably, and securely supply power.

The term "Smart Grid" does not have a broadly agreed definition. It can be explained in straightforward words or more complicated terms.

3.3.1 What is smart grid?

The term "smart grid technology" refers to a technology that has been created to maximize the advantages of utilities and their customers and to deliver affordable and dependable electrical services via the optimal use of available resources and smart tools. A smart grid is an intelligent network that integrates information technology and the existing power system network [31]. Due to its fault-finding and self-healing capabilities, the smart grid (SG) is viewed as a possible future network of power. Sustaining energy, integrating renewable energy sources, and developing a control system that is effective are critical factors that are considered for building an SG system. The smart grid is an opportunity to leverage modern communication and information technologies to transform the conventional electrical power grid. Smart refers to something that is clever, tidy, trim, fashionable, or automated; a grid is a network of electrical cables that supply power to specific spots. One may get a sense of what the smart grid is by reading [32].

The notion of the smart grid has developed from a vision to a goal that is gradually being achieved. As technology advances, gadgets and systems become capable of facilitating the development of a more intelligent grid. Concretized energy regulations enable governments to implement smart grid programs. Smart grid practices across areas show no rivalry but rather an ad hoc community of shared objectives and lessons learned.

3.3.2 Why are smart grid technologies attractive?

Ever since the implementation of smart grids began, the power sector has worked with a distinct separation of its generation, transmission, and distribution subsystems. In the conventional electric grid, around 20% to 30% of power is lost, owing to inadequate generating, transmission, and distribution operations. The smart grid (SG) represents a technological advancement over the traditional electric grid. Electromechanical control of the conventional grid results in a smart, intelligent system dubbed the "Smart Grid" (SG) [32]. A smart grid is a critical idea for providing consumers with an easy-to-integrate and dependable service. Smart grid is a critical concept for delivering reliable and easily integral services to consumers. It can rapidly detect and address issues within the power system, reducing the need for manual intervention. Moreover, it enables targeted responses to system faults or inefficiencies. Automatic metering systems within the distribution system enable utilities to remotely monitor the condition of consumers' premises, consumption records, and alarms [33].

The source does not have real-time knowledge about the service parameters at the terminating point; the system is a pipeline that is only one-way. As a result, the electric system is overengineered in order to resist the greatest predicted peak demand over its whole load. Because peak demand is seldom, a system structured around it is wasteful. Additionally, system stability has deteriorated as a result of the massive increase in demand for electricity and the lack of infrastructure investment. With safe boundaries depleted, any disruption in the distribution network or unanticipated increase in demand that results in component failures could result in catastrophic blackouts. Utility companies have implemented a variety of command and control functions to facilitate troubleshooting and maintenance of their high-cost upstream assets [34]. SCADA is a frequent and widely used example. Additionally, the utilities' (utility companies') inability to extend their capacity in response to rising electricity demand and the rapid rise in fossil fuel prices has accelerated the need to modernize the distribution network through the introduction of new technologies that aid in revenue protection [35]. However, the aging infrastructure, changes in climatic conditions, greenhouse gas emissions, carbon footprint, inefficient electrical networks, and higher power costs compelled the need for a smart grid.

The smart grid's promising characteristics include intelligent decentralized control, resilience, flexibility, sustainability, digitalization, intelligence, consumer empowerment, green energy, and smart infrastructure [36].

Additionally, researchers explore and uncover important enabler technologies in order to have a better understanding of the present state of smart grid development. This review study concentrated on architectural models aimed at demand response programs (DRPs), consumer empowerment (CE), and demand side management (DSM) [37]. Additionally, it covered communication technologies such as phasor measurement units, wirelessadvanced metering infrastructure, machine-to-machine communication, and supervisory control and data acquisition (SCADA). Additionally, applications in the field of power systems such as intergrid, micro, nano, and pico grid, distributed generation, and virtual power plants have been thoroughly reviewed [38]. Integration of renewable energy resources (RERs) as distributed generation (DG) is critical [39]. Additionally, this study analyzes the smart grid's architectural model, with a particular emphasis on consumer empowerment and prosumer engagement.

Recent years have seen a dramatic growth in the use of RERs inside the smart grid. Power electronics advancements enabled a regulated and straightforward RER's integration with the SG. By incorporating the RER idea into the SG, we can get more sustainable and dependable power at a lower cost than we can with a normal electric system. The RER has the ability to significantly contribute to meeting the rising energy demand [40]. The renewable energy sources interfaced models with the smart grids during their design are subject to vary depending on the installation position and other considerations. The RER is inherently unpredictable and extremely reliant on meteorological conditions. The needed variability and unpredictability of RERs add to the difficulty in its integration [41]. The aforementioned challenges in the RERs' integration process drew academics from throughout the world to this research.

Solar photovoltaic (PV) energy, biomass energy, and wind energy are the primary components of RER power generation technology. Among other renewable sources, solar PV is the strongly suggested unit for integration with the smart grid. Wind power, due to its decentralized attractiveness, is also an ideal option for SG integration [42]. Despite the fact that only about 20% of biomass can be converted, biomass remains an attractive option for grid electrification due to its low environmental impact [43]. Wind energy is the fastest-growing energy source on the planet.

The RER interface is achieved by the use of power converter technology. To generate energy from RER, a structure that is cohesive and well-managed is required [44]. In comparison to a centralized system, a decentralized system is safer and more trustworthy. Due to the fact that RERs are highly dynamic and complex systems, the key challenge with RER integration becomes optimization [45]. The primary problems and variables impacting RER integration are dependability, quality of power, voltage profile, and efficiency. Voltage fluctuations occur as a result of RER's unpredictable nature, which is the primary impediment to renewable energy's integration with the smart grid [46]. The authors demonstrated the integration difficulties inherent in storage devices [47]. Each component of the storage system had its own control mechanism, which resulted in dumps and sluggish reaction times. Significant aspects, including economy, quality, and dependability, were also stressed while discussing smart grid storage technology. The authors explored the security difficulties surrounding RER in SG applications in [48, 49] and discussed integration difficulties with an emphasis on solar photovoltaic and wind energy. Solar photovoltaic integration requires power conditioning equipment because of the fluctuation on the DC output side. Researchers in [50] propose an AC–DC hybrid microgrid that employs interfaces for power electronics. [51] offered a structured hybrid with storage technologies along with forecasting based on neural networks. The meteorological department provided the data, and the multiagent system was used to perform the plans.

Thus, by developing a more effective controller and interface technology for the integration of renewable sources, the efficacy of RER may be increased.

However, several unresolved issues remain in SG technology, including a lack of knowledge, customer adoption, data gathering systems, energy metering, energy control, cyber terrorism, and dynamic optimization. Considering the foregoing, a full examination of the SG's development, technologies, and procedures should be conducted. Additionally, the vision for the future power grid necessitates a novel approach to power system protection. This requirement stems from the dynamic nature of the future smart grid, which has necessitated the development and implementation of a vast variety of protective approaches and processes over the years. Particularly, research into adaptive power protection approaches is accelerating since it is very desired for the power system to adjust its protection settings in real time in response to system operating states and situations.

3.3.3 Smart grid as a better future technology

Any substantial modification to the system of traditional power requires a thorough explanation and is overpriced due to the magnitude of investment done over the years and the power system's enormous size. Climate experts unanimously agree that greenhouse gases produced by humans are causing hazardous climate change. In terms of climate change, the capabilities of the smart grid are numerous. It supports both climate change adaptation (CCA) and climate change mitigation (CCM) [52] through institutional and behavioural measures, such as energy conservation and demand management, as well as through technological means, including the integration of renewable energy sources.

The existing method of transmitting and distributing power has been demonstrated to be wasteful and unreliable. A smart grid has the potential to reduce greenhouse gas emissions by up to 211 million metric tonnes and is significantly more reliable than a traditional grid [53]. With the numerous manifestations of climate change and the ever-increasing need for energy, energy sustainability and environmental preservation have become global problems. As cities and countries improve technologically, power consumption increases to levels that may become unmanageable if left unmanaged. The smart grid enables the transition to more environmentally friendly technologies such as distributed generation and microgrids. It is critical that the general public be aware of the issue and that future researchers and policymakers pay due attention. The move to the smart grid creates new regulatory challenges, that may cross jurisdictional lines and necessitate increasing collaboration among state, federal, and municipal legislators and supervisors.

Population growth necessitates a high demand for vast resources, including energy and material consumption. As a result of this suffocating situation, many developing countries are forced to confront additional issues such as waste generation and climate change. To overcome this perilous scenario, a smart alternative known as Waste-to-Energy (WtE) technology emerges as a viable solution. While this may appear to be a promising solution at the moment, the availability of these resources and the technology's longevity are both unknown. A brief review of the contemporary nexus of waste and energy with the added value of climate change may help to clarify this connection and forecast its long-term viability. In this regard, smart grids are expected to play a vital role.

The future smart power grid, in which distributed generations (DGs) are strongly integrated and self-healing capabilities are required, is designed to supply stable and high-quality electricity in an economically and ecologically responsible way. The many domains of the smart grid conceptual model include but are not limited to the customer domain, the market domain, the utility domain, the operation domain, and the generation, transmission, and distribution domains.

3.3.3.1 Consumer domain

Finally, the customer is the party to whom the entire power infrastructure was built. Consumer domain actors enable customers to take control of their energy usage and generation. The customer domain is often defined by the utility meter and the energy services interface (ESI) [54]. The ESI streamlines the process of establishing secure utility-to- consumer connections. Additionally, the ESI can serve as a connection to facility-based systems, for example, a consumer's energy management system (EMS) or building automation system (BAS). The consumer domain is often subdivided into residential, building/commercial, and industrial subdomains.

3.3.3.2 Market domain

Interactions within the market's domain necessitate auditable, dependable, and traceable communications. They must adhere to the rules governing electronic commerce and integrity. The critical challenges in the market's domain are prices to each of the consumer subdomains by extending the range of DER signals, the expansion of aggregator capabilities, interoperability of market information throughout all utility companies and consumers, the simplification of market rules, and the evolution of price and energy characteristic communication mechanisms between and among markets and consumers.

3.3.3.3 Utility domain

The utility province offers services that aid electricity consumers, distributors, and producers in their business operations. These business operations span from standard utility services like invoicing and account administration to advanced consumer facilities like energy management and home energy generation. The utility domain interacts with the domains of operations, markets, and consumers [55]. While it is necessary to communicate with the domain of operations for situational awareness and control of the system, communication relating to the consumer and market domains is necessary to enable economic growth through the creation of "smart" services.

3.3.3.4 Operation domain

Today, the majority of these duties are performed by a regulated utility. The smart grid enables utilities to outsource a greater number of them; others may change with time. Regardless of the evolution of markets and utility domains, fundamental functions for planning and operating a "wires" company's service delivery points will remain [56]. EMS is used in transmission operations to analyze and run the transmission power system effectively and reliably, whereas equivalent DMS is used in distribution operations to analyze and operate the distribution system. The bulk generating domain should convey critical performance and service quality problems, such as generator shortage and failure. These communications may result in the channeling of power from other sources into the transmission system. Inadequate supply can be handled directly (as a result of operations) or inadvertently (via markets). The bulk generation domain now requires greenhouse gas emission reductions, increased use of renewable energy sources (RESs), as well as storage to mitigate RESs' variability [57].

3.3.3.5 Transmission domain

Typically, the SCADA system monitors and manages the transmission network, in which all components of the system are integrated, including monitoring devices, communication networks, and control devices. Transmission is a domain populated by actors such as power quality and sag monitors and remote terminal units (RTUs), fault recorders substation meters, phasor measurement units, (PMUs), protection relays, and substation user interfaces. Additionally, it comprises distributed energy resources (DERs) such as units of electrical storage or peak production [57].

3.3.3.6 Distribution domain

Historically, distribution systems were radial in layout, lacked telemetry, and practically all communication within the domain was conducted manually. The first sensor base in this area was a consumer with a telephone, whose contact triggered the field's dispatch to carry out the restoration of electricity [58]. Historically, numerous communication interfaces were hierarchical and unidirectional within this domain, but they can now be considered bidirectional, similar to what electrical networks are starting to do. The distribution domain of the smart grid operates in real time and connects closer ties to the operational domain. This manages power flows associated with more markets that are constantly changing, as well as other eco-friendly and safety concerns. The market field interacts with the distribution area in such a way that localized consumption and generation are impacted. As a result, market-induced behavioral changes may have implications for the structural and electrical systems for both the distribution and broader utility grids.

3.3.4 Technological components of smart grids

Smart grid components are essentially an architecture comprised of multiple levels of technology, innovation, intelligence, and new tools.

3.3.4.1 Interface for smart devices component

Smart devices are electronic gadgets that are typically devices or networks that are connected to another device or network via numerous wireless protocol variants and are capable of operating interactively and independently. Monitoring and control devices that are intelligent are involved in the development of mechanisms [59] for real-time data processing. These sources must be seamlessly integrated into the operation of both centralized and Distributed Energy Resources (DERs). Several prominent examples of smart devices include intelligent automobiles, intelligent doorbells, intelligent refrigerators, intelligent bands, intelligent thermostats, intelligent locks, intelligent phablets and tablets, intelligent watches, intelligent key chains, intelligent smartphones, and intelligent speakers [60].

3.3.4.2 Component of storage

Due to the unpredictable nature of renewable energy sources and the incompatibility of peak demand and supply, it is critical to develop ways for storing future-use energy. The storage component enhances the utility grid's dependability and resilience, as well as that of power users. Flow batteries, ultracapacitors, flywheels, pumped hydroelectricity, superconducting magnetic fields energy storage, and compressed air are all examples of energy storage technologies [61].

3.3.4.3 Component of technology for monitoring and control

Among the components of monitoring and control technology, self-monitor, self-heal, and flexibility certainty of the generation, as well as a smart intelligent network, are the well-known devices. Devices are capable of coping with dependability concerns, congestion, and instability. This new flexible grid system must be able to withstand shock (durability and reliability) and be sufficiently dependable to adapt to changes in its use in real time [62]. Intelligent energy-efficient devices and intelligent distributed energy resources (DERs) include monitoring and control capabilities. These technologies are self-aware and capable of acting autonomously in response to their context awareness.

3.3.4.4 Component for intelligent grid distribution

The network of distribution is the final stage in the process of delivering electricity to customers. Small industrial, commercial, and residential clients are served through primary and secondary distribution feeders. Intelligent assistance schemes will incorporate capabilities for distribution-level monitoring. Smart meters are used to automate processes by establishing utility companies' communication channels with their customers, energy management components, and AMI [63]. Self-learning capabilities will be built into the automation function, which will include modules for fault detection, automatic billing, feeder reconfiguration, restoration, and voltage and load optimization, based on real-time pricing (RTP).

3.3.4.5 Intelligent metering component

A smart meter is an energy or gas meter that is capable of both metering and communication. It collects data on energy use and enables remote reading and display on a device within the home or secure transmission [32]. Additionally, the meter may be updated remotely, for example, to move from a credit-based to a prepayment system or to change pricing information. It serves two critical functions, such as (i) giving energy consumption data to customers to aid in cost and consumption control; and (ii) data transmission to the utility in order to ensure that peak-load requirements are met, control the load factor, and develop pricing strategies based on consumption data. Smart meters must have automated data collection and communication between utilities and consumers in both directions [64]. Smart meters are being developed to track data on energy, gas, and water use. Smart meters offer users information on how and when they spend energy, as well as the cost per kilowatt hour of electricity.

3.3.4.6 Demand management component

The authors develop options for DSM and energy efficiency for adjusting consumer demand. This is done in order to reduce operating costs by avoiding the utilization of costly generators and deferring capacity expansion [65]. DSM alternatives improve generating dependability and help cut emissions. These alternatives have a cumulative effect on the utility load curve. It is critical to have a standardized protocol for customer distribution via two-way information highway technology. There will be energy-efficient buildings and smart homes, as well as requirements for clean air, demand-side meters, plug-and-play, and enhanced energy efficiency through consumer interfaces [66].

The deployment of wireless objects or devices is a critical component of a large number of network-based applications. The items might be smart meters embedded in a smart grid, sensors or actuators in a wireless sensor network, or Internet of Things devices [67]. Following the physical deployment of such devices, another critical operation is network address assignment, which enables network connections among the devices for device control or message delivery reasons.

3.4 AI in Smart Distributed Generation

Artificial intelligence and information technology are emerging technologies in the power sector. Electrical networks are extremely complex and extensive; the data generated by generation, transmission, and distribution is abundant and must be processed for increased efficiency and reliability. To improve their efficiency, power grids are incorporating and adapting to new technologies [68]. The Internet of Things (IoT) is a modernized technology that enables the collection of data and the monitoring of physical components that are not visible to grids. Numerous works [69] discuss the use of modern technology in smart grids to improve their reliability and efficiency of operation. Numerous researchers have discussed the concept of distributed generation and the application of IOT in such a scenario [69].

The exponential growth of distributed generation in a variety of methods and capacities is undergoing the transformation of traditional distribution network scheduling. Despite the numerous benefits associated with renewable distributed generation technologies, when distributed generation is not properly integrated into already-established distribution networks, it can result in a number of economic and technical challenges. Thus, optimal distributed generation planning is critical to ensuring that the distribution of the network performs as expected in terms of voltage stability, power quality, reliability, reduction in power losses, and profitability. Many authors discussed numerous traditional and meta-heuristic techniques for the problem of optimal planning of distributed generation [70]. Frequently, metaheuristic methods are used because they provide greater flexibility, especially, when dealing with multiobjective planning problems where a globally optimal solution is not required. Analytical methods are deemed suitable for modeling the mechanisms underlying power systems and validating numerical methods.

3.4.1 Role of AI in smart distributed generation

Renewable energy's maturity and increasing availability are effectively relocating power generation closer to where it is required, resulting in the development of microgrids, decentralized power grids on a small scale, or in conjunction with the primary electrical grid of an area [71]. Microgrids are typically self-sufficient in terms of energy, metering, generation, and loads, in addition to definable boundaries that effectively scale them down to a small localized power station that is disconnected from, but connected to, the primary electrical grid. Cities that are intelligent leverage cutting-edge technological advancements in existing services, those based on cloud computing, edge computing, the Internet of Things, smart systems, and smart grids [72].

Numerous benefits associated with various renewable energy-based DG technologies are classified according to their economic, technical, and environmental characteristics. Grid reinforcement, reliability, enhanced power quality, reduction in power losses, and voltage security and stability are the technical advantages. Economic benefits include cost savings associated with transmission and distribution operations and a decrease in emissions with savings in fossil fuel prices. This, finally, leads to environmental aids including greenhouse gas reduction and natural resource conservation. Despite the numerous benefits of distributed renewable energy resources, various technical and economic challenges continue to exist in integrating distributed energy generation into existing power distribution networks [73]. Adequate placement of DG and their sizing can lead to significant voltage instability, degraded power quality power losses, and inadequate shielding in electricity distribution networks.

Distributed generation's optimal sizing and location are required to ensure the distribution network's operational effectiveness in terms of voltage stability, power quality, profitability, and reliability. The optimal decision-making problem for DGs has been studied in terms of a variety of economic and technological objectives, including minimization of loss, optimization of the voltage profile, and cost-benefit analysis [74]. There are a variety of available solution techniques, including traditional methods (e.g., optimal load flow, analytical techniques, probabilistic approaches, and exhaustive analysis) and metaheuristic techniques (e.g. GA, PSO, and AN optimization). In essence, optimal distributed generation planning entails determining a set of distributed generation variables in decision-making, such as location, size, and type, in order to minimize or maximize an objective function. These objectives are determined in accordance with a number of equality and inequality constraints. Technical objectives, such as optimizing voltage profile and restricting losses with economic objectives of maximizing profit [75], are used. These techniques fall into two broad categories: traditional methods and metaheuristic techniques.

3.4.1.1 Conventional techniques

Analytical techniques analyze the power network of distribution mathematically, generating a collection of numerical equations formed through which an objective function is derived. These techniques are frequently simple to implement and ensure that the DG planning solution is convergent [76]. These techniques can be used to achieve a variety of goals, including reduction of power loss and enhancement of voltage profile [75]. Additionally, several authors proposed a logical procedure for determining the optimal location and size of distribution gateways within the primary distribution network to minimize power losses. The authors developed a set of analytical equations for determining and minimizing power loss through the optimal size and placement of different types of direct current generators. In [76], an analytical method based on multiple objective indices was proposed to determine the photovoltaic DG units' optimal capacity while reducing the voltage deviation along with active and reactive power losses. An exhaustive study optimizes a sole objective, such as minimization of losses, by thoroughly searching the solution population space for all candidates corresponding to DG capacities and distribution network locations. The authors of [77] conducted analysis by employing a multiobjective index, to determine the optimum DG sizes and locations. This further helps to evaluate and minimize the active, as well as reactive power losses, along with voltage drops in the distribution network. Extensive analysis was found to be computationally savvy when considering the location of a single DG unit in response to a particular load-generation scenario. However, when numerous DG units operating in a variety of load generation scenarios are considered, the intricacy of this strategy can be computationally prohibitively expensive. When multiple DG units with various scenarios for load generation are considered, the complexity of this approach can be prohibitively expensive computationally. Probabilistic techniques are used to address the indeterminate factors inherent in the DG planning process such as irregular generation, electricity consumption, and availability of renewable DG. [78] proposed optimal DG planning via a probabilistic-based multiobjective method in which a balance was struck between economic costs and pollutant emissions while taking load consumption and electricity price uncertainties into account. However, such probabilistic techniques frequently require a large volume of operational data and a high level of capability for data processing.

3.4.1.2 Metaheuristic algorithms

An algorithm that employs metaheuristics is an iterative investigation process that manipulates candidate solutions using a variety of concepts to maintain command of a subservient heuristic [79]. These methods are capable of taking into account various types of objectives, either single or multiple. However, their efficiency and effectiveness are strongly dependent on the parameters of optimization being adjusted. A heuristic's objective is to generate a solution in a reasonable amount of time that is adequate for solving the problem at hand. Heuristics are at the heart of the entire field of artificial intelligence and computer simulations of thought, as they can be used in situations where no known algorithms exist. In general, metaheuristics do not make use of explicit knowledge discovered through advanced machine learning (ML) models during the search process. During the search process, meta-heuristics produce a huge amount of data. When it comes to the target problem and the instance characteristics to solve, the data can be static. However, with enhanced capabilities, artificial intelligence (AI) is capable of performing or controlling tasks that are typically performed by humans due to their intelligence and discernment requirements.

3.4.2 How AI work

Interest in demand response (DR) has grown in recent years as a cost-effective means of enhancing the adaptability and thus the reliability of energy systems. Since the tasks associated with DR are extremely complex in combination with the frequent requirement for decisions made in near-real time, means that machine learning (ML), a subcategory of artificial intelligence (AI), has established itself as a critical technology for facilitating demand-side responsiveness. AI techniques are used to address a variety of issues, including determining the most appropriate set of consumers to respond to gaining knowledge of their characteristics and preferences, pricing, scheduling, and device control that are dynamic. Further, they are used in learning how to incentivize and reward DR participants in a fair and economically efficient manner.

As the modern power system evolves at a breakneck pace, more distributed smart grid components are required. These are communication infrastructure, smart metering infrastructure, distributed resources, and electric vehicles. These are now tightly coupled to the power system, encompassing a massive number of electrical networks and the communication system that underpins it. These components generate enormous amounts of data to automate and enhance the performance of the smart grid by enabling a variety of applications, including distributed energy management [80], system state forecasting [81], FD, and cyberattack security [82].

Artificial intelligence (AI) is a term that refers to any device that is capable of perceiving its environment and further maximize its chances of success. The ideal state of artificial intelligence is one in which it thinks humanely and rationally, and acts humanely and rationally [83]. AI serves as a knowledge and database of rules, a decision-maker, an executor of an action, and a controller for appliances [83].

The qualitative inductive method consisted of several steps, each of which addressed one of the five fundamental functions of smart homes including the following: management of devices, health care, energy, interaction with intelligence, and security. Tang explained that intelligent buildings use artificial neural networks as intelligent decision-making forming expert systems [84]. According to this classification, intelligent buildings are capable of managing devices, optimizing energy use, supporting healthcare functions, enabling intelligent human-building interaction, and ensuring security. The term "data processing" encompasses a variety of techniques, which include semantic analysis, data mining, and rule-based technologies. In terms of product evaluation, there are six functions that are powered by artificial intelligence for smart homes. These are entertainment systems, energy management, health care, intelligent interaction, personal robots, and security [85]. The vastness of AI encompasses a variety of modes of operation, facilitated by a variety of learning techniques. These include the following: energy management systems, intelligent interaction, health care, personnel robots, entertainment, and security.

3.4.2.1 Machine learning (ML)

It is a substantial subfield of AI concerned with the discovery of existing data patterns. By analyzing the characteristics of input, machine learning algorithms can be used to construct models for specialized errands such as prediction, cataloging, and bunching. During the machine learning model's training phase, experiences are gathered from the statistical properties of the data. Numerous machine learning algorithms have been proposed thus far, and several of them have found widespread application. The authors advanced the concept of the support vector machine (SVM) [86], a machine learning technique that performs well on sorting and regression problems. SVM seeks to determine the hyperplane that is optimal, the decision limit is called, for categorizing the input figure points. SVM can process nonlinear division due to the kernel trick's introduction, which converts single-dimensional data to multidimensional data. The decision tree (DT) algorithm [87] is another well-known and interpretable machine learning method that makes decisions using a tree-like structure. It makes predictions through the use of a hierarchy of rules. To obtain a purer division while training DT, it is necessary to determine the optimal characteristics and thresholds for each node of decision. Additionally, the depth of the tree should be considered in relation to the overfitting problem.

3.4.2.2 Deep learning (DL)

DL is a subset of machine learning that focuses on directly processing large amounts of data and developing models for performing tasks from one end to another. It is not reliant on domain-specific knowledge or techniques for feature engineering; rather it includes extraction of useful features for data preprocessing [88]. In comparison to traditional ANNs used in shallow machine learning, deep neural networks (DNNs) have more hidden layers, which enhances the capacity of DL to represent intricate relationships. However, in the absence of adequate initial weights, the back-propagation algorithm cannot be used to train a DNN. Hinton et al. did not propose a greedy layer-wise pretraining method until 2006, which was based on deep belief networks (DBN) to solve such kinds of problems. This was a watershed moment in the growth of deep learning methods (DL) [89]. The convolutional neural network's success (CNN) in computer vision exemplifies the utility of deep learning (DL). Convolution and pooling layers in multiple pairs serve as filters and allow the extraction of structural features of varying resolution in image recognition input images. However, when a CNN with multiple layers is trained, the overfitting problem becomes apparent. The dropout technique has been shown to be effective of regularization that aids in avoiding overfitting [90]. In the field of natural language processing, recurrent neural networks (RNNs) are frequently preferred. The long short-term memory (LSTM) RNN is a widely used type of RNN, which contains memory cells for storing history. These are continually updated in accordance with the gating mechanism. The memory cell makes an attempt to deduce data's long-range dependencies, which enables LSTM to outperform conventional RNNs in the processing of large time series. A human-inspired attention mechanism [91] is frequently advantageous when dealing with extremely long time series. Critical information will be prioritized which is directly related to output and helps improve performance.

3.4.2.3 Reinforcement learning

In contrast to supervised learning, which requires labeled training samples, reinforcement learning (RL) is capable of training an agent without requiring labeled training samples. This is done via the agent's information interaction with the environment. Typically, Markov decision processes (MDPs) are used to model RL problems. This consists of the following elements such as action at, policy p, state st, and reward R. In general, reinforcement learning algorithms can be classified as model-based types and model-free types. The distinction is whether the agent must be aware of or learn the model of the environment [92]. World Models, MBMF, and Alpha Zero are all popular model-based reinforcement learning algorithms. However, inconsistencies exist between reality and the learned model. The resulting error may be used by a trained agent along with the learned model to limit the failure in the desired results of the real world. Rather than needing a model of the environment, through direct information interaction, model-free reinforcement learning algorithms optimize agent parameters. This is done in the real world, which is easier to apply and is more frequently utilized in various applications.

Due to the inability of conventional computational techniques to handle the massive amounts of data generated by intelligent grid systems, artificial methods have garnered considerable attention. Numerous research efforts have been devoted to studying these artificial intelligence techniques to address the challenges, as they utilize large-scale data to enhance the performance of the smart grid. The artificial intelligence techniques used in the smart grid can be broadly classified into the following categories:

(i) ES: A human proficient in the loop technique that is used to solve specific problems.
(ii) Supervised learning: A subset of artificial intelligence models that examines the mapping between inputs and outputs in order to forecast the outputs of new contributions.
(iii) Unsupervised learning: Unlabeled data are used to represent the similarity and dissimilarity among the data in a machine learning class.
(iv) Reinforcement learning (RL): The intelligent agent's strategy distinguishes it from administered and unsupervised learning, which is oriented around the concept of cumulative reward maximization.
(v) Ensemble methods: It combines the results of several artificial intelligence algorithms in order to overcome the limitations of a single algorithm and achieve a higher overall performance.

3.4.3 Artificial intelligence in practical application

AI applications are now the most critical and rapidly evolving concept in significant manufacturing sectors on the environment and sustainability, renewable energy sources (solar, hydropower, wind, and biomass), the catalysis industry and smart grids, and energy storage and distribution. Due to the generalization capabilities of AI networks, they are preferable to other machine learning algorithms. Demand for machine learning techniques in the energy sector will increase significantly in the coming years, as academic programs related to artificial intelligence in science, math, and engineering continue to grow in popularity. Data generation, management, and security are expected to be critical to the successful implementation of machine learning algorithms that can be shared among major energy stakeholders, thereby promoting the development of ambitious energy management projects [93]. New algorithms will facilitate the exchange of data between machine learning and systems for the generation of reliable data and the addition of additional sources of data.

While artificial intelligence (AI) approaches have been applied to a variety of power system applications, they have only recently begun to generate significant research interest in the field of demand-side response. Demand response (DR) has been identified as one of the most promising approaches for providing demand flexibility to the power system; consequently, many system operators place a premium on expanding the scale and scope of DR programs [94]. This enhanced functionality of DR schemes necessitates an automated framework that is adaptable and capable of learning in a dynamic environment.

Numerous optimization techniques are available in the literature for energy problems in general, including methods of conventional and intelligent search. The quest for the DG's optimal capacity and location is typically modeled mathematically as a problem of nonlinear optimization. To begin, a variety of constraints and objective functions are defined by generating a single optimal solution or a collection of superior solutions or variables utilizing a condensed initial set of input variables. The optimization technique aids in decision-making.

By integrating microgrids into smart city processes via renewable energy, efficient generation, smart metering, and optimal load management, the platform's capabilities are enhanced while incurring minimal additional costs. Artificial intelligence (AI) has emerged as the most promising concept for resolving these issues. When solving real-world simulation scenarios, scalable Bluewave-ai generally collects a list of data sources required to enhance AI capabilities [95]. This is accomplished through the use of deep learning in the data center and AI inference at edge computing nodes and Internet of Things sensors.

Underprovisioning of power, overprovisioning of power, virtual stored charge value, energy usage prediction, and savings estimation of net savings or losses associated with discharging available charge in batteries based on the price of energy stored initially, and compared to future energy generation and usage predictions are all addressed through the application of AI to a microgrid power system. Six theoretical energy pricing levels ranging from the cheapest to the most expensive can be realized with on-site renewable energy and on-site battery storage. Six theoretical energy pricing levels, ranging from the cheapest to the most expensive, can be realized with on-site renewable energy and on-site battery storage, followed by community grid energy, community battery storage, off-peak main grid energy, and finally, main grid energy during peak times. BluWave-ai Edge and data simulation have been studied and implemented using AI.

Several authors have also proposed and implemented artificial intelligence-based optimal distributed generation placement in smart power distribution systems. Prior to implementation, the interfacing of DGs with power distribution grids was considered to ensure optimal operation with stability, control, and efficient energy management. Additional constraints such as line loss reduction, voltage stability enhancement, reliability enhancement, economic benefits, enhanced security, and environmental pollution optimization were addressed [96]. Tabu search, ant colony method, and particle swarm optimization techniques were used to achieve the objectives of minimizing investment and operation costs, as well as minimizing power loss and customer energy demand loss. Minimizing losses associated with line loading and the total reactive power capacity requires the relationship between line loading and reactive power losses, investment costs associated with DG sources, costs associated with the operation and maintenance of DG sources, and cost of purchasing energy from the transmission grid loss costs. The objective function is constructed by combining indices indicating the effect of the DG on real and reactive power losses, voltage profile, and MV. The objective function is formulated by incorporating indices that reflect the impact of distributed generation (DG) on real and reactive power losses, voltage profile, and minimum voltage (MV), while also considering conductor capacity and the system's short-circuit level to ensure reliable and efficient network operation.

Uncertainty and changeability are the key challenges related to renewable energy sources, particularly with the intermittent availability of solar, wind, and hydropower. To enable the integration of a significant portion of renewable energy sources, it is critical to have suitable planning tools capable of optimizing the integrated renewable energy sources that are variable in nature.

GA was used to reduce the cost of system expansion and increase system reliability [97]. Due to the fact that these objectives are incompatible, the appropriate solution was found in a single run using pareto-optimal models.

3.4.3.1 Sizing and optimal allocation of DG

In the DG sizing and siting problem, novel improved techniques are proposed in which GA is used in conjunction with the multiattribute decision-making (MADM) to account for the power system's various parameters. Other enhanced GA methods include the adaptive genetic algorithm and quantum genetic algorithm mentioned in [98] found to be more tenacious and possess a greater capacity for searching. Simulated Annealing was used in [99] to identify and characterize the capabilities of DGs while requiring less computation time than GA or TS. TS was instrumental in resolving the problem of locating and sizing DGs. Chauhan et al. concentrated on optimizing the DG planning process with the goal of reducing loading and line losses. The PSO method is used to ascertain the optimal type, site, and size of distributed generators. Harmonic limits and protection constraints are taken into account during DG integration [100]. Additionally, a PSO was used to improve the voltage profile while reducing total harmonic distortion (THD), cost, and losses. In the problem of sizing and siting of DGs, new enhanced PSO methods are being proposed, including improved PSO, binary PSO, PSO with inertia, PSO through constriction factor, and social learning-based PSO [101]. Authors in [102] proposed ACO to address the issue of sizing and siting DGs derived from renewable energy sources in systems of radial distribution while minimizing system-wide losses. In [103], HS when combined with the factor of loss sensitivity, then the technique was used to calculate the optimum position of DG. The CSA (Cuckoo search algorithm) was applied in [104] to optimize the voltage distribution and minimize power losses in solar-thermal DGs and direct-drive biomass units. The SFLA (shuffled frog leaping algorithm) was successfully applied to the issues regarding the size and allocation of DG. The shuffled bat Algorithm is based on microbats' behavior involving echolocation. This proposed algorithm was implemented [105] considering a radial distributed network, first at 100 percent base load and then at 120 percent. The plant growth simulation algorithm (PGSA) was effectively used to enhance the voltage profile and reduce power losses. The primary advantage of PGSA is that it can operate regardless of actual or reactive energy losses or loading. The imperialist competitive algorithm was implemented to ensure the proper operation of sensitive loads in a distribution network's islanding mode [106].

3.4.3.2 Forecasting

Load forecasting is a critical component of smart grids because it uses artificial intelligence to forecast the required electrical power (AI). Deep learning is widely used in the smart grid for load forecasting via an artificial neural network (ANN). Generally, deep learning requires massive resources for the aggregation or centralization of data and a significant computational cost in terms of time of computation in the smart grid.

The machine learning-based smart meter system is effective at detecting daily living activities in the ambient assistive living (AAL) space [107]. Machine learning has been applied to smart meters in order to improve end-user load modeling [108]. In smart power meters, AI is also combined with edge computing and analytics [109]. At the moment, the success of the smart grid, smart home, and smart meter is dependent on artificial intelligence and communication security [110]. Swarm intelligence, deep learning, and machine learning are all examples of common artificial intelligence approaches that find widespread use in smart grids. Machine learning is being used in the smart grid to analyze large amounts of data and to address security concerns regarding the Internet of Things (IoT).

Accurate forecasting of solar power/irradiance is critical for ensuring the economic operation of the smart grid [111]. To forecast load, appliance power consumption data has been incorporated, using the fuzzy logic method [112]. Long short-term memory recurrent neural networks (LSTM-RNN) models were developed using smart meter data in forecasting residential load in Australia [113]. Others have made strides toward comprehending and forecasting customer behavior in demand response. The nearest neighbor and Markov chain algorithms were used in [114] to forecast user behavior and energy management. Recent studies on load forecasting have shifted their focus to deep learning techniques [114]. A time-dependent CNN along with cycle-based LSTM was used for load forecasting. The LSTM-RNN method has been found to be robust for forecasting load data. The forecast of the medium-term load has been found to be complex, which necessitates the use of superior techniques such as deep learning to low ANN. Regardless of this recommendation, an ANN was implemented in a recent SA load forecasting [115]. Semero et al study focused on forecasting of short-term load, of about 30 minutes fast using ANFIS [116]. In another recent study, Motepe et al.

discovered that DBN models performed better for the same load profile in [117] considering the temperature.

3.4.3.3 Stability

Evolving artificial intelligence methods offer influential and promising gears for analyzing and controlling the stability and operation of smart grids, and have gained increasing attention. In smart grids, security, fault diagnosis, and stability control are all assessed. As stable operation is a critical criterion for SG, hence these usages have accomplished remarkable feats. The addition of new structures has resulted in the smart grid's stability characteristics being distinct and significantly more complex than those of traditional energy grids. Under these conditions, stability analysis in the conventional sense and control methods became ineffective. Alternatively, AI is capable of adapting to these changes and makes promising tools available for meeting the smart grid's stringent security and stability requirements.

The voltage stability of power systems has been assessed using a support vector machine (SVM) and biogeography-based optimization (MOBBO). The optimal allocation of capacitor banks (CBs) with the goal of minimizing active power loss and optimizing bus voltages was accomplished using the salp swarm algorithm SSA. IEEE 33 and 69 bus radial distribution systems are used to validate the proposed algorithm [118]. The simulation results demonstrate the proposed algorithm's efficiency in comparison to other existing optimization algorithms.

A technique called multidirectional long short-term memory (MLSTM) is being proposed for the purpose of forecasting the stability of intelligent grid networks [119]. The obtained results are compared to those obtained using other well-known deep learning techniques such as traditional LSTM, gated recurrent units (GRU), and recurrent neural networks (RNN). Experiments demonstrate that the MLSTM approach outperforms all other machine learning approaches [120].

AI and machine learning have the potential to enable the smart grid to make sound judgments and to react intelligently to unexpected changes in customer demand, power outages, abrupt drops and increases in renewable energy output, or any other catastrophic event. Additionally, machine learning can assist in capturing customer consumption patterns, forecasting energy demand and intermittent energy generation, and predicting equipment failures.

Using feature extraction [121], a machine learning algorithm is used to forecast the stability of a smart grid. The authors here employ three distinct methods for selecting features. Selection of Binary Particle Swarm Optimization Features (BPSOFS), Binary Kangaroo Mob Optimization Features (BKMOFS), and Multivariate Adaptive Regression Splines (MARS). Four classifiers are used to forecast the grid's stability: logistic regression (LR), random forest (RF), gradient-boosted trees (GBT), and multilayer perceptron classifier (MPC) [122]. Several studies [123] used data-matching learning networks to predict the stability of decentralized power generation.

3.4.3.4 Fault diagnosis

Numerous factors contribute to the vulnerability of distributed generation systems to faults and errors. Among them, the most common are component failures in the power system and lightning strikes. Typically, power system failures are caused by component failure, human error, or aging equipment. The aforementioned events jeopardize the reliability of the power system, resulting in costly repairs, lost productivity, and dissatisfied customers. Due to the unpredictable nature of faults, rapid fault location and isolation are required to mitigate the impact of faults in distribution systems. As a result, numerous methods for locating and detecting these faults in distribution power systems have been developed previously by researchers. Artificial intelligence techniques such as artificial neural networks, fuzzy logic, genetic algorithms, support vector machines, and matching approaches are all examples of artificial intelligence techniques. However, fault detection using intelligent methods is challenging to implement because it requires training data and is time-consuming.

According to [124], the resistance of the fault and the distance between the occurrence of the fault, and the point of measurement have an effect on impedance-based methods. The estimation error increases as resistance and fault distance increase. The accuracy of impedance-based techniques is generally dependent on the line's parameters, the value of load, and its characteristics. Localization error is also dependent on the network's complexity. This is characterized by systems that are out of balance and varying degrees of resistance to faults. When multiple distances between faults are estimated, the protection devices are frequently used to ascertain the precise location [125].

To locate faults in EDS with DG's, the authors employ the AI technique with a standard backpropagation approach [126]. The training data for this method is derived from the current fed to each distributed generator under the fault condition. As a result, the method's accuracy is extremely reliant on the system's total number of DG sources. Further, SVMs have also been used to detect the presence of faults in EDS. These are based on measurements of voltage and current taken at the substation and from DG sources. When the network's number of DG sources is increased, this technique becomes more robust [127]. In [128], the authors demonstrate how to diagnose faults in the presence of DG, using the SVM in EDS.

The methodology proposed is demonstrated on a 132/11 kV substation's distribution feeder in India with loads distributed throughout and multiple DG resources. The proposed scheme for locating faults is capable of determining the type of fault, its location, and its impedance. The simulation result demonstrates that the classification and regression algorithms perform satisfactorily. The fault line's classification accuracy was 100, 99.95, and 92.06 percent, respectively, with three DG sources, two DG sources, and one DG source.

If effective restorations and identifications cannot be completed quickly, unknown faults may result in unnecessary costs. The identification of these faults has been studied using soft computing techniques such as artificial neural networks, fuzzy logic, support vector machines, genetic algorithms, and hybrid combinations [129].

3.4.3.5 Integrating renewable sources

Uncertainty and variability are the primary challenges associated with renewable energy sources, particularly with the intermittent obtainability of solar, wind, and hydropower. To facilitate the integration of a large proportion of renewable energy sources, it is critical to possess effective planning means. Numerous optimization techniques, such as conventional and smart search methods, exist in the literature for energy problems in general. The search for DG's optimum siting and volume can be modeled mathematically as a problem of nonlinear optimization problem.

DG sizing and siting problem, as demonstrated in [130], where multi-attribute decision-making (MADM) method was combined with a genetic algorithm considering various power system parameters. Other enhanced GA methods, including the adaptive genetic algorithm (AGA), have been portrayed in [131] as a dynamic searching ability same as quantum genetic algorithm [132].

Particle swarm optimization is used in [133] to estimate the optimum site, size, and type of distributed generator units in order to attain the best possible amalgamation of DGs considering harmonic constraints. Additionally, a PSO was used in [134] to improve the voltage profile while reducing the losses, total harmonic distortion (THD), and prices.

Numerous authors have investigated several meta-heuristic approaches for the optimum allocation of DGs and planning in the network of distribution. The techniques are the cuckoo search method, the symbiotic organism search algorithm, the artificial bee colony, the harmony search algorithm, bacteria foraging, the bat algorithm, the whale optimization algorithm, the grey wolf optimization algorithm, and the firefly algorithm [135].

3.5 Future Prospects

The new generation of artificial intelligence (AI), dubbed AI 2.0, has emerged as a research focus in recent years [136]. The development of smart energy and electric power systems will be accelerated by data-driven artificial intelligence 2.0. (Smart EEPS). Machine learning (ML) is a representative algorithm category in AI 2.0 that is used to make predictions and judgments by analyzing and learning from massive amounts of historical and synthetic data in order to assist humans in making optimal choices. Preliminary applications of machine learning have been made in the smart grid (SG) and energy internet (EI) domains, which are significant representatives of Smart EEPS. AI 2.0, particularly machine learning, is undergoing a critical stage of rapid development on a global scale and will be critical to Smart EEPS.

Additionally, with a focus on greener technology, it is necessary to comprehend weather parameters in order to enhance the performance of AI models in load forecasting. Further research is required to determine the effect of temperature inclusion on the performance of these techniques.

References

[1] Kabalci, Ersan, Aydn Boyar, and Yasin Kabalci. "Centralized power generation." Hybrid Renewable Energy Systems and Microgrids. Academic Press, 2021. 47-72.

[2] Willis, H. Lee, and Walter G. Scott. Distributed energy generation: planning and assessment. Crc Press, 2018

[3] Willis, H. Lee, and Walter G. Scott. Distributed energy generation: planning and assessment. Crc Press, 2018

[4] Sorin, Etienne, Lucien Bobo, and Pierre Pinson. "Consensus-based approach to peer-to-peer electricity markets with product differentiation." IEEE Transactions on Power Systems 34.2 (2018): 994-1004

[5] Carvallo, Juan-Pablo, et al., "Distributed resources alter paradigms in power system design, planning, and operation: An application of the GAP model." IEEE Transactions on Industrial Electronics 107.9 (2019): 1906-1922

[6] Santos, Lael N., et al. "A Distributed Generation Manager with Support for Distributed Network Operator Commands." International Conference on Industrial Technology (IEEE) 2020 (ICIT). 2020, IEEE.

[7] Obi, Manasseh, Tylor Slay, and Robert Bass. "Distributed energy resource aggregation using customer- owned equipment: A review of the literature and standards." Energy Reports 6 (2020): 2358-2369

[8] Alkaabi, Sultan S., Hatem H. Zeineldin, and Vinod Khadkikar. "Adaptive planning technique for customer distributed generating installations in smart distribution networks." IET Renewable Energy Generation, Vol. 12.1 (2018), pp. 81-89

[9] El Khaouat, Atimad, and Laila Benhlima. "Distributed integrated control system architecture for microgrids." 2018 19th IEEE Mediterranean Electrotechnical Conference (MELECON). IEEE, 2018

[10] Razavi, Seyed-Ehsan, et al., "Impact of distributed generation on distribution system protection and voltage regulation: A review," Renewable and Sustainable Energy Reviews 105 (2019): 157-167

[11] Wolsink, Maarten. "Co-production in distributed generation: renewable energy and the creation of space for infrastructure integration within landscapes." Landscape Research 43.4 (2018): 542-561

[12] Manfren, Massimiliano, Paola Caputo, and Gaia Costa. "Paradigm shift in urban energy systems via distributed generation: Methods and models." Applied energy 88.4 (2011): 1032-1048

[13] Roberts, Mike B., Anna Bruce, and Iain MacGill. "A comparison of arrangements for increasing self- consumption and maximising the value of distributed photovoltaics on apartment buildings." *Solar Energy* 193 (2019): 372-386.

[14] Freitas, S., C. Reinhart, and M. C. Brito. "Minimizing storage needs for large scale photovoltaics in the urban environment." *Solar Energy* 159 (2018): 375-389.

[15] Schreurs, Twan, et al. "Techno-economic analysis of combined heat pump and solar PV system for multi- family houses: An Austrian case study." *Energy Strategy Reviews* 36 (2021): 100666.

[16] Copping, Andrea, et al. "Maritime renewable energy markets: power from the sea." *Marine Technology Society Journal* 52.5 (2018): 99-109.

[17] Raj, Utkarsh, and Ravi Shankar. "Deregulated automatic generation control using novel opposition-based interactive search algorithm cascade controller including distributed generation and electric vehicle."

Iranian Journal of Science and Technology, Transactions on Electrical Engineering 44.3 (2020): 1233-1251.

[18] Distributed IEA. Generation in liberalised electricity markets. Paris; 2002. p. 128

[19] [Garau, Michele, et al. "Co-simulation of smart distribution network fault management and reconfiguration with lte communication." *Energies* 11.6 (2018): 1332.]

[20] Celli, Gianni, et al. "Multi-objective Modelling and Optimization for DG-Owner and Distribution Network Operator in Smart Distribution Networks." *Operation of Distributed Energy Resources in Smart Distribution Networks*. Academic Press, 2018. 249-283.

[21] Kashin, A. M. "Distributed generation based on cutting-edge technology and digital systems." естник оссиско академии наук 89.4 (2019): 326-330.

[22] Batarseh, Issa, and Khalil Alluhaybi. "Emerging potential in distributed power electronics and battery integration: Preparing for an energy storage revolution." IEEE Power Electronics Magazine, Volume 7.2 (2020), pages 22-32.

[23] B. K. Bose, "Power electronics smart grid and renewable energy systems," IEEE Proc., vol. 105, no. 11, pp. 2011-2018, 2017.

[24] F. Blaabjerg, Y. Yang, K. Ma, and X. Wang, "Power electronics: The critical technology for renewable energy system integration," Proc. International Conference on Renewable Energy Research and Applications (ICRERA), pp. 1618-1626, 2015.

[25] Kakran, Sandeep, and Saurabh Chanana. "Smart grid operations with distributed generation: A review." Renewable and Sustainable Energy Reviews 81 (2018): 524-535.

[26] Keles, Cemal, and Yusuf Yigit. "Implementation of demand side load management using a home load shedding algorithm for traditional energy metres." 2019 IEEE International Conference on Artificial Intelligence and Data Processing (IDAP). IEEE, 2019.

[27] Ourahou, Meriem, et al., "Review on smart grid control and dependability in the presence of renewable energy sources: Challenges and opportunities," Mathematics and Computers in Simulation 167 (2020): 19-31.

[28] Abdmouleh, Zeineb, Adel Gastli, and Lazhar Ben-Brahim. "Survey about public perception regarding smart grid, energy efficiency & renewable energies applications in Qatar." *Renewable and Sustainable Energy Reviews* 82 (2018): 168-175.

[29] Zame, Kenneth K., et al., "Smart grid and energy storage: Policy suggestions," Renewable and Sustainable Energy Reviews, vol. 82 (2018), pp. 1646-1654.

[30] Wang, Yongli, et al., "Energy management of a smart microgrid with response loads and distributed generation considering demand response," Journal of cleaner production 197 (2018): 1069-1083.

[31] Al-Badi, Abdullah Hamed, et al. "Survey of smart grid concepts and technological demonstrations worldwide emphasizing on the Oman perspective." *Applied system innovation* 3.1 (2020):

[32] Dileep, G. "A survey of smart grid technology and applications." Renewable energy 146 (2020): 2589-2625.

[33] Butt, Osama Majeed, Muhammad Zulqarnain, and Tallal Majeed Butt. "Recent advancements in smart grid technology: Future prospects in the electrical power network." Ain Shams Engineering Journal 12.1 (2021): 687-695.

[34] Leszczyna, Rafa. "Standards on cyber security assessment of smart grid." International Journal of Critical Infrastructure Protection 22 (2018): 70-89

[35] Zame, Kenneth K., et al. "Smart grid and energy storage: Policy recommendations." Renewable and Sustainable Energy Reviews 82 (2018): 1646-1654.

[36] Kim, Seong Cheol, Papia Ray, and S. Surender Reddy. "Features of smart grid technologies: An overview." ECTI Transactions on Electrical Engineering, Electronics, and Communications 17.2 (2019): 169-180.

[37] Nawaz, Abdullah, et al., "An intelligent integrated approach for efficient demand side management with forecaster and advanced metering infrastructure frameworks in smart grid." IEEE Access 8 (2020): 132551-132581.

[38] Shaukat, N., et al. "A survey on consumer empowerment, communication technologies, and renewable generation penetration within the Smart Grid." Renewable and Sustainable Energy Reviews 81 (2018): 1453-1475.

[39] Shaukat, N., et al., "A survey on consumer empowerment, communication technologies, and renewable generation penetration within the smart grid." Renewable and Sustainable Energy Reviews 81 (2018): 1453-1475.

[40] Vijayan, Vineeth, Abheejeet Mohapatra, and S. N. Singh. "Demand Response with Volt/Var Optimization for unbalanced active distribution systems." *Applied Energy* 300 (2021): 117361.

[41] Lokeshgupta, B., and S. Sivasubramani. "Multi-objective residential demand side management with solar and wind energy integration." 2019 8th International Conference on Power Systems (ICPS). IEEE, April 2019.

[42] Ahmed, Shakir D., et al., "Grid integration issues of wind energy: A study." IEEE Access 8 (2020): 10857-10878.

[43] Zheng, Yingying, et al. "Optimization of a biomass-integrated renewable energy microgrid with demand side management under uncertainty." Applied Energy 230 (2018): 836-844.

[44] Worighi, Imane, et al., "Integrating renewable energy in smart grid systems: Architecture, virtualization, and analysis." Sustainable Energy, Grids, and Networks 18 (2019): 100226.

[45] Worighi, Imane, et al., "Integrating renewable energy into smart grid systems: Architecture, virtualization, and analysis." Sustainable Energy, Grids, and Networks 18 (2019): 100226 .

[46] Ourahou, Meriem, et al., "Review on smart grid control and reliability in the presence of renewable energies: Challenges and prospects." Mathematics and computers in simulation 167 (2020): 19-31.

[47] Zame, Kenneth K., et al. "Smart grid and energy storage: Policy suggestions." Renewable and Sustainable Energy Reviews 82 (2018): 1646-1654.

[48] El Mrabet, Zakaria, et al. "Cyber-security in smart grid: Survey and challenges." Computers & Electrical Engineering 67 (2018): 469-482.

[49] Alam, Md Shafiul, et al. "High-level penetration of renewable energy sources into grid utilities: Challenges and solutions." IEEE Access 8 (2020): 190277-190299.

[50] Kazerani, Mehrdad, and Kambiz Tehrani. "Grid of Hybrid AC/DC Microgrids: A New Paradigm for Tomorrow's Smart Cities." 2020 IEEE 15th International Conference on System of Systems Engineering (SoSE). IEEE, 2020.

[51] Aktas, Ahmet, et al "Dynamic energy management for solar power systems in smart grid applications, incorporating hybrid energy storage. "Environment 162 (2018): 72–82.

[52] Chen, A. A., et al. "Pathways to climate change mitigation and stable energy from 100 percent renewable energy for a small island: Jamaica as an example." Renewable and Sustainable Energy Reviews 121 (2020): 109671.

[53] Ortiz, Paul, et al., "Greenhouse gas emission reduction system in photovoltaic nano grid with battery and thermal storage reservoirs," Journal of Cleaner Production 310 (2021): 127347.

[54] Ahene, Emmanuel, et al., "Efficient signcryption with proxy re-encryption and its use in the smart grid," IEEE Internet of Things Journal 6.6 (2019): 9722-9737.
[55] Shakeri, Mohammad, et al., "An overview of the building energy management system taking demand response programmes, smart strategies, and the smart grid into account." Energies 13.13 (2020): 3299.
[56] Ghosal, Amrita, and Mauro Conti. "Key management systems for smart grid advanced metering infrastructure: A survey." IEEE Communications Surveys & Tutorials 21.3 (2019): 2831-2848.
[57] Baleentis, Tomas, and Dalia treimikien. "Sustainability in the electricity sector through advanced technologies: Energy mix transition and smart grid technology in China." Energies 12.6 (2019): 1142.
[58] Chaves, Tiago R., et al. "IoT Device to Detect Fuse Status in Overhead Distribution Grids." IEEE 4th International Electrical and Energy Conference, 2021 IEEE (CIEEC). IEEE, c. 2021.
[59] Criado, Javier, et al., "Integrating cyber-physical systems in a component-based strategy for smart homes," Sensors 18.7 (2018): 2156.
[60] Dumitru, Andreea-Larisa Boboc, and Daniel Moise. "Marketing Study: Online Shopping Trends for Smart Devices." Bucharest 2018 (2018): 106.
[61] Nalamati, Chandra Sekhar. "Grid Energy Storage Technologies." Integration of Renewable Energy Sources with the Smart Grid (2021): 119-139.
[62] Attia, Shady. "Challenges and Future Directions of Smart Sensing and Control Technology for Adaptive Façades Monitoring." Lucerne University of Applied Sciences and Arts (2018): 505-514.
[63] Jenssen, Robert, and Davide Roverso, "Intelligent monitoring and inspection of power line components powered by UAVs and deep learning," IEEE Power and energy technology systems journal 6.1 (2019): 11-21.
[64] Huang, Can, et al., "Smart Meter Pinging and Reading Through AMI Two-Way Communication Networks to Monitor Grid Edge Devices and DERs." IEEE Transactions on Smart Grid (2021).
[65] Guelpa, Elisa, and Vittorio Verda. "Demand response and other demand side management techniques for district heating: A review." Energy 219 (2021): 119440.

[66] Noor, Sana, et al. "Energy demand side management within microgrid networks enhanced by blockchain." Applied energy 228 (2018): 1385-1398.

[67] Shahriar, S., et al., "Bridging Internet of Things and Wireless Sensor Networks: Applications and Challenges," Indonesian Journal of Computing, Engineering, and Design (IJoCED), 2.1 (2020), pp. 13-23.

[68] Cheng, Lefeng, and Tao Yu. "A new generation of AI: A review and perspective on machine learning technologies applied to smart energy and electric power systems." *International Journal of Energy Research* 43.6 (2019): 1928-1973.

[69] Kankanhalli, Atreyi, Yannis Charalabidis, and Sehl Mellouli. "IoT and AI for smart government: A research agenda." *Government Information Quarterly* 36.2 (2019): 304-309.

[70] Coelho, Francisco Carlos Rodrigues, et al. "Optimal distributed generation allocation in unbalanced radial distribution networks via empirical discrete metaheuristic and steepest descent method." *Electrical Engineering* 103.1 (2021): 633-646.

[71] Smith, Collin, Alfred K. Hill, and Laura Torrente-Murciano. "Current and future role of Haber–Bosch ammonia in a carbon-free energy landscape." *Energy & Environmental Science* 13.2 (2020): 331-344.

[72] Huang, Yutao, et al. "An edge computing framework for real-time monitoring in smart grid." 2018 IEEE International Conference on Industrial Internet (ICII). IEEE, 2018.

[73] Bajaj, Mohit, and Amit Kumar Singh. "Grid integrated renewable DG systems: A review of power quality challenges and state-of-the-art mitigation techniques." International Journal of Energy Research 44.1 (2020): 26-69.

[74] Mahmoud, Ahmed M., et al. "A Cost-Benefit Analysis of Optimal Active and Reactive Power Compensators and Voltage Conditioners Allocation in a Real Egyptian Distribution System." 2021 22nd International Middle East Power Systems Conference (MEPCON). IEEE, 2021.

[75] Haider, Waseem, et al. "Voltage profile enhancement and loss minimization using optimal placement and sizing of distributed generation in reconfigured network." *Machines* 9.1 (2021): 20.

[76] Ravindran, S., and T. Aruldoss Albert Victoire. "A bio-geography-based algorithm for optimal siting and sizing of distributed generators

with an effective power factor model." *Computers & Electrical Engineering* 72 (2018): 482-501.

[77] Bouhouras, Aggelos S., Paschalis A. Gkaidatzis, and Dimitris P. Labridis. "Network reconfiguration in modern power distribution networks." *Handbook of optimization in electric power distribution systems*. Springer, Cham, 2020. 219-255.

[78] Contreras, Sergio F., Camilo A. Cortés, and Johanna MA Myrzik. "Probabilistic multi-objective microgrid planning methodology for optimizing the ancillary services provision." *Electric Power Systems Research* 189 (2020): 106633.

[79] Houam, Yehya. Control and Optimization of an Autonomous Photovoltaic System using Metaheuristic Methods. Diss. Université Mohamed Khider–Biskra, 2021.

[80] Kumar, Nallapaneni Manoj, et al. "Distributed energy resources and the application of AI, IoT, and blockchain in smart grids." *Energies* 13.21 (2020): 5739.

[81] Qadir, Zakria, et al. "Predicting the energy output of hybrid PV–wind renewable energy system using feature selection technique for smart grids." Energy Reports 7 (2021): 8465-8475.

[82] De Dutta, Sarmistha, and Ramjee Prasad. "Security for smart grid in 5G and beyond networks." Wireless Personal Communications 106.1 (2019): 261-273.

[83] Day, Min-Yuh. "Introduction to Artificial Intelligence for Text Analytics." (2022).

[84] Tang, Kai-Yu, Ching-Yi Chang, and Gwo-Jen Hwang. "Trends in artificial intelligence-supported e- learning: A systematic review and co-citation network analysis (1998–2019)." *Interactive Learning Environments* (2021): 1-19.

[85] Criado, Javier, et al. "Integrating cyber-physical systems in a component-based approach for smart homes." *Sensors* 18.7 (2018): 2156.

[86] Sadeghi, Mohsen, Hossein Gholamalinejad, and Maaruf Ali. "A new Database for Underwater Sound Recognition using a Nonlinear Support Vector Machine." 2019 International Conference on Computing, Electronics & Communications Engineering (ICCECE). IEEE, 2019.

[87] Meng, Xiangfei, et al. "Construction of decision tree based on C4. 5 algorithm for online voltage stability assessment." International Journal of Electrical Power & Energy Systems 118 (2020): 105793.

[88] Fan, Cheng, et al. "Deep learning-based feature engineering methods for improved building energy prediction." Applied energy 240 (2019): 35-45.

[89] Geng, Zhiqiang, Zhongkun Li, and Yongming Han. "A new deep belief network based on RBM with glial chains." Information Sciences 463 (2018): 294-306

[90] Kong, Xiangtao, et al. "Reflash Dropout in Image Super-Resolution." arXiv preprint arXiv:2112.12089 (2021).

[91] Zhou, Hangxia, et al. "Short-term photovoltaic power forecasting based on long short term memory neural network and attention mechanism." Ieee Access 7 (2019): 78063-78074.

[92] Zhang, Weitong, Dongruo Zhou, and Quanquan Gu. "Reward-free model-based reinforcement learning with linear function approximation." Advances in Neural Information Processing Systems 34 (2021).

[93] Rangel-Martinez, Daniel, K. D. P. Nigam, and Luis A. Ricardez-Sandoval. "Machine learning on sustainable energy: A review and outlook on renewable energy systems, catalysis, smart grid and energy storage." Chemical Engineering Research and Design 174 (2021): 414-441.

[94] Kirchem, Dana, et al. "Modelling demand response with process models and energy systems models: Potential applications for wastewater treatment within the energy-water nexus." Applied Energy 260 (2020): 114321.

[95] Sahabiev, Ilnas, Elena Smirnova, and Kamil Giniyatullin. "Spatial Prediction of Agrochemical Properties on the Scale of a Single Field Using Machine Learning Methods Based on Remote Sensing Data." Agronomy 11.11 (2021): 2266.

[96] Gholami, Khalil, and Mohammad Hasan Parvaneh. "A mutated salp swarm algorithm for optimum allocation of active and reactive power sources in radial distribution systems." Applied Soft Computing 85 (2019): 105833.

[97] Ghorbani, Narges, et al. "Optimizing a hybrid wind-PV-battery system using GA-PSO and MOPSO for reducing cost and increasing reliability." Energy 154 (2018): 581-591.

[98] Zhang, Junxi, and Shiru Qu. "Optimization of Backpropagation Neural Network under the Adaptive Genetic Algorithm." Complexity 2021 (2021).

[99] Nazari-Heris, Morteza, et al. "Optimal distributed generation allocation using quantum inspired particle swarm optimization." Quantum

computing: An environment for intelligent large scale real application. Springer, Cham, 2018. 419-432.

[100] Chauhan, Vineeta, and Jaydeep Chakravorty. "Investigations On Diversified Multi-Objective Optimal Power Flow Techniques with Integration Of Renewable Based Distributed Generation

[101] Essallah, Sirine, and Adel Khedher. "Optimization of distribution system operation by network reconfiguration and DG integration using MPSO algorithm." *Renewable Energy Focus* 34 (2020): 37-46.

[102] Eid, Ahmad. "Allocation of distributed generations in radial distribution systems using adaptive PSO and modified GSA multi-objective optimizations." *Alexandria Engineering Journal* 59.6 (2020): 4771-4786.

[103] Quadri, Imran Ahmad, Suman Bhowmick, and Dheeraj Joshi. "A hybrid teaching–learning-based optimization technique for optimal DG sizing and placement in radial distribution systems." *Soft Computing* 23.20 (2019): 9899-9917.

[104] Nguyen, Thuan Thanh, and Thang Trung Nguyen. "An improved cuckoo search algorithm for the problem of electric distribution network reconfiguration." *Applied Soft Computing* 84 (2019): 105720.

[105] Ajeigbe, Olusayo A., Josiah L. Munda, and Yskandar Hamam. "Towards maximising the integration of renewable energy hybrid distributed generations for small signal stability enhancement: A review." *International Journal of Energy Research* 44.4 (2020): 2379-2425.

[106] Kumar, Nagendra, Barjeev Tyagi, and Vishal Kumar. "Multiarea deregulated automatic generation control scheme of power system using imperialist competitive algorithm based robust controller." *IETE journal of Research* 64.4 (2018): 528-537.

[107] Ramachandran, Anita, et al. "Performance Analysis of Machine Learning Algorithms for Fall Detection." *2019 IEEE International Conference on E-health Networking, Application & Services (HealthCom).* IEEE, 2019.

[108] Ghazal, Taher M., et al. "Energy demand forecasting using fused machine learning approaches." (2022).

[109] Fujimoto, Yu, Megumi Fujita, and Yasuhiro Hayashi. "Deep reservoir architecture for short-term residential load forecasting: An online learning scheme for edge computing." *Applied Energy* 298 (2021): 117176.

[110] Gholizadeh, Nastaran, and Petr Musilek. "Federated learning with hyperparameter-based clustering for electrical load forecasting." *Internet of Things* 17 (2022): 100470.

[111] Akhtaruzzaman, Md, et al. "HSIC bottleneck based distributed deep learning model for load forecasting in smart grid with a comprehensive survey." *IEEE Access* 8 (2020): 222977-223008.

[112] Sajjad, Muhammad, et al. "A novel CNN-GRU-based hybrid approach for short-term residential load forecasting." *Ieee Access* 8 (2020): 143759-143768.

[113] Agrawal, Rahul Kumar, Frankle Muchahary, and Madan Mohan Tripathi. "Long term load forecasting with hourly predictions based on long-short-term-memory networks." *2018 IEEE Texas Power and Energy Conference (TPEC)*. IEEE, 2018.

[114] Dong, Yunxuan, Xuejiao Ma, and Tonglin Fu. "Electrical load forecasting: A deep learning approach based on K-nearest neighbors." *Applied Soft Computing* 99 (2021): 106900.

[115] Tao, Yuxin, et al. "Application of SVR optimized by modified simulated annealing (MSA-SVR) air conditioning load prediction model." *Journal of Industrial Information Integration* 15 (2019): 247-251.

[116] Semero, Yordanos Kassa, et al. "An accurate very short-term electric load forecasting model with binary genetic algorithm based feature selection for microgrid applications." *Electric Power Components and Systems* 46.14-15 (2018): 1570-1579.

[117] Motepe, Sibonelo, et al. "Using Deep Learning Techniques for South African Power Distribution Networks Load Forecasting." *2019 International Aegean Conference on Electrical Machines and Power Electronics (ACEMP) & 2019 International Conference on Optimization of Electrical and Electronic Equipment (OPTIM)*. IEEE, 2019.

[118] Gholami, Khalil, and Mohammad Hasan Parvaneh. "A mutated salp swarm algorithm for optimum allocation of active and reactive power sources in radial distribution systems." *Applied Soft Computing* 85 (2019): 105833.

[119] Alazab, Mamoun, et al. "A multidirectional LSTM model for predicting the stability of a smart grid." *IEEE Access* 8 (2020): 85454-85463.

[120] Zhou, Yuhao, and Yude Yang. "Transient stability assessment using gated recurrent unit." *2020 5th Asia Conference on Power and Electrical Engineering (ACPEE)*. IEEE, 2020.

[121] Shamisa, Abdolah, Babak Majidi, and Jagdish C. Patra. "Sliding-window-based real-time model order reduction for stability prediction in smart grid." *IEEE Transactions on Power Systems* 34.1 (2018): 326-337.

[122] Moldovan, Dorin. "Improved Kangaroo Mob Optimization and Logistic Regression for Smart Grid Stability Classification." *Computer Science On-line Conference*. Springer, Cham, 2021.

[123] Shi, Zhongtuo, et al. "Artificial intelligence techniques for stability analysis and control in smart grids: Methodologies, applications, challenges and future directions." *Applied Energy* 278 (2020): 115733.

[124] Ji, Liang, et al. "A new single ended fault location method for transmission line based on positive sequence superimposed network during auto-reclosing." *IEEE Transactions on Power Delivery* 34.3 (2019): 1019-1029.

[125] Ding, Wenbo, and Hongxin Hu. "On the safety of iot device physical interaction control." *Proceedings of the 2018 ACM SIGSAC Conference on Computer and Communications Security*. 2018.

[126] Santolamazza, Annalisa, Daniele Dadi, and Vito Introna. "A data-mining approach for wind turbine fault detection based on SCADA data analysis using artificial neural networks." *Energies* 14.7 (2021): 1845.

[127] Perez, Ramón, Carmen Vásquez, and Amelec Viloria. "An intelligent strategy for faults location in distribution networks with distributed generation." *Journal of Intelligent & Fuzzy Systems* 36.2 (2019): 1627-1637.

[128] Srivastava, Adhishree, and S. K. Parida. "A Robust Fault Detection and Location Prediction Module Using Support Vector Machine and Gaussian Process Regression for AC Microgrid." *IEEE Transactions on Industry Applications* 58.1 (2021): 930-939.

[129] Mishra, Manohar. "Power quality disturbance detection and classification using signal processing and soft computing techniques: A comprehensive review." *International transactions on electrical energy systems* 29.8 (2019): e12008.

[130] Bozorg-Haddad, Omid, Babak Zolghadr-Asli, and Hugo A. Loaiciga. *A handbook on multi-attribute decision-making methods*. John Wiley & Sons, 2021.

[131] Ramadas, Sandeep Segu, and Rudranna Nandihalli. "Optimal system designing for hybrid renewable energy system with the aid of adaptive genetic algorithm incorporates Cauchy mutation (AGA- Cauchy)." *Journal of Green Engineering* 8.1 (2018): 1-16.

[132] Wang, Yuxing, and Chunyu Wei. "Design optimization of office building envelope based on quantum genetic algorithm for energy conservation." *Journal of Building Engineering* 35 (2021): 102048.

[133] Abdel-Salam, Mazen, Mohamed Th El-Mohandes, and Lubna Mahmoud. "A PSO-based Multi-objective Method for Optimal Weight Factors, Placement and Sizing of Multiple DG Units in a Distribution System." *2019 21st International Middle East Power Systems Conference (MEPCON)*. IEEE, 2019.

[134] Keypour, Reza, et al. "A new population-based optimization method for online minimization of voltage harmonics in islanded microgrids." *IEEE Transactions on Circuits and Systems II: Express Briefs* 67.6 (2019): 1084-1088.

[135] Tarraq, Ali, et al. "Meta-heuristic optimization methods applied to renewable distributed generation planning: A review." *E3S Web of Conferences*. Vol. 234. EDP Sciences, 2021.

[136] Wu, Fei, et al. "Towards a new generation of artificial intelligence in China." *Nature Machine Intelligence* 2.6 (2020): 312-316.

4

Critical Analysis of the Role of Solar PV and Advocating the Need for a Circular Economy for Solar PV Recycling

Lalit Agarwal[1], Vikalp Gupta[2], and Purujit Mohapatra[3]

[1]MAIT, GGSIPU, India
[2]Coventry University, United Kingdom
[3]University of Glasgow, Scotland
E-mail: l.agarwal15@gmail.com; vikalp13199@gmail.com; purujit97@gmail.com

Abstract

The study predicts that the capacity of India's solar energy sources will grow from 120 GW in 2024 to 304.9 GW in 2036 and that solar energy has surpassed wind and hydro-power. It is vital to evaluate the influence of solar energy on India's future and see if it can replace coal power. With a predicted lifespan of 25 years, the growth of the solar industry will result in a significant amount of waste, up to 16%. Although the existing solar study focuses on efficiency, the removal of wrecked solar panels is also a concern. Recycling is a solution, but only a few companies can handle it effectively. In the approaching periods, disposing of solar panels will be a major ecological subject. The study focuses on the recycling of solar panels, including environmental protection, recycling technology, recycling legislation, financial implications, and waste management. It also provides suggestions for improving technology and making political decisions to ensure a smooth rollout of solar energy. A forecasting model predicts the volume of waste produced by solar panels and its balance of the system employing a reliability function. The investigation also evaluates the quantity of raw materials that can be recovered after recycling. Numerous nations are considering expanding the accountability of solar material traders by

DOI: 10.1201/9788743809852-4

integrating reuse or disposal as a fundamental component of their recycling management tactics. The solar sector should persistently enhance its practicality, efficient recovery rate, environmental impact, and economic viability to contribute to a more sustainable and cleaner future.

Keywords: Solar energy, solar PV waste, extended producer responsibility, renewable energy regulation, solar panel degradation analysis.

4.1 Introduction

India ranked third in the world in terms of primary power demand in 2018, with 809.2 metric tons, although its per-person consumption is lower (0.6 tons) than the worldwide average (1.8 tons). India is susceptible to energy shortages and environmental degradation due to its reliance on imported energy and significant CO_2 emissions from its electricity industry (40% in 2018). The issue has gotten worse as a result of the increased output of coal-fired energy facilities to meet demand. India needs to switch from using fossil fuels to renewable energy to address this issue, and solar energy is the most economical choice given its potential for widespread deployment and technological developments. India has a large potential for solar energy production, with 3.28 million km^2 of land and an average of 3 to 7 kWh/m^2 of sunshine every day.

A reliable and nonpolluting renewable energy source that satisfies climate policy objectives is solar energy. Although solar photovoltaic (SPV) technology is widely employed, there has not been much research on how to dispose of solar photovoltaic modules (SPMs) that are no longer functional. This is a serious problem because SPMs include more precious, dangerous, and basic materials than other e-waste do. In accordance with the utilization of waste electrical and electronic equipment (WEEE) regulations, this study focuses on SPM in India and its effect on environmental sustainability. Despite its expanding energy needs, coal, oil, and biomass still provide 80% of India's energy. With more than 70% of all energy production coming from coal in 2020, together, SPV and wind produced only about 18% capacity and less than 10% of energy production (Anam, Z. et al., 2022, Psomopoulos, C. S. et al., 2022, Jolly, K., et al.).

India has experienced significant repercussions due to the COVID-19 pandemic, both in terms of fatalities and business closures. Since India's economy and population are expanding quickly, there are questions over how long the country will be able to continue its current levels of energy use. In India, solar energy is anticipated to increase quickly over the next 20 years and eventually replace coal as the country's primary energy source. Due to the cost-competitiveness of solar and India's objective of acquiring 420 GW of renewable capacity by 2040, SPV capacity has increased by 60% over the last five years as shown in Figure 4.1. India is a driving driver behind this transition (IEA "India Energy Outlook" 2021).

As of December 31, 2020, India had 37.39 GW of operating solar energy facilities, with an additional 36.69 GW of planned installations still waiting to be commissioned and 18.47 GW of proposed projects that have been publicly announced. By 2022, the target of 100 GW is anticipated to be attained. Solar power facilities with a combined capacity of over 40 GW are anticipated to be online by March 2021. The estimated capacity for solar energy production in India is believed to be 751 GW. It is expected that by the fiscal year of 2020–2021, the various solar initiatives will have resulted in the creation of a combined extent of 40,000 MW.

Unfortunately, as there are more SPVs deployed, there will be more trash produced when they approach the conclusion of their 25-year operational period. This causes hazardous waste to increase, endangering the environment. Studies predict that by 2030, around 2 million tons of SPM waste will be generated, and that number will rise to 60 million tons by 2050. In order to appropriately dispose of this garbage, recycling must be addressed with a

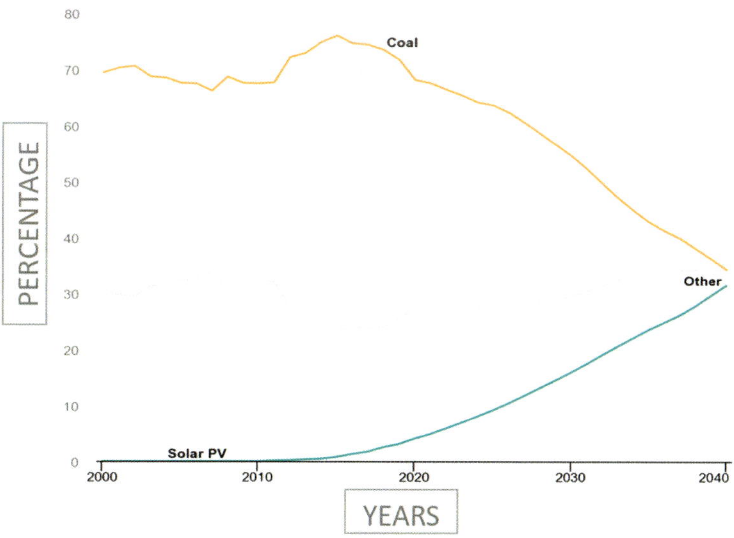

Figure 4.1 Variations in the distribution of electricity production in India from the year 2000 to 2040 (IEA "India Energy Outlook" 2021).

policy framework. The government must create a strategy for all parties to follow in order to recycle the waste, and the industry must produce low-cost recycling technologies. In the following 20–25 years, the majority of solar panels installed over the past 5–6 years will be considered garbage (Rathore, N. et al 2022, Fiandra et al., 2019; Deng, R. et al., 2022).

The study by IRENA in 2016 highlights the main reasons for solar panel failure, which include recurring and early losses. Due to the limited supply and recovery methods of the raw materials used in solar panels, they may become scarce relative to their consumption in the future. Recycling of solar panel waste is not a priority in India, and as the quantity of deployments surges, the amount of waste generated will also increase, posing a significant challenge in the future. The NCPRE and NISE led a study to evaluate the dependability and sturdiness of SPM in various portions of India, but a comprehensive analysis of solar panel waste management and disposal has yet to be conducted. Only Japan and the United States have established recycling processes for solar panel waste, while only the European Union has legislation in place. Improper disposal of solar panels can lead to environmental problems such as waste of materials, leaching of toxic elements, and deprivation of natural supplies (Ganti, K., et al 2022, Gervais, E et al., 2021, Yu, F. et al., 2022, Weckend, S et al, 2016).

In India, recycling obsolete solar panels is favored over disposal. For recovered minerals, recycling can offer a new resource distribution network. The evaluation and estimation of the amount of recyclable materials that can be obtained from SPM waste and the required time frame is necessary. Due to India's reliance on imported commodities, particularly gold, silver, and copper, the SPM sector is susceptible to financial crises. A material flow study of the resources utilized in SPM and EOL solar e-waste is necessary for approaches and methods for dealing with this type of trash. Even though they only make up a small fraction of the waste, valuable and necessary metals considerably increase their financial value. Establishing a circular supply chain is crucial for the proper disposal of end-of-life waste and maintaining a sustainable financial system (Hagelüken, C. et al., Ansanelli, G. et al., 2021, 2022, Mahmoudi, S. et al., 2019).

India needs to investigate solar energy's potential and set up a reverse supply chain for resources. The study will build on CE and enhance sustainable waste management for solar panels. We will examine the rise in solar panel installations and forecast the likelihood of solar panel waste using publications from The Energy Resource Institute (TERI), International Renewable Energy Agency (IRENA), National Institution for Transforming

India (NITI Aayog), Jawaharlal Nehru National Solar Mission (JNNSM), International Energy Agency (IEA), and Bridge to India (BTI).

The study's model calculates the total energy production, carbon emissions, and financial benefits from solar photovoltaic (SPV) projects in India by aiming to construct 305 GW of solar photovoltaic modules (SPM) by 2035 and implementing a 3Rs (reuse, recycle, and reduce) strategy for waste management. The model also calculates the potential recovery of recyclable materials from the SPM and the associated balance of system (BOS) waste. The benefits are estimated by analyzing the rate of energy traded to the power grid (de Sousa Jabbour, A. B. L. et al., 2019). The study first utilizes the Homer Pro program to determine the power output, financial advantages, and carbon emissions of a 1 GW rated setup equipped with a 500 KW inverter. Next, the Weibull distribution model, a flexible and well-established method for assessing the likelihood of solar photovoltaic module failure, was utilized. The study takes into account early and frequent SPM breakdowns through the application of the Weibull distribution model. The quantity of leftovers produced by the SPM and its balance of system (BOS) is also estimated using the Weibull distribution model and fixed loss calculation based on India's specific criterion for initial levels of fragility (Weckend, S. et al., 2016; Heath, G. et al., 2022).

Studies indicate that as the capacity of solar panel installations increases, a substantial volume of end-of-life SPM trash will enter India around 2030 and continue to increase rapidly. The calculation is based on the projected deployment of SPM by the year 2035 and the resultant trash created by 2050. The study considers the financial value of hazardous, critical, and valuable metals to ensure a balanced recovery of these elements, as well as annual recycling targets and the demand for raw materials. The analysis focuses on recycling 21 elements commonly found in EOL solar panel waste.

4.2 Overview of the Modeling Process

A solar photovoltaic (SPV) system includes SPM (solar panels), transformers, inverters, mounting structures, cabling, and sometimes, batteries for off-grid installations, and monitoring devices to improve sun exposure. The software "Homer Pro" was used to determine the total power generation and discharges from a 1 GW-rated solar plant (Yang, Q. et al 2022). The peak capacity for the system was determined after connecting it to a 500 KW inverter and integrating it with the power grid. The study selected three sites with differing latitudes but similar longitudes for estimation (Mahmoudi S, et al. 2019).

The study considers three categories of loss in energy and trash triggered by the breakdown of solar modules: fixed loss, regular loss, and early loss. The fixed loss represents the destruction to SPM that occurs after its typical lifespan. The regular loss is the destruction that occurs at the typical lifespan of SPM, and the early loss is the destruction that occurs before its typical lifespan. The study uses the Weibull reliability function, a proven method

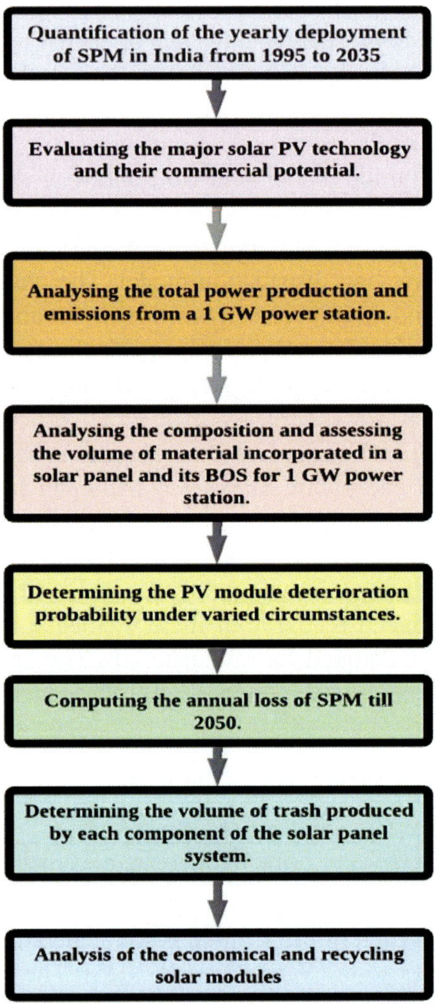

Figure 4.2 The methodology used to identify the materials recovered from SPV systems is depicted.

for calculating SPV breakdown frequency, to account for early and regular failures of solar panels (Urbina, A. 2022, Weckend S, et al., 2016).

The study is divided into seven sections as shown in Figure 4.2. Sections 4.2.1 to 4.2.4 focus on organizing, classifying, and transforming the variables needed to predict the resource inventory of EOL SPV debris and the resources needed for SPV deployment. Sections 4.2.5–4.2.7 determine the amount of debris that must be handled, recycled, or disposed of each year. The yearly and accumulated trash production in India is calculated in tons/GW of installed solar capacity, based on the market share. The study evaluates the viability and financial feasibility of recuperating resources from EOL SPM and its BOS components in Section 4.2.8.

This study predicts the expanse of trash engendered by SPV panels in India up to 2050, based on the NITI Aayog's energy forecast. The majority of the waste will be produced by generations one and two of SPV panels until 2050, accounting for 97% of the total. After 2050, generation three of SPV panels will also contribute to the waste. The study will compare its projections with the NITI Aayog's 2050 energy forecasts and the 25-year panel life set by the JNNSM policy, which is scheduled to end in 2022. The calculations are considered realistic due to the focus on the first two generations of SPV panels until 2050 (Islam M. T. et al., 2020).

4.2.1 The quantification of annual installation of SPM in India from 1995 up to 2035

Table 4.1 shows the solar energy production in India from 2010 to 2022, alongside the forecasts up until 2035 from various sources. The adoption of solar power systems in India started around 1995 but at a slow pace. If India implements a higher green energy strategy, it is projected that operational size will reach 305 GW by 2035. However, if the moderate green energy scenario is chosen, the deployed capacity will reach 161 GW, according to a study by TERI (NCPRE and NISE, 2017). The actual SPV deployment data up to 2022 was obtained from the MNRE website. The reports from IEA, TERI, IRENA, NITI Aayog, JNNSM, and BTI are used as a reference for this research's presumption, as shown in Table 4.1. Graph 4.1 illustrates the solar installation's explosive expansion in India from 1995 until 2040 as an S-shaped curve (D'Adamo et al., 2020). The manufacturing of solar modules will endure to require materials like precious, foundation, and others in order to be produced at affordable prices. Therefore, resource sustainability challenges and understanding the CE is crucial (Wang. X et al., 2022).

Table 4.1 Overview and projection of solar PV installed in India (GW) (MNRE, 2018, 2021, NCPRE and NISE, 2017, Physical Progress, Solar Current-Status, BTI India Solar Compass, TERI, 2015, 2021, NITI Ayog, 2015, India Renewables Outlook 2024, IEA Outlook, (2021)).

Actual installed capacity by (GW)									
Year	IEA	MNRE	NITI Ayog	Bridge to India (BTI)	Target (JNNSM)	IRENA	TERI		This study
							HERS	LERS	
2010	0.23	0.31	1		1	0.5		0.011	0.19
2015	3	4.9		4.8	10	5.36		3.74	4.79
2020	34	34.6		49	62	38.98	65	34.6	38.1
2022	--	59.3	26	66.8	100		100	53.9	58.4
2025	118			82		134	147	93	121
2030	186		66			200	228	189	200
2035	300					310	--	--	305

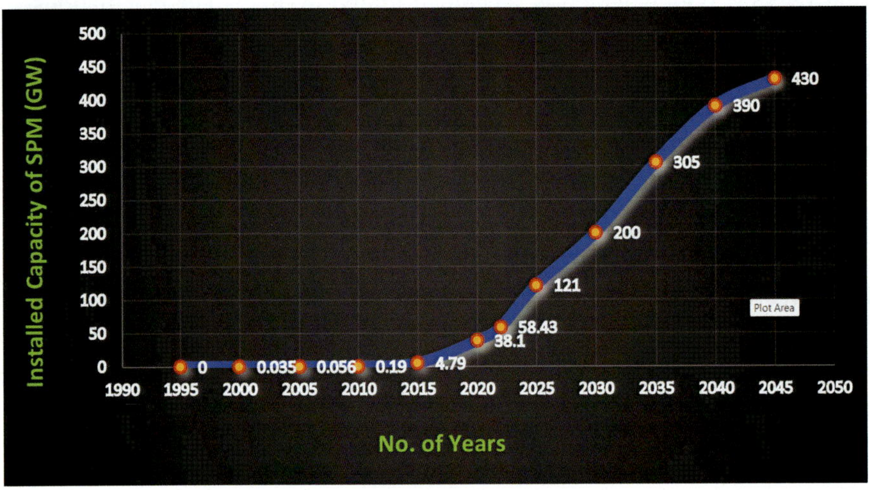

Graph 4.1 Estimates for the installation of SPV capacity (GW) in India until 2045.

4.2.2 Assessing the primary SPV technologies and their commercial viability

The design of solar panels can vary, as shown in Table 4.2. Different types of solar panels use different materials and have varying levels of hazardous and safe substances. Table 4.3 shows the changing market share of different solar panel technologies over time, reflecting advancements. This study focuses on three main types of solar panels made from Si, CdTe, and CiGs, as supplementary tech has a minor commercial value and lies beneath the categories of Si-based and thin-film tech. By 2022, it is predicted that CdTe will have a 6.3% market share, Si will have 87%, and CiGs will have 6.7%. The regular lifespan of SPM amplified from 25 to 30 years in 2014, according to Table 4.3 (Arora N. et al., 2018; ITRPV, 2020, NISE, 2018; Contreras-Lisperguer, R. et al., 2017, Sharma, S. et al., 2015).

Table 4.2 Share in the market by tech and annual PV generation of trash (Domínguez, A. et al 2017).

SPM technology classification(Bagher, A. M. et al., 2015;Arora N. et al., 2018M)	
Devices	Type of Cell
SPM centered on Silicon	· Monocrystalline· Poly or multicrystalline · Ribbon · a-Si (amorph/micromorph)
SPM centered on Thin flim	· Copper indium gallium (di)selenide (CIGS) · Cadmium telluride (CdTe)
Others	· Concentrating solar PV (CPV) · Organic PV/dye-sensitised cells (OPV) · Crystalline silicon (advanced c-Si) · CIGS alternatives, heavy metals (e.g. perovskite), · advanced III-V

Table 4.3 Share in the market by tech and annual PV generation of trash (Domínguez, A. et al 2017).

Deplyoment Year (i)	Year of SPV Debris Creation(j)	(%) Share In the market(%)		
		CdTe	CiGs	Si-based
2000	2025	0	0	100
2005	2030	0	2	98
2010	2035	17	1	82
2014	2044	5	2	93
2020	2050	5.2	5.8	89
2022	2052	6.3	6.7	87
2025	2055	5.9	10.5	83.6
2030	2060	4.7	15.7	79.7
2035	2065	4.2	22.8	74

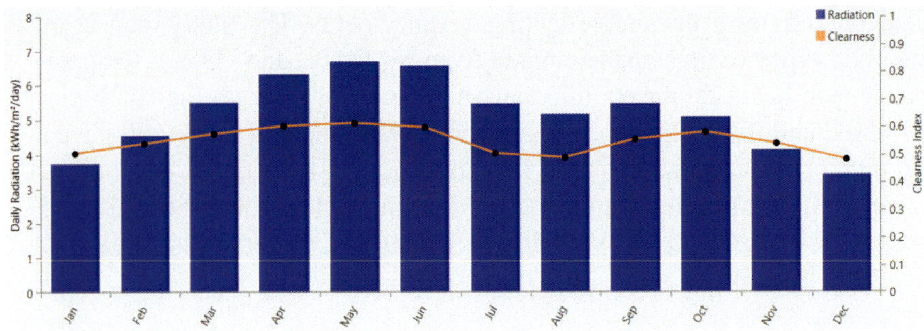

Figure 4.3 The clearness index and daily radiation was obtained from the Homer Pro program.

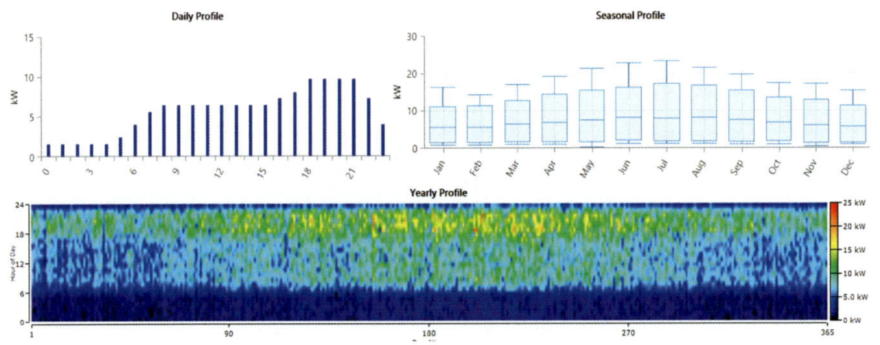

Figure 4.4 Load demand as per Homer pro.

4.2.3 Using the Homer Pro program for determining the emissions, financial benefits, and total energy output (by examining the rate of energy traded to the grid) from SPV systems in India

The emissions, financial benefits, and power creation of a 1 GW-rated SPV system in India were calculated using the Homer Pro software. As shown in the Figures 4.3 and 4.4 the system was designed with 5.18 KWh/m^2 as the average daily global radiation and was linked to a 500 KW inverter and the grid. During optimal conditions, the surplus electricity produced by the system was sent to the utility grid.

Three locations with the same longitude but different latitudes were chosen for the calculation. (a) Maharashtra, India (20°14.7'N, 76°53.1'E), (b) Tamil Nadu, India (10°33.7'N, 76°53.1'E), and (c) Delhi, India (28°36.1'N)

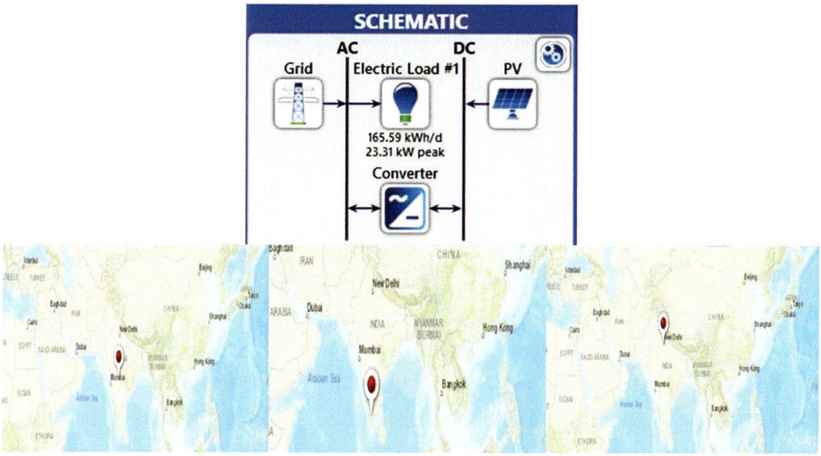

Figure 4.5 A schematic design of the solar module and its simulated location. (a) Maharashtra (20°14.7'N, 76°53.1'E), (b) Tamil Nadu (10°33.7'N, 76°53.1'E), and (c) Delhi (28°36.1'N, 76°53.1'E).

min), 76°53.1'E), as shown in Figure 4.5. The cost of a 1 GW photovoltaic (SPV) system is estimated at $6 billion in capital costs and $5 billion in replacement costs. Maintenance costs are estimated at $10 million per year at 95% efficiency. The cost of the converter was estimated at $5.5 billion in capital costs, $2.5 billion in replacement costs, and $1 billion annually in operation and maintenance.

Power purchase and resale rates were estimated at $0.44/kWh and $0.05/kWh, respectively, based on the assumption that Si-SPV technology continues to dominate his SPV market and commodity prices remain constant. The SPV system design is shown in Figure 4.5.

A 1 GW plant's total energy production, emissions, and financial benefits were calculated, and the findings were calculated along with the total prediction for that specific period using Section 4.1 as a guide.

4.2.4 Evaluating the amount and composition of substances fused into the SPM and its BOS of a 1 GW power plant

The majority of SPV deployments worldwide are based on crystalline silicon technology. This technology generates various types of waste, including valuable, basic, critical debris, and hazardous, and has specific reprocessing (Nain et al., 2020). The primary components of these PV systems include

Table 4.4 SPM specification (Domínguez, A. et al 2017; Mahmoudi, S. et al., 2019).

SPV tech.	Area(m2) /nominal power	Area(m2) /GW	Nominal power (Wp)	Weighs(tons) /GW	Commercial Value(%)
CiGs	0.72	9000000	80	157500	5.8
CdTe	0.72	11076923	65	184615.4	5.2
Si based	1.46	6517857	224	102678.5	89

silicon, aluminum, glass, ethylene vinyl acetate (EVA), silver, gold, and copper (Arora et al., 2018). Table 4.4 presents the features of every PV tech, including the amount of waste generated per ton for a 1 GW installation, its nominal area, mass, nominal power, and market share. The component concentrations of each SPM tech are transformed into tons per GW for ease of estimation (Nain, P. et al., 2020, Arora, N. et al 2018).

In Table 4.5, the components of photovoltaic (PV) systems are classified based on valuable, special, hazardous, critical, and other criteria (ITRPV (2019), Domnguez et al. (2019), Mahmoudi et al. (2019), Xu et al. (2018)).

For this analysis, the components of the BOS were mainly considered to be the mountings, inverter, and cable. The battery and transformer were considered as external components. The whole system is highly dependent on the inverter since most PV systems are connected to the grid, which helps in the conversion of DC to AC and vice versa.

The investigation focused on a 500 kW string converter with a ratio of 1.15, utilizing information from the Ecoinvent database version 3.4 (Martulli, A. et al., 2022). The PV system is connected to a combiner box, a converter, and the transformer with the help of cables, which are made of copper or aluminum. For a 1 GW solar deployment, the total volume of copper is estimated to be 4712.07 tons or 0.64 Kg per square meter of the solar plant (Mahmoudi et al., 2019).

The SPV is anchored to the ground using anchoring frames, which are usually made of copper or aluminum. The components of the balance of system (BOS) are listed in Table 4.5 based on data from the Ecoinvent database 3.4. The transformer is a crucial component of the PV system and characteristically 0.8 as the power factor (D'Adamo et al., 2020).

Iron and aluminum or copper are the main materials used for the construction of a transformer. An Al transformer weighs 45% less than a Cu transformer. The balance of system (BOS) may also contain additional components that are not included in this research, as their recycling process is already established, such as tracking systems and batteries. The weight of

Table 4.5 Mass Composition and Recycling Yield for Various Components

Components	**Metal**	**Mass composition SPV (tonne/GW)**			**Mass composition SPV BOS (tonne/GW)**			**Recycling Yield (%)**	**Expense USD/ton**
		Si	CdTe	CiGs	Inverter	Cabling	Mounting		
Base & Special	Sn	0.06	0.01	90	0.06	0	0	32	19340
	Cr	0	33.24	0	0	0	0	20	9900
	Cu	736.64	5538.47	450	1949.25	4712.07	0	100	6460
	Al	16557.9	166.16	135	753.25	0	25423.5	100	2000
	Fe	0	0	0	0.29	0	48891.4	90	600
	Mn	0	0	0	0.01	0	0	37	5
	Ni	1.07	0	0	0.92	0	0	41	14520
	Zn	0.01	0.01	90	2.3	0	1760.09	27	2610
	Ti	0.01	0.01	0	0	0	0	52	640
	Mo	0	0	90	0	0	0	18	17800
	Ta	0	0	0	0.12	0	0	21	194000
Precious	Ag	57.96	0	0	2.13	0	0	95	651000
	Au	0	0	0	2.94	0	0	36	49375000
Other Metals	Si	795.31	553.85	0	0	0	0	100	29998
Hazardous	Se	0	0	90	0	0	0	89	50300
	Pb	4.7	7.76	0	10.35	0	0	98	2320
	Cd	0	221.54	270	0	0	0	95	1100
Other Materials	EVA	6518.86	6646.16	8100	0	0	0	0	890
	Glass	65840.46	168369.2	135000	0	0	0	95	790
Critical Metals	Mg	522.82	0	423	0.06	0	0	33	2010
	In	0	0	45	0	0	0	90	460000
	Te	0	221.54	0	0	0	0	95	89000
	Ga	0	0	90	0	0	0	90	295000

all components in a 1 GW photovoltaic (PV) system is calculated (Pamponet et al., 2022). The lifespan of the inverter is estimated somewhat around 10 years on average, while that of other BOS components is estimated to be 25 years on average (TERI, 2015).

The number of SPM can be calculated using the equation

$$\text{No. of SPM req.} = PV \frac{\text{Capacity}(W)}{\sum(\text{ Commerical Value }(\%) * \text{ Nominal Power }(W_p))}$$

$$\text{while material contain of SPV}(\text{Kg}) = \text{ Material Composition }\left(\frac{\text{kg}}{\text{m}^2}\right) * \text{Area}\left(\text{ m}^2\right)$$

$$\text{and, NO. of inverter required} = \frac{\text{SPV capacity }(W) * \text{Inverter Sizing Ratio}}{\text{Inverter capacity }(W)}$$

4.2.5 Investigating the likelihood of degradation of the solar photovoltaic module under different conditions

The fixed loss approach simulates that the solar module will be replaced after a predetermined period of time, referred to as its end-of-life (EOL), which was estimated to be 25 years prior to 2014 and 30 years after that, according to a 2016 analysis by IRENA (Weckend, et al., 2016). Once the EOL is reached, it is assumed that the SPM will be immediately replaced, as represented by eqn (4.1)

$$u_{\text{fA}}{=}x_{\text{P}}, \tag{4.1}$$

where the deployment time (*A*) ranges from 1995 to 2035 and the debris production period (*P*) ranges from 1995 to 2050. For fixed loss

$$D{=}P{+}A; A{=} \text{ 30 if } D{>} \text{ 2014}; A{=} \text{ 25 if } D{\leq}2014.$$

Calculating the Weibull distribution of probabilities is essential to ascertain the SPM's dependability due to the event's unpredictability and the transient nature of the SPM fall. In the IRENA 2016 study, the basic causes and analysis of SPM failure are covered; regular loss and early loss SPV breakdowns are distinguished.

Eqn (4.2) states that the Weibull distribution of probabilities can be used to calculate the prospect of damage in the respective situation. The representation's suitability for modeling the collapse of the solar module was first proved in the linked articles. The form factor *Y* determines how the SPM failure develops with time *A*. The benchmark rate, which assumes existence harm with a possibility of 99.99% in 40 years and average lifecycle anticipation of 30 years, is speculated to be 5.3749 for the regular loss set of conditions and 2.4928 for the early loss set of environments. The IEA's PV energy systems scheme and various other papers were used to determine the SPV value (Mahmoudi, S., et al. 2021; Weckend, S. et al., 2016; D. Oteng et al., 2022).

The Weibull distribution of probabilities is portrayed as

$$F(A) = 1 - e^{-\left(\frac{A}{T}\right)^{\Upsilon}}, \tag{4.2}$$

where "A" is a time period that can be anything between 0 and 40 decades. The shape factor "Υ" determines the distinctive S form of the Weibull distribution of the probability curve. Graph 4.2 displays the slope of $F(t)$, the parameter describing the Weibull distribution of probability.

Consequently, in case of regular losses

$$u_{\mathrm{RA}} = F(A) \ \text{with} \ \Upsilon = 5.37. \tag{4.3}$$

In case of early losses

$$u_{\mathrm{EA}} = F(A) \ \text{with} \ \Upsilon = 2.4928. \tag{4.4}$$

925 SPM were evaluated by NCPRE, IIT Bombay, and NISE in India at 37 sites in all six climatic categories, with an estimated early loss failure rate of 1.55% yearly (NCPRE and NISE, 2017). The u_{EA} for India is stated without taking into account how the rate of SPM failure and degradation varies by country and location.

$$u_{\mathrm{EA}} = x_{\mathrm{D}}(1 - 0.0155)^{\mathrm{A}}. \tag{4.5}$$

The multiplication of the eqn (4.5) and (4.4) is used to determine the annual loss, whereas eqn (4.1) and (4.3) are used to calculate the regular loss.

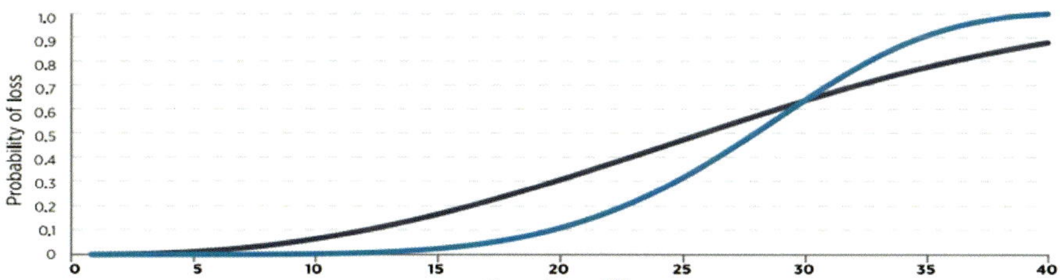

Graph 4.2 The cumulative distribution characteristics of the Weibull function for early and regular loss scenarios (Weckend, S. et al., 2016).

4.2.6 Calculating the yearly losses in SPM until the year 2050

This subsection reflects the goals of the examination. Eqn (4.5) shows how the adjusted loss for each period was estimated using the mutual influence of regular and early loss for the relevant time. Corrected loss would increase greatly in size and be comparable to the annual demand for repowering. The corrected loss is shown as

$$U_{\mathrm{CD}}=U_{\mathrm{ED}}+U_{\mathrm{RD}}, \tag{4.6}$$

where "*P*" ranges from 1995 to 2030, and $D=P+A$.

4.2.7 Estimating the amount of waste generated by each part of the SPM

Once the above-mentioned six phases have been successfully implemented, eqn (4.7) is employed to compute the total power loss subjected in each year, taking into account the volume of every substance and the market share of respective technology, as shown in segments 4.2.4 and 4.2.6, can be used to calculate the total mass of SPV debris as well as the years' worth of resources that are needed and rebounded from trash.

$$W_{\mathrm{D}\alpha}=U_{\mathrm{Cd}}w_{\alpha\mathrm{P}}p_{\alpha} \tag{4.7}$$

The volume of components listed in Table 4.5 was multiplied by the adjusted loss for each time to determine the percentage of every metal and resource in the SPV EOL trash.

4.2.8 Analysis of the economical and recycling solar modules

The structure for a CE is laid out in this section. Globally, the recycling of waste from SPV is still in its early stages, predominantly regarding physical substructure and technology regulations. To ensure the proper disposal of this waste from consumers and reduce costs from both financial and cultural perspectives, policies must be established and long-standing agendas and amenities for managing the waste must be established (Jain, S. et al., 2022, Kastanaki, E. et al 2022).

India currently lacks any official guidelines for managing SPM trash. According to the MNRE guidelines, designers must ensure that all SPM collected from their power plants once they reach EOL are disposed of in accordance with "E-waste (Management and Handling) Regulations," which were created by the federal government and are updated periodically. The information in this publication relates to recycling for solar thermal systems.

Additionally, these regulations do not mention the treatment of solid particulate matter waste or provide specific methods for it. Despite the Ministry of Environment, Climate Change, and Forecast having established rules and guidelines for e-trash organization and dispensation in 2011 and updating them in 2016 and 2018, India still lacks expertise in this area (Suresh, S., et al., 2019; Sheoran, M. et al., 2022). In April 2019, the MNRE created a timeline for the importation, manufacture, employment, and recycling of SPV. Records kept by the CPCB show that the country's actual capacity is much minor than the notional perimeter of 22% of the total volume of e-trash it produces.

Although the waste produced by EOL SPV is costly, the bulk of it may be recycled and used again (IRENA, 2017). In order to analyze the total quantity of additional resources recuperated from garbage, the volume of remaining resources is multiplied by the output of the material's recycling process while taking into account both present and emerging technologies. Numerous publications were checked, as shown in Table 4.7, to regulate the recovering return of all metals used in SPV systems.

Despite the waste generated from EOL SPV systems, a significant portion of it can be recycled and reused, according to a 2017 study by IRENA. The sum of leftover resources is multiplied by the efficiency of the recycling process, considering both existing and newly developed technologies, to determine the total amount of additional material recovered from the waste. Table 4.5 offers the consequences of the investigation on the amount of each metal used in SPV systems that was recycled.

4.3 Results

An investigation was conducted to forecast overall power generation, financial profits, emissions, and EOL SPV trash in India from 1995 to 2050. The calculation took into account the substantial growth of SPV in the country, which was based on predictions of the total installed power generation capacity and EOL waste until 2050. The yearly energy output, financial benefits, greenhouse gas emissions, and waste generated at the end of a product's life cycle were forecasted for the same period. The yearly projections of SPV installations, on the other hand, were based on various sources and estimated until 2035 (Karali, N. et al., 2022). The design of a circular economy supplier network for EOL waste was grounded on the valuation of the amount and duration of waste disposed of each year (Duran, A. S., et al., 2022). The predictions were based on two assumptions: the Si SPV technology would continue to dominate the industry and the raw material composition in SPV installations would remain unchanged over time. However, the sensitivity analysis highlights that the volume of trash and retrieved materials can be affected by changes in metal compositions, deployment pace, recycling efficiency, SPV technology, and commercial share (Ndzibah, E. et al., 2022).

4.3.1 The total energy output, financial gains, and emissions resulting from the expected SPV installations in India are being calculated

This section used Homer Pro software to examine the overall power generation, greenhouse gas emissions, and financial advantages (the price of energy traded to the grid) of a 1 GW-rated SPV system in India. SPV output power for all locations over a year and the total monthly energy production are displayed in Figures 4.6 and 4.7. According to Table 4.6, the combined annual energy output for the three sites was roughly 1582.379 GW. The total annual emissions from all sites are shown in Table 4.7, with carbon dioxide coming in at the highest level at 15994.33333 kg/year. The monthly monetary gains from each site are displayed in Table 4.8. Using the constant net metering price of 0.044 $ per KWh, the gain in money over a year was determined to be 106819.4467 $.

The findings of the Homer Pro analysis for a 1 GW-rated power plant are shown in Table 4.9. They include total energy production, maximum output, mean output, mean output daily, financial advantages, and emissions of carbon dioxide, sulfur dioxide, and nitrogen dioxide from 1995 to 2035. If the projected growth rate holds, total annual energy production is predicted to be 482618 GW, with a daily production of 1322.24149 GW, by 2035. If net metering is possible, this might lead to financial gains of $32579931.24. Over time, it is projected that there were emissions of carbon dioxide, sulfur

Table 4.6 The greenhouse gas emissions from all locations are being calculated.

Site/Pollutant	A	B	C	Average	Units
Carbon Dioxide	15,862	16,020	16,101	15,994.33	kg/yr
Sulphur Dioxide	68.8	69.5	69.8	69.37	kg/yr
Nitrogen Oxides	33.6	34	34.1	33.9	kg/yr

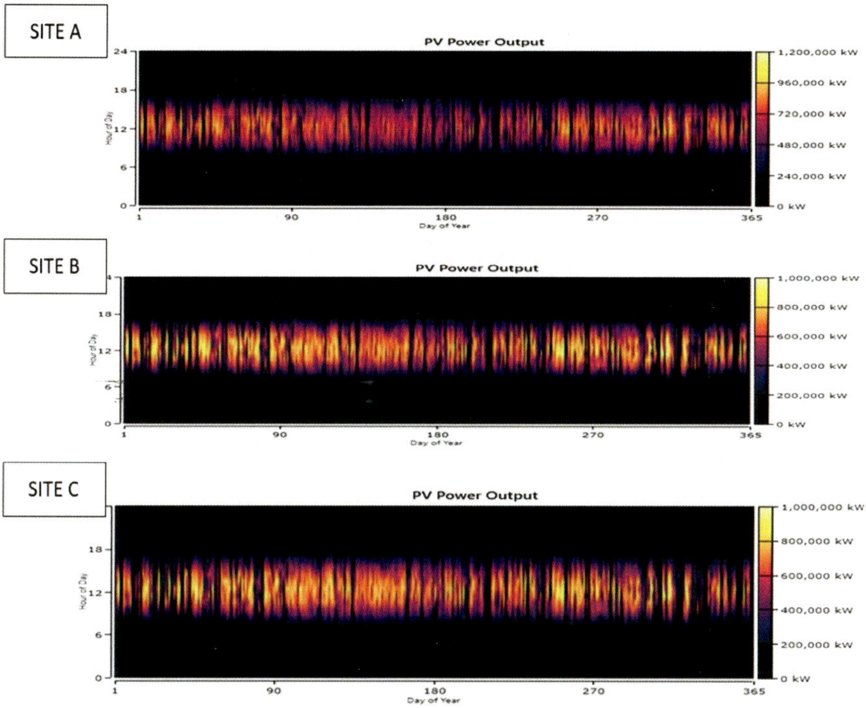

Figure 4.6 The SPV energy output for all sites over the course of one year is being determined.

Table 4.7 The energy productions generated by Homer Pro from all locations are being measured.

Site	A	B	C	Average	Units
Maximum Output	9,93,366	9,93,366	10,02,423	9,96,385	kW
Hours of Operation	4,408	4,408	4,367	4,394	hrs/yr
Capacity Factor	17.9	17.2	19.1	18	%
Mean Output	1,78,632	1,72,285	1,90,985	1,80,634	kW
Mean Energy Output	42,87,166	41,34,851	45,83,636	43,35,218	kWh/d
Total Production	1,56,48,15,413	1,50,92,20,513	1,67,30,27,001	1,58,23,54,309	kWh/yr

dioxide, and nitrogen dioxide totaling 4878170, 21156.935, and 10339.5 kg, respectively.

Table 4.8 Energy Data and Financial Benefits Analysis

	Energy Purchased from Grid (KWH)				**Energy Sold (KWh)**				**Monetary Benefits $**			
Site/Months	**A**	**B**	**C**	**Average**	**A**	**B**	**C**	**Average**	**A**	**B**	**C**	**Average**
January	1,948	1,948	1,948	1,948	1,68,381	1,68,381	1,68,381	1,68,381	8,321.63	8,321.63	8,321.63	8,321.63
February	1,781	1,781	1,781	1,781	1,52,040	1,52,040	1,52,040	1,52,040	7,512.95	7,512.95	7,512.95	7,512.95
March	2,069	2,074	2,069	2,071	1,76,063	1,76,002	1,76,255	1,76,107	8,699.71	8,696.37	8,709.32	8,701.80
April	1,851	2,133	1,851	1,945	1,91,865	1,81,147	1,91,865	1,88,292	9,500.71	8,950.71	9,500.71	9,317.38
May	2,092	2,092	2,085	2,090	1,98,031	1,98,031	1,99,524	1,98,529	9,796.92	9,796.92	9,871.92	9,821.92
June	2,191	2,191	2,104	2,162	1,91,332	1,91,332	2,03,745	1,95,470	9,457.07	9,457.07	10,082.07	9,665.40
July	2,264	2,264	2,264	2,264	1,97,677	1,97,677	1,97,677	1,97,677	9,770.68	9,770.68	9,770.68	9,770.68
August	2,255	2,271	2,255	2,260	1,97,543	1,97,059	1,97,543	1,97,382	9,764.40	9,739.40	9,764.40	9,756.07
September	2,339	2,409	2,287	2,345	1,79,841	1,76,911	1,81,789	1,79,514	8,875.09	8,725.09	8,975.09	8,858.42
October	2,209	2,209	2,278	2,232	1,82,775	1,82,750	1,76,229	1,80,585	9,028.29	9,027.06	8,697.51	8,917.62
November	2,098	1,982	2,326	2,135	1,64,175	1,77,559	1,49,903	1,63,879	8,103.86	8,778.86	7,378.86	8,087.19
December	2,001	1,993	2,228	2,074	1,68,269	1,69,761	1,53,496	1,63,842	8,313.39	8,388.39	7,563.39	8,088.39
Annual	25,099	25,348	25,476	25,308	21,67,992	21,68,650	21,48,447	21,61,696	1,07,144.69	1,07,165.13	1,06,148.52	1,06,819.45

Table 4.9 A comprehensive analysis of the overall results over the period from 1995 to 2035 is being conducted.

Year (i)	Anuual Rated Installed Capac-ity (GW)	Total Amount of Rated Solar Power (xi)(GW)	Maximum Output (KW)	Mean Output (KW)	Mean Output per Day (KW/day)	Total Production (KWh)	Monetary Benefits $	Emissions (kg/year) Carbon Dioxide (kg)	Emissions (kg/year) Sulphur Dioxide (kg)	Emissions (kg/year) Nitrogen Oxides (kg)
1995	0	0	0	0	0	0	0	0	0	0
2000	0.035	0.035	34,873.475	6,322.19	151,732.63	55,382,400.82	3,738.68	559.79	2.4278	1.1865
2005	0.021	0.056	20,924.085	3,793.314	91,039.578	33,229,440.49	2,243.21	335.87	1.4567	0.7119
2010	0.134	0.19	133,515.59	24,204.956	580,919.212	212,035,477.4	14,313.81	2,143.196	9.295	4.5426
2014	4.6	4.79	4,583,371	830,916.4	19,942,002.8	7,278,829,821	491,369.45	73,572.4	319.088	155.94
2020	33.31	38.1	33,189,584.35	6,016,918.54	144,406,111.6	52,708,222,033	3,558,155.77	532,760.14	2,310.61	1,129.209
2022	20.3	58.4	20,226,615.5	3,666,870.2	88,004,925.4	32,121,792,473	2,168,434.77	324,678.2	1,408.15	688.17
2030	141.6	200	141,088,116	25,577,774.4	613,866,868.8	2.24061E+11	15,125,633.65	2,264,750.4	9,822.37	4,800.24
2035	105	305	104,620,425	18,966,570	455,197,890	1.66147E+11	11,216,041.9	1,679,370	7,283.53	3,559.5
Total			303,897,425	55,093,370	1,322,241,490	4.82618E+11	32,579,931.24	4,878,170	21,156.94	10,339.5

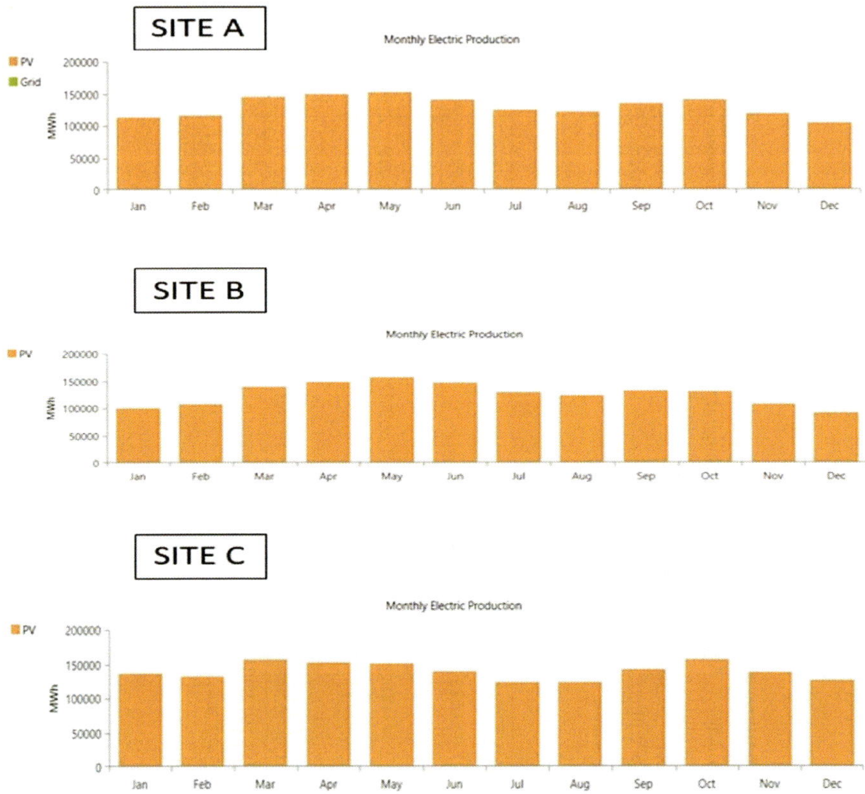

Figure 4.7 The energy generated each month from all locations is being depicted.

4.3.2 The resources required to meet the expected demand for SPV are being determined

Table 4.10 presents a comprehensive overview of the materials needed for SPV deployment in the coming decade. It calculates the quantity of SPVs, yearly installed capacity, valuable metals (gold and silver), and the amount of inverters needed each year grounded on the predicted cumulative installation of 305 GW by 2035.

The forecasted installation of 305 GW of SPV systems by 2035 is depicted in Graphs 4.3, 4.4, and 4.5, which illustrate the inclination in the average number of panels, the amount of gold (Au) required, and the amount of silver (Ag) required per year, correspondingly. The quantity of essential materials necessary is presented in Graph 4.16. The demand for each SPV component follows a similar pattern. By 2035, India will require

Table 4.10 The expected accumulation of materials necessary for the deployment of SPV until 2035 is being predicted.

Year (i)	**Annual Installed Capacity**	**Cumulative Installed Capacity**	**No of req SPM**	**No of req Inverter**	**Ag req**	**Au Req**
1995	0	0	0		0	0
2000	35,000,000	0.035	156,250	805	805.13	0.15778
2005	21,000,000	0.056	95,664.9	483	483.08	0.09467
2010	134,000,000	0.19	729,532	3,082	3,082.51	0.60407
2014	4,600,000,000	4.79	22,081,418.06	105,800	105,817.36	20.7368
2020	3.331E+10	38.1	167,084,679	766,130	766,255.71	150.16148
2022	2.03E+10	58.4	104,166,676.1	466,900	466,976.61	91.5124
2030	1.416E+11	200	793,152,910.4	3,256,800	3,257,334.4	638.3328
2035	1.05E+11	305	633,445,966.9	2,415,000	2,415,396.3	473.34
Total					2,645,216	1,374.94

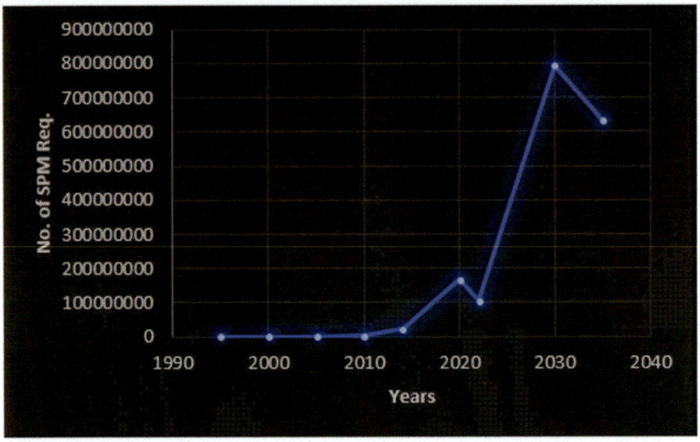

Graph 4.3 As per the evaluation, the deployment of SPM (solar photovoltaic modules) is deemed essential by 2035.

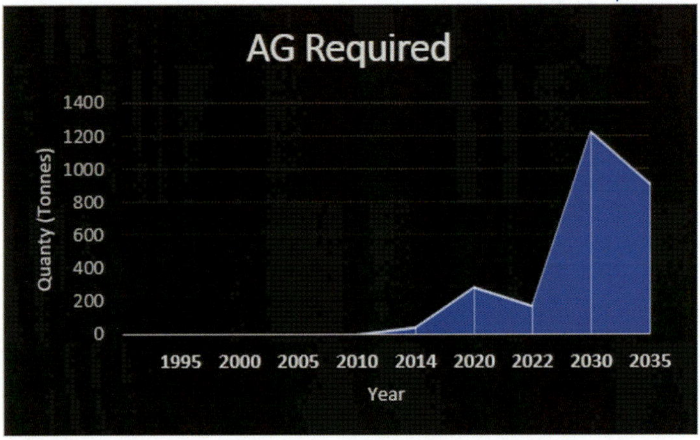

Graph 4.4 The demand for Ag (silver) is projected to persist until 2050.

approximately 7.01 million inverters and 1.74 billion solar panels, as well as 1374.94 tons of gold and 2645.216 tons of silver. 70% of the waste generated comes from solar panels, while only 30% is from BOS components such as inverters, cables, and mounting structures. This is why the waste generated from BOS components has been included in the research alongside the waste from SPVs.

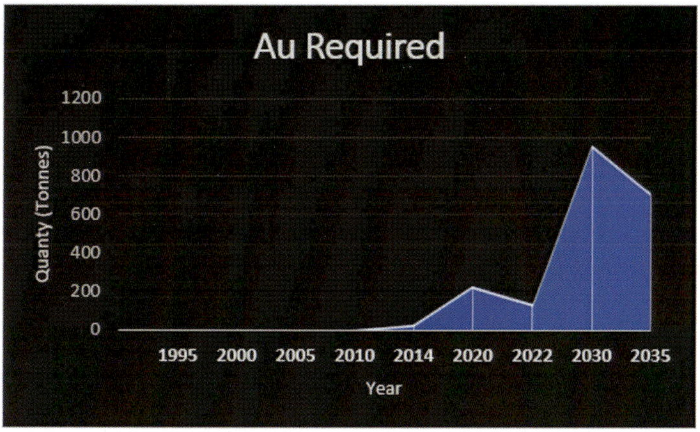

Graph 4.5 The demand for Au (gold) is projected to persist until 2050.

4.3.3 An inventory of waste and resources is being conducted

The predicted EOL energy loss from SPV in India is contrasted with the loss of solar power indicated in a 2016 IRENA analysis in Graph 4.7. It is obvious that the volume of trash in the forthcoming years will be significantly more than what was forecasted by IRENA as India boosted its SPV deployment goals under the JNNSM program from 21 to 101 GW in 2014.

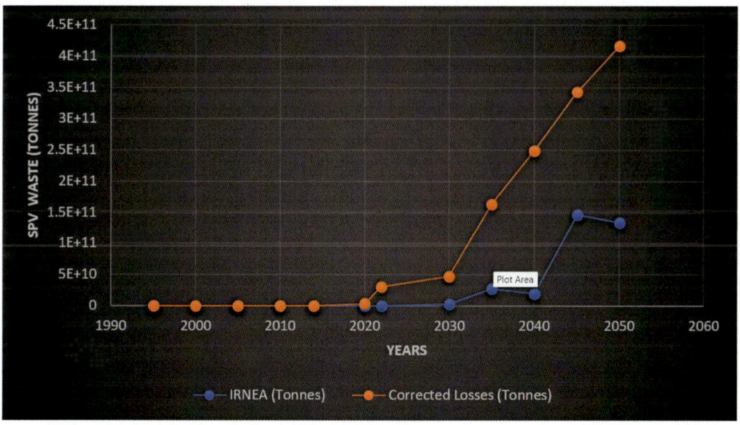

Graph 4.6 A comparison between the estimated waste generated by the IREA and the end-of-life solar panel trash by 2050 is being quantified.

India's tendency for early breakdown due to poor management is significant. India has a consistent overall breakdown percentage, as recorded in section 4.2, while the IRENA description emphasizes its EOL SPV waste estimate on the Weibull probability function.

Graph 4.6 shows the solar photovoltaic (SPV) systems' need for recharging each year due to various forms of degradation being analyzed between 1995 and 2050, including regular, early, and fixed losses. Losses and waste production are inversely related, which highlights the importance of repowering. The data shows that early losses generate more waste than regular losses, but fixed losses are inversely proportional to the deployment rate. The quantity of e-waste from solar photovoltaic systems until 2050 is shown

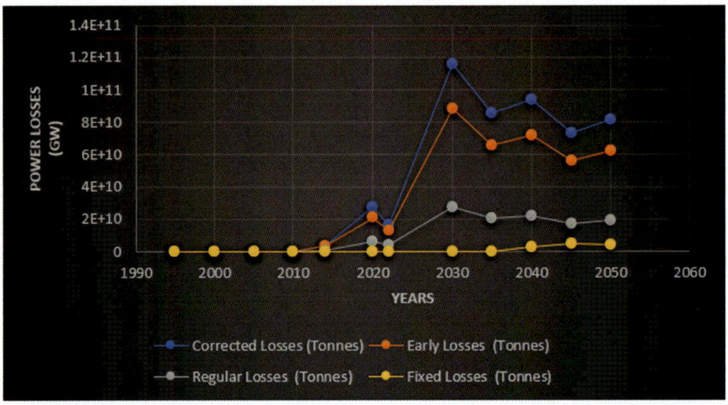

Graph 4.7 The SPV systems' need for recharging each year due to various forms of degradation is being analyzed.

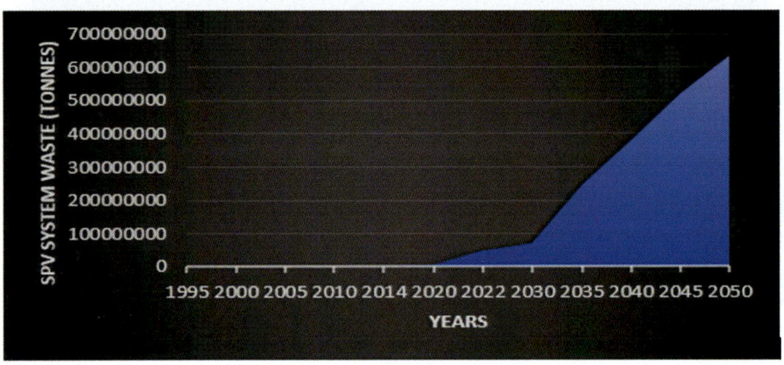

Graph 4.8 The total waste generated by SPV systems on a yearly basis from 1995 to 2050 is being calculated.

Table 4.11 The predicted weight and financial contribution of metals present in end-of-life PV system waste generated in India between 2020 and 2047 is being estimated.

Metal	Mass Share (to	Mass Share (%	Price (US	Commodity Value (USD)	Economic Share (%)	Mass Recovered (tonne	Economyrecovered (USD	MASS loss (tonne)	Money loss (USD
Ag	72687.63297	0.222643638	615000	9952815004	0.415229187	69053.25132	9455174254	3634.381649	497640750.2
Au	5667.275854	0.017358977	49375000	4857419101	0.202650425	2040.219307	1748670876	3627.056547	3108748225
Fe	9407172.749	28.81435365	600	1.62637E+11	6.785177191	8466455.474	1.46373E+11	940717.2749	16263696148
Mg	324001.8151	0.99242388	2010	646309748.5	0.026963896	106920.599	213282217	217081.2161	433027531.5
Ni	2357.937812	0.00722241	14520	247275.4911	1.03163E-05	966.754503	101382.9513	1391.183309	145892.5397
Pb	22639.48937	0.069345198	2320	3642260.483	0.000151954	22186.69958	3569415.274	452.7897873	72845.20967
Si	435144.343	1.332855612	2998	1738793770	0.072542082	435144.343	1738793770	0	0
Sn	7969.130219	0.024409601	19340	3762080.459	0.000156953	2550.12167	1203865.747	5419.008549	2558214.712
Ta	231.3173818	0.00070853	194000	31795.67351	1.32651E-06	48.57665018	6677.091438	182.7407316	25118.58207
Al	1.65E+07	50.66309308	2000	1.67596E+12	69.92066916	16540244.99	1.67596E+12	0	0
Ti	5.618478029	1.72095E-05	640	0.061882395	2.58172E-12	2.921608575	0.032178845	2.696869454	0.029703549
Zn	461812.3408	1.414540209	2610	1704988047	0.071131715	124689.332	460346772.7	337123.0088	1244641274
Cd	37380.03613	0.11449578	1100	4707842.029	0.00019641	35511.03433	4472449.928	1869.001807	235392.1015
Cu	5192732.868	15.90544208	6460	5.33549E+11	22.25954002	5192732.868	5.33549E+11	0	0
Cr	5179.412503	0.015864642	9900	813478.2752	3.39381E-05	1035.882501	162695.655	4143.530002	650782.6201
GA	78205.39753	0.239544658	295000	5526437143	0.230561706	70384.85778	4973793429	7820.539753	552643714.3
In	3910.269877	0.011977233	460000	21543738.02	0.0008988	3519.242889	19389364.21	391.0269877	2154373.802
Mn	19.27644848	5.90441E-05	5	0.005690806	2.37419E-13	7.132285939	0.002105598	12.14416254	0.003585208
MO	7820.539753	0.023954466	17800	3334595.971	0.000139119	1407.697156	600227.2748	6412.842598	2734368.696
Se	7820.539753	0.023954466	50300	9423043.671	0.000393127	6960.28038	8386508.867	860.2593729	1036534.804
Te	34520.06757	0.10573564	89000	324850129.1	0.013552674	32794.06419	308607622.6	1726.003378	16242506.45
Total	32647523.05	100		2.39694E+12	100	31114656.34	2.37482E+12	1532866.705	22126253673

in Graph 4.8 and Table 4.11, considering the necessary materials including precious metals, inverters, and solar photovoltaic modules, along with the resources used and commercial share. The projected amount of waste from 2022 until 2050 is 1.909 billion tons. Graph 4.9 and Table 4.12 categorize the waste according to whether it came from the solar photovoltaic deployment or balance of system components and show the annual amounts of gold and silver available for recycling and disposal.

By 2050, e-waste from solar photovoltaic (SPV) systems is expected to reach 35 million tons, making it crucial for recycling businesses to take action soon. This waste will greatly impact India in the next 5 years and will become even more severe in the future. Graph 4.9 shows the forecasting of proportion of waste generated by different components and technologies in end-of-life solar panels. Other components not considered in this study may also be present in SPV waste. It is also expected that waste from CdTe and CiGs technologies will increase due to advancements in technology after 2045.

Table 4.12 predicts that 1.909 billion tons of metal waste will be generated by SPV and BOS from 2022 to 2050, linked to the expected SPV deployment of 305 GW by 2035. Graph 4.10 shows the expected proportion of major materials used in EOL SPV waste, including silicon for solar panels (1%), aluminum for EVA, copper for wiring (6%), panel laminate (5%), glass (51%), and panel frames (31%).

Table 4.12 The estimated decrease in power generation (in GW), material waste, and loss of precious metals (in tons) between 2020 and 2047 is being calculated.

Year (i)	Corrected Losses	Total Year Wise Wasete	Year Wise Waste Genrated SPM	Year Wise Waste for BOS	Ag Waste	Ag Recovered	Au Waste	Au Recovered
1995	0	0			0	0	0	0
2000	2637993.536	4013.208876	26756063.58	13228539.76	2.287889397	2.173494927	0.178381123	0.064217204
2005	4220789.657	8026.417753	53512127.16	26457079.52	4.575911343	4.347115776	0.356762246	0.128434408
2010	15375733.75	10434.34308	69565765.31	34394203.37	5.948644483	5.651212259	0.463790919	0.166964731
2014	353566505	25799.19992	172003265.9	85040612.74	14.7082417	13.97282962	1.14673579	0.412824884
2020	2741627266	553249.5094	3688514479	1823648695	315.4024988	299.6323739	24.59111194	8.852800298
2022	2024151852	4698318.332	31323688317	15486831800	2678.463996	2544.540797	208.8332119	75.17995628
2030	14670835530	7224433.525	48165298310	23813539006	4118.576828	3912.647986	321.1152482	115.6014893
2035	13337539210	24845020.29	1.65642E+11	81895398138	14163.89366	13455.69898	1104.323935	397.5566165
2040	25738012954	37911133.36	2.52754E+11	1.24965E+11	21612.75133	20532.11376	1685.093088	606.6335117
2045	32281798571	52221638.15	3.48162E+11	1.72136E+11	29771.02397	28282.47277	2321.173589	835.6224921
2050	97104048470	63421163.64	4.22829E+11	2.09052E+11	15732.5677	14945.93931	0	0
TOTAL		190923230	1.27289E+12	6.29331E+11	88420.20067	83999.19064	5667.275854	2040.219307

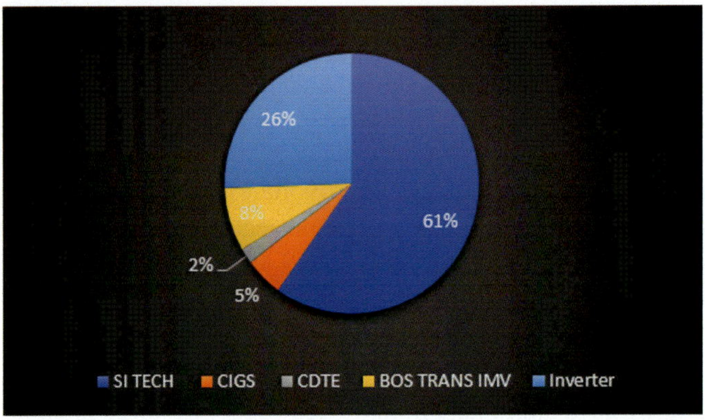

Graph 4.9 The proportion of waste generated by different components and technologies in end-of-life solar panels is predicted to vary between 2022 and 2050.

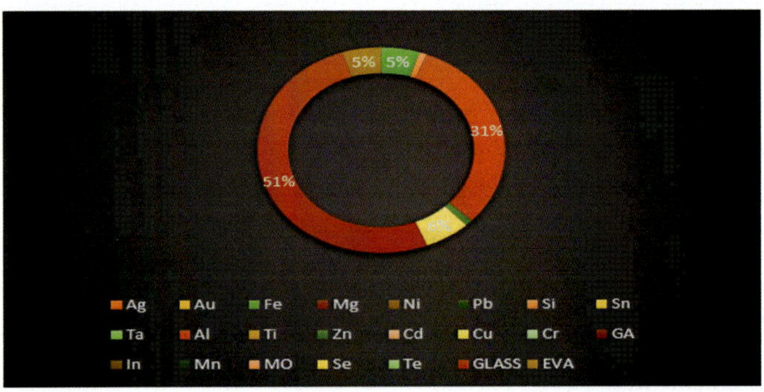

Graph 4.10 The estimated proportion of SPV system components that will become obsolete by 2050 is being calculated.

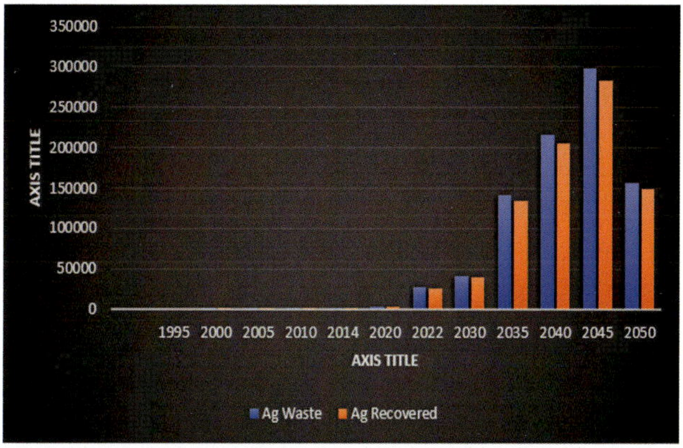

Graph 4.11 The trend in the amount of silver (Ag) produced as waste and recovered by 2050 is being analyzed.

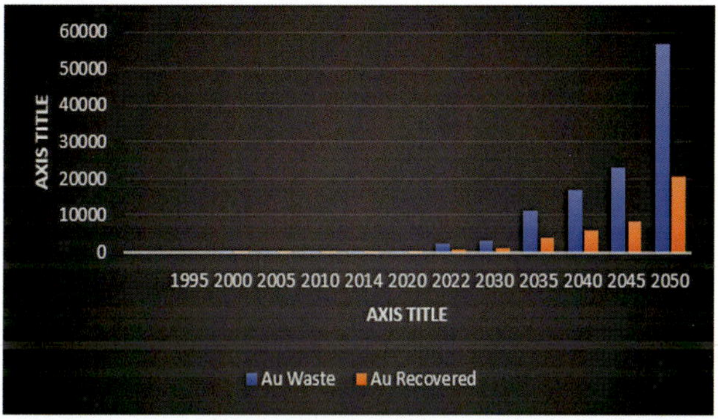

Graph 4.12 The trend in the amount of gold (Au) produced as waste and recovered by 2050 is being analyzed.

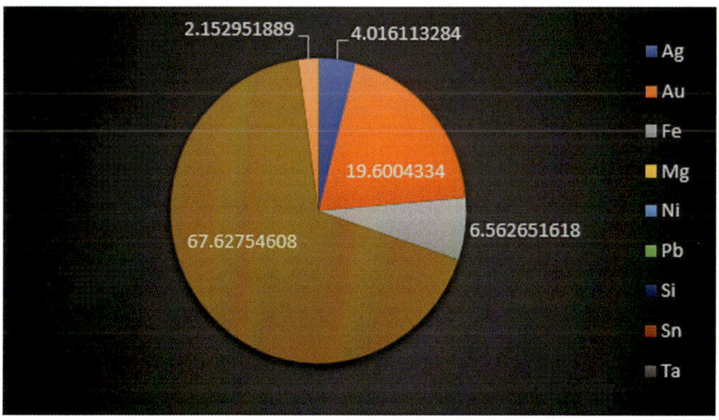

Graph 4.13 The expected financial contribution of metals in waste generated from solar photovoltaic (SPV) systems between 2022 and 2050 is being estimated.

Graphs 4.11–4.13 show the amount of waste formed and the amount of valuable resources (Ag and Au) recovered by 2050. Graph 11 demonstrates that due to high silver recycling, approximately 95% of the Ag can be recovered from EOL, SPM, and converter, which will be enough to meet a substantial share of future silver demand. India's dependence on imports of scarce materials highlights the need for further attention. Graphs 4.14–4.16 show the trash situation for critical and valuable materials.

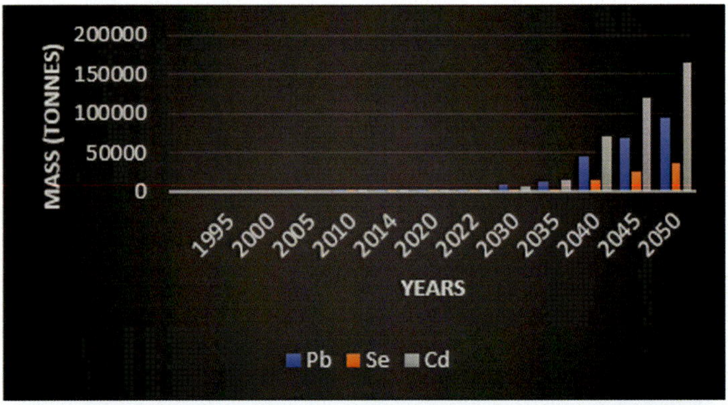

Graph 4.14 The number of hazardous metals produced by the end-of-life solar photovoltaic (SPV) system by 2050 is being estimated.

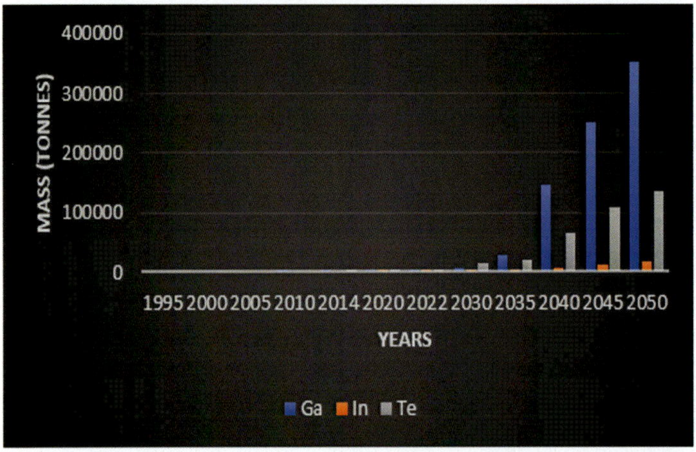

Graph 4.15 The required amount of crucial materials by 2035 is being determined.

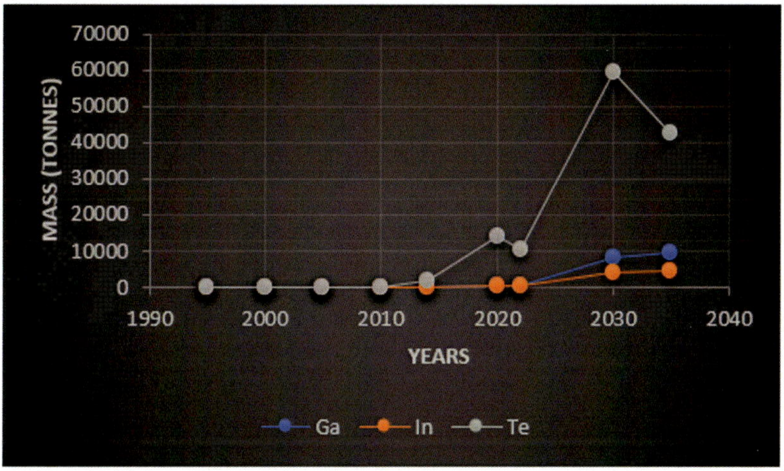

Graph 4.16 The trend in the output of critical materials as waste by 2050 is being analyzed.

4.3.4 An economic analysis of recycling and metal inventory is being conducted

The study evaluated the recycling yields of 21 elements, with an overall return rate of 60%–97% for the entire SPM. BOS has a 90% yield in recycling. Recycling is critical for environmental protection and preserving natural resources. Table 4.10 shows the metal inventory of SPV based on mass and commercial value and the amount of elements recovered after recycling.

Graph 4.13 presents the economic analysis of SPV waste elements, valued at 247 trillion US dollars between 2022 and 2050, with a potential 70% recovery (214 trillion US dollars) through advanced recycling processes. Recycling is crucial and economically reasonable for EOL SPV waste. Table 4.11 and Graph 4.13 show that the main valuable materials in EOL SPV waste are iron (6.5%), copper (2.1%), aluminum (67%), silver (4%), and gold (19.6%). The majority of commercial benefits come from precious and base metals, while other important metals like magnesium, gallium, tellurium, and indium also have potential commercial value.

Silver and gold are used in C–Si cells and inverters, respectively, and are commercially important due to their scarcity. This study analyzed the potential silver waste of 726,878.266 tons and gold waste of 56,672 tons generated from SPV systems. Table 4.11 depicts the mass and monetary losses of all materials if inadequate recycling techniques are employed.

4.3.5 Sensitivity analysis of EOL SPV systems

As SPM technologies advance, the amount of waste and recovered materials will change. This also holds true for the size of the total market, the proportion of metal components, the rate of local deployment, and the effectiveness of recycling. The study makes use of an accurate prediction algorithm that takes these changes into consideration. Despite Si-based SPM's current market dominance, Table 14 demonstrates that by 2035, advanced Si-based and thin-film SPM will hold a sizable market share. With a projected market share of 22.8% for CiGs, 4.2% for CdTe, and 74% for Si-based technologies in 2035, the market is split among three primary technologies (Domínguez, A. et al 2017).

The report states that advancements in SPM technology result in an upsurge in waste creation by 22% and a reduction in silver production by 10%. A shift toward thin-film SPM leads to a rise in demand for critical elements like indium, selenium, and gallium, but a reduction in the need for magnesium, aluminum, and silicon. The recycling yield was calculated using advanced US equipment, and India's recycling network is noted to be insufficient, only able to handle 22% of e-waste generated. The retrieved material volume in the report represents the highest yield possible, but definite retrieval could be inferior. The specific resources recovered may also be subordinate since the reprocessing process for most components is still new (D. Osteng et al., 2022).

The transformer is a vital component of SPV systems and most use copper–iron transformers due to copper's availability and cost-effectiveness compared to aluminum. However, using aluminum conductors instead of copper can reduce copper usage and waste while increasing the proportion of aluminum used. Technological advancements can also reduce the amount of equipment used, requiring further advancements in recycling technology and changing the components in EOL SPV waste (Ganesan, K. et al., 2022). The extraction of metals may also be affected by the collection, transportation, and disposal methods of EOL SPV waste (Wang, C. et al., 2022).

4.4 Discussion and Results

It is estimated that by 2035, 482,618 GW of energy will be produced annually, with a daily production of 1322.24149 GW. This could bring in annual financial gains of around $32 million if net metering and a relentless reimbursement amount of 0.044$/KWh are available. SPV is vital for India's power requirements and coal power plants may be substituted by SPV if there is reliable growth in solar power production and administration support. However, for this to be achieved, the electrical grid system must be upgraded with hybrid smart grids and backup energy tech like pumped hydro or lithium-ion batteries. The power manufacture at site (a) is not as consistent as sites (b) and (c) due to India's weather patterns and the Monsoon season. Delhi (site c) has the highest energy production, with peaks in March, April, May, and October, as shown in Tables 4.5 and 4.7 and Figures 4.6 and 4.7 (Islam, M. R. et al., 2018, Yang, D., et al., 2022, Mustafa, J. et al., 2022).

According to Homer Pro simulations, even though renewable energy is perceived as a green source, all sites analyzed had considerable heights of sulfur dioxide, nitrogen oxide, and carbon dioxide discharges. These discharges are associated with the distribution, transmission, and manufacturing of properties or electricity. The production of SPV systems encompasses two stages that produce noteworthy discharges, such as the manufacturing of SPMs. The total carbon dioxide discharges from the production of SPMs are approximately 3,312,106 kg. Moreover, the waste disposal of SPVs is also a concern and will release a significant amount of CO_2, similar to coal-based thermal generation stations. Thus, SPVs are not emission-free, and policies for the integration of end-of-life SPVs are essential to make it a clean energy system (Chandrasekharam, D. & Ranjith Pathegama, G., 2020).

Due to the recent surge in SPV deployment, which has increased fivefold, India is likely to have a higher rate of early losses, as shown in Graphs 4.1 and 4.6. This will outcome in a bulky sum of EOL SPV waste, compared to the estimates made by IRENA. The production of SPVs and their associated components, such as the balance of system, requires a large amount of materials and other assets, as indicated in Tables 5 and 7. The projected waste from this investigation can be compared to the EOL solar trash conveyed by organizations like IEA, IRENA, BTI, and MREN. The harm to SPM is calculated using the Weibull probability distribution and taking into account the market share and reliability of the technology. From 2022 until 2050, India is estimated to create about 1.909 billion tons of electric trash, including rare, vital, and valuable metals from SPV systems, as shown in Graph 4.9. The balance of system accounts for about 30% of the total waste produced by solar systems. According to Graph 4.8 and Table 4.12, EOL solar waste will have a noteworthy impact on India's economy starting in 2022, and its growth is expected to be substantial.

According to data from Table 4.11 and Graphs 4.4, 4.5, 4.11, and 4.12, it seems that a large amount of electronic waste will be produced across

India. The valuation of the trash engendered by Indian solar panel deployment is much larger than most other electronic waste, making it problematic to dispose of because India lacks the organization to accomplish it. This information helps policymakers understand the opportunities and challenges in managing end-of-life solar panel waste. Table 4.11 and Graphs 4.13 and 4.10 demonstrate the monetary and weight portion of each of the 21 materials and other materials contained in end-of-life solar panels. India's target to increase its JNNSM solar execution goals from 22 to 101 GW in 2014 and the higher frequency of initial failure due to poor management have resulted in a greater volume of e-waste being produced compared to other studies. The financial value of all waste expected to be produced between 2022 and 2050 is estimated to be around 247 trillion \$, and up to 70% (214 trillion \$) of it can be recovered with advanced recycling processes. Recycling end-of-life SPV trash is not only desirable but also economically feasible. Although there are challenges in implementing advanced recycling technologies and using reclaimed materials from SPV e-waste for production, there is a need for such a framework to reduce dependence on foreign materials and address the disposal of toxic materials like cadmium and lead (Weckend, S et al., 2016). India currently lacks the infrastructure to properly deal with end-of-life SPV waste (Nain, P. et al. 2021). The current e-trash regulations from 2011 are insufficient, and the lack of specific regulations for SPV waste makes it difficult to establish proper recycling methods. Only a small percentage of India's e-trash can be recycled in the formal economy, and the handling of toxic materials like cadmium, lead, and others poses additional challenges (Mahmoudi, S. et al., 2021, Priyadarshini, P. et al., 2020). The study found that while basic metals like copper, aluminum, iron, silicon, and zinc make up a large part of SPV waste, valuable, toxic, and hazardous elements also need to be considered. The procurement of these elements is affected by economic and political factors, with toxic materials primarily used in thin-film SPVs and crucial components primarily used in silicon SPVs (Rathore, N. et al., 2022, Daniela-Abigail, H. L. et al., 2022).

Graphs 4.14–4.16 show India's usage of these possessions from 1995 to 2035. India is heavily dependent on imports for critical elements such as gallium as it lacks the necessary natural resources (Ueberschaar, M. et al., 2017). This analysis does not consider repowering, which would result in a significant increase in both the quantity of raw materials required and the amount of EOL SPV WEEE in 2050 (de Oliveira, R. P. et al., 2021).

The report evaluates the final power manufactured, emissions, and economic benefits from solar power in India until 2035, including the EOL SPV waste. It considers the use of different technologies, their economic value, and the impact of repowering. The study finds that silicon-based SPV technology will likely dominate the industry, while the volume and type of raw materials used in other technologies will not change much (Veleva, V. et al., 2018). The report highlights the importance of monitoring the quantity and quality of EOL SPV waste, as well as implementing relevant legislation and a proper collection and recycling system. The waste can be seen as a supplementary resource sector, and a well-functioning collection and recycling infrastructure is necessary to make use of these valuable materials while minimizing harm to the environment (Ndzibah, E. et al., 2022).

4.4.1 Key administrative insights gathered from the current study

This section highlights the administrative considerations derived from the current study and focuses on their role in supporting the smooth growth of the SPV industry in India.

The use of solar tech can lead to cleaner power commerce, bringing benefits such as energy security, safety, environmental protection, and economic growth. Quick action and a strong export market could result in job creation and profitable growth. However, the high cost of manufacturing is the biggest obstacle to sustainable growth, and the country must improve its R&D operations to lower these costs and make SPV accessible. To ensure the growth of the solar sector, simplifying governance is crucial to attract developers and increase public interest.

Governments, organizations, and regulations all play a character in promoting the embracing of new tech in the household and industrial sectors in India. Previous studies show that countries with a strong SPV support system and efficient regulations have seen a significant increase in SPV usage. However, adopting SPV requires significant financial resources, and regulatory institutions must effectively manage and provide funding for sustainable implementation in rural areas. The next section discusses recommended actions.

To promote the adoption of solar photovoltaic (SPV) energy in India, there are several suggestions made based on current research. These include

1. Promoting SPV energy through demand-side reductions, increased manufacturing productivity, and replacement of coal power plants with SPV energy. These initiatives should be tailored to meet India's sustainable development goals and involve local communities.

2. Modernizing the Indian electrical grid by utilizing SPV power systems for cost-effective and environmentally friendly energy. This requires modernizing transmission lines, distribution facilities, and investing in energy storage systems.
3. Updating policies to increase local capacity and SPV production. The government is working to develop a comprehensive energy policy to leverage India's SPV potential.
4. Careful preparation, such as finding suitable land for SPV systems and avoiding impacts on agriculture.
5. Simplifying administration by promoting cooperation between different organizations, providing accurate information, and simplifying policies and certification processes (Sindhu, S. et al., 2017).

4.4.1.1 Policy and administration guidance

This section highlights the administrative considerations related to managing EOL solar waste in India, with a particular emphasis on the actions and inputs required to establish a framework for effectively handling such waste.

The study offers significant functioning and legislative suggestions for the solar recycling division and government tangled in creating a background for electric waste management. Key findings include

The need for the industry to establish a proper framework for recycling SPV components such as solar panels and other BOS elements, as projections estimate a large volume of SPV waste in 2022 (0.29 million tons).

A forward-thinking approach is needed to manage the anticipated 1.909 billion tons of SPV waste generated by 2050, with the creation of a specialized reverse supply chain based on circular economy principles to track the e-waste.

Economic analysis shows that all PV e-waste produced between 2022 and 2050 is valued at over 247 trillion $, with the potential to recover 70% (214 trillion US dollars) through advanced recycling methods. Regular evaluations and material flow analyses are crucial to improve recycling capabilities and innovations.

4.4.2 Employing a CE for EOL SPV in India is advocated

The proper disposal of EOL equipment has both ecological and financial paybacks. However, many companies have yet to fully utilize their potential for managing EOL waste. To address this, a framework for the commercial management of EOL waste is needed (Palaniappan, S. K., et al., 2021). The

PV division is required to create a supply chain strategy for EOL recovery, which will facilitate CE certifications. Techniques such as life-cycle analysis, alternative materials, tax enticements, and inter-sector collaboration can help achieve a CE for solar (Wang, X.,et al 2022). A CE in solar manufacturing can be achieved through efficient production processes, material utilization, and power efficiency. A CE for solar enterprise can be accomplished through design for modularity, disassembly, versatility, and recycling (Xu, Y., et al., 2018).

Efforts such as extended producer responsibility (EPR), promoting recycling, and creating reverse logistics systems can help in creating a closed-loop for the collection and disposal of SPV waste (Trypolska, G., et al., 2022, JV, R. L. et al., 2019). Techniques like down-cycling, cascading, energy recovery, and substance restoration can aid in the cyclical recycling of SPV (Heath, G. A., et al., 2020, Korniejenko, K., et al., 2021). A regulated and easily disassembled architecture of SPM and BOS can revolutionize recycling. All parties involved in the SPV supply chain must understand their waste management obligations (Curtis, T. L., et al., 2021). Optimizing producers responsibility organizations (PRO) and EPR programs is crucial to addressing EOL SPV waste (Di Foggia, G. et al. 2022, Lee, J. et al. 2022).

4.5 Conclusion

India is facing significant environmental issues due to air pollution caused by the energy division which primarily relies on nonrenewable power sources. SPV power, which is both affordable and environmentally friendly, is crucial to India's sustainable development. India is well-suited for large-scale SPV projects due to its favorable physical location and market, including 300 days of sun per year, abundant barren land, and low external costs for SPV electricity generation.

Despite the favorable conditions in India for the growth of solar power, the industry faces challenges such as a lack of regulations, inadequate infrastructure, limited funding, low innovation, and transparency issues. However, the government is taking steps to address these issues and make it easier for solar energy companies to operate in India. Efforts are being made to boost capacity, allow for full foreign investment, provide financial assistance, and establish a favorable atmosphere for partnerships and technology transfers, with the goal of enticing significant solar energy companies globally to establish India as a center for solar power ventures.

The study advocates for the implementation of solar tech and associated schemes for a sustainable imminent in India. It aims to identify the challenges and find solutions to overcome them for the wider adoption of SPV energy. However, it is important to note that no human-made effort, including SPV energy, is completely safe for the environment. The impact of the environment depends on the project's scale, type, and location. The study's findings can help decision-makers improve regulations and reduce emissions through the adoption of SPV energy leading to a CE.

This study also sheds light on the potential problems posed by the EOL solar waste in India, which is not well equipped to handle it due to the lack of recycling regulations and facilities. By 2050, the research findings predict that photovoltaic systems and their balance of system components will generate around 1.909 billion tons of waste, mostly made up of aluminum frames (31%), glass covers (51%), EVA for module lamination (5%), copper cables (6%), and silicon for photovoltaic modules (1%). The major contributors to this waste are 21 metals, with a total worth of $247 trillion, including gold (19%), aluminum (67%), copper (2%), and silver (4%). These findings indicate that by using advanced recycling methods, $214 trillion worth of resources can be recaptured from the EOL waste.

The study proposes the creation of a monitoring system to manage the waste generated by SPV technology in India. It is crucial to establish a specialized reverse supply chain for effective recycling and retrieval of valuable materials from end-of-life SPV waste. A financial framework is necessary to support this ecosystem. The study serves as a resource for future evaluations of the financial viability, life cycle analysis, and improvement of the closed-loop waste management system. It can aid policymakers and researchers in their work. The study includes categorizing the metals involved, material flow analysis, and economic feasibility analysis for efficient collection, deconstruction, and recycling of the waste. SPV technology is a clean source of energy that can help India address its energy needs, environmental concerns, and strive for a sustainable future. The study is positive and highlights the importance of promoting a circular economy for SPV waste management.

References

[1] Anam, M. Z., Bari, A. M., Paul, S. K., Ali, S. M., & Kabir, G. (2022). Modelling the drivers of solar energy development in an emerging

economy: Implications for sustainable development goals. Resources, Conservation & Recycling Advances, 13, 200068.

[2] Ansanelli, G., Fiorentino, G., Tammaro, M., & Zucaro, A. (2021). A life cycle assessment of a recovery process from end-of-life photovoltaic panels. Applied Energy, 290, 116727.

[3] Arora, N., Bhattacharjya, S., Baksh, S. K., & Anand, M. (2018). Greening the Solar PowerSPV value chain. European Union's Resource Efficiency Initiative (EU-REI) Project. www.resourceefficiencyindia.c om.

[4] Chandrasekharam, D., & Ranjith Pathegama, G. (2020). CO2 emissions from renewables: solar pv, hydrothermal and EGS sources. Geomechanics and Geophysics for Geo-Energy and Geo-Resources, 6(1), 1-17.

[5] Contreras-Lisperguer, R., Mu noz-Cerón, E., Aguilera, J., & de la Casa, J. (2017). Cradle-to-cradle approach in the life cycle of silicon solar photovoltaic panels. Journal of Cleaner Production, 168, 51–59.

[6] Curtis, T. L., Buchanan, H., Smith, L., & Heath, G. (2021). A Circular Economy for Solar Photovoltaic System Materials: Drivers, Barriers, Enablers, and US Policy Considerations (No. NREL/TP-6A20-74550). National Renewable Energy Lab.(NREL), Golden, CO (United States).

[7] D'Adamo, I., Gastaldi, M., & Rosa, P. (2020). Recycling of end-of-life vehicles: Assessing trends and performances in Europe. Technological Forecasting and Social Change, 152, 119887.

[8] Daniela-Abigail, H. L., Tariq, R., El Mekaoui, A., Bassam, A., De Lille, M. V., Ricalde, L. J., & Riech, I. (2022). Does recycling solar panels make this renewable resource sustainable? Evidence supported by environmental, economic, and social dimensions. Sustainable Cities and Society, 77, 103539.

[9] de Oliveira, R. P., Benvenuti, J., & Espinosa, D. C. R. (2021). A review of the current progress in recycling technologies for gallium and rare earth elements from light-emitting diodes. Renewable and Sustainable Energy Reviews, 145, 111090.

[10] de Sousa Jabbour, A. B. L., Luiz, J. V. R., Luiz, O. R., Jabbour, C. J. C., Ndubisi, N. O., de Oliveira, J. H. C., & Junior, F. H. (2019). Circular economy business models and operations management. Journal of cleaner production, 235, 1525–1539.

[11] Di Foggia, G., & Beccarello, M. (2022). An overview of packaging waste models in some European countries. Recycling, 7(3), 38.
[12] Domínguez, A., & Geyer, R. (2017). Photovoltaic waste assessment in Mexico. Resources, Conservation and Recycling, 127, 29–41.
[13] Duran, A. S., Atasu, A., & Van Wassenhove, L. N. (2022). Cleaning after solar panels: applying a circular outlook to clean energy research. International Journal of Production Research, 60(1), 211–230.
[14] Fiandra, V., Sannino, L., Andreozzi, C., Corcelli, F., & Graditi, G. (2019). Silicon photovoltaic modules at end-of-life: Removal of polymeric layers and separation of materials. Waste Management, 87, 97–107.
[15] Ganesan, K., & Valderrama, C. (2022). Anticipatory life cycle analysis framework for sustainable management of end-of-life crystalline silicon photovoltaic panels. Energy, 245, 123207.
[16] Ganti, P. K., Naik, H., & Barada, M. K. (2022). Environmental impact analysis and enhancement of factors affecting the photovoltaic (PV) energy utilization in mining industry by sparrow search optimization based gradient boosting decision tree approach. Energy, 244, 122561.
[17] Gervais, E., Shammugam, S., Friedrich, L., & Schlegl, T. (2021). Raw material needs for the large-scale deployment of photovoltaics–Effects of innovation-driven roadmaps on material constraints until 2050. Renewable and Sustainable Energy Reviews, 137, 110589.
[18] Hagelüken, C., & Goldmann, D. (2022). Recycling and circular economy—towards a closed loop for metals in emerging clean technologies. Mineral Economics, 35(3-4), 539–562.
[19] Heath, G. A., Silverman, T. J., Kempe, M., Deceglie, M., Ravikumar, D., Remo, T., ... & Wade, A. (2020). Research and development priorities for silicon photovoltaic module recycling to support a circular economy. Nature Energy, 5(7), 502–510.
[20] Heath, G. A., Silverman, T. J., Kempe, M., Deceglie, M., Ravikumar, D., Remo, T., ... & Wade, A. (2020). Research and development priorities for silicon photovoltaic module recycling to support a circular economy. Nature Energy, 5(7), 502–510.
[21] International Energy Agency. (2021). India energy outlook 2021. OECD Publishing.
[22] Islam, M. R., Mahfuz-Ur-Rahman, A. M., Muttaqi, K. M., & Sutanto, D. (2018). State-of-the-art of the medium-voltage power converter technologies for grid integration of solar photovoltaic power plants. IEEE Transactions on Energy Conversion, 34(1), 372–384.

[23] Islam, M. T., & Huda, N. (2020). Assessing the recycling potential of "unregulated" e-waste in Australia. Resources, Conservation and Recycling, 152, 104526.

[24] ITRP, V. (2019). International Technology Roadmap for Photovoltaic (ITRPV): Results 2018. Itrpv, no. March. http://www.itrpv.net/.cm4all/iproc. php/ITRPV Seventh Edition including maturity report, 20161026.

[25] ITRPV. (2020). International Technology Roadmap for Photovoltaic (ITRPV): 2019 Results.

[26] Jain, S., Sharma, T., & Gupta, A. K. (2022). End-of-life management of solar PV waste in India: Situation analysis and proposed policy framework. Renewable and Sustainable Energy Reviews, 153, 111774.

[27] Jolly, K., Modibbo, U. M., Chauhan, J., & Ansari, M. (2022). A Cost-Based Optimization Modelling of Solar Power Generation in India for Sustainable Development. In Computational Modelling in Industry 4.0 (pp. 35–47). Springer, Singapore.

[28] JV, R. L., Jabbour, C. J. C., Ndubisi, N. O., Caldeira de Oliveira, J. H., & Junior, F. H. (2019). Circular economy business models and operations management.

[29] Karali, N., & Shah, N. (2022). Bolstering supplies of critical raw materials for low-carbon technologies through circular economy strategies. Energy Research & Social Science, 88, 102534.

[30] Kastanaki, E., & Giannis, A. (2022). Energy decarbonisation in the European Union: Assessment of photovoltaic waste recycling potential. Renewable Energy, 192, 1–13.

[31] Korniejenko, K., Kozub, B., Bąk, A., Balamurugan, P., Uthayakumar, M., & Furtos, G. (2021). Tackling the circular economy challenges—Composites recycling: Used tyres, wind turbine blades, and solar panels. Journal of Composites Science, 5(9), 243.

[32] Lee, J., Lin, K. Y. A., Jung, S., & Kwon, E. E. (2022). Hybrid Renewable Energy Systems Involving Thermochemical Conversion Process for Waste-to-Energy Strategy. Chemical Engineering Journal, 139218.

[33] Mahmoudi, S., Huda, N., & Behnia, M. (2021). Multi-levels of photovoltaic waste management: A holistic framework. Journal of Cleaner Production, 294, 126252.

[34] Mahmoudi, S., Huda, N., & Behnia, M. (2021). Multi-levels of photovoltaic waste management: A holistic framework. Journal of Cleaner Production, 294, 126252.

[35] Mahmoudi, S., Huda, N., Alavi, Z., Islam, M. T., & Behnia, M. (2019). End-of-life photovoltaic modules: A systematic quantitative literature review. Resources, Conservation and Recycling, 146, 1–16.
[36] Mustafa, J., Husain, S., Khan, U. A., & Akhtar, M. M. (2022). Prediction of diffuse solar radiation using machine learning models based on sunshine period and sky-clearness index for the humid-subtropical climate of India. Environmental Progress & Sustainable Energy, e13973.
[37] Nain, P., & Kumar, A. (2020). Identifying issues in assessing environmental implications of solar PVs-related waste. Recent Developments in Waste Management, 71–90.
[38] NCPRE, NISE, 2017. All-India Survey of Photovoltaic Module Reliability: 2016
[39] Ndzibah, E., Pinilla-De La Cruz, G. A., & Shamsuzzoha, A. (2022). Collaboration towards value creation for end-of-life solar photovoltaic panel in Ghana. Journal of Cleaner Production, 333, 129969.
[40] Ndzibah, E., Pinilla-De La Cruz, G. A., & Shamsuzzoha, A. (2022). Collaboration towards value creation for end-of-life solar photovoltaic panel in Ghana. Journal of Cleaner Production, 333, 129969.
[41] NISE, 2018. Annual report 2017-18. National Institute of Solar Energy.
[42] NITI Ayog, 2015. Report on India's Renewable Electricity Roadmap 2030.
[43] Oteng, D., Zuo, J., & Sharifi, E. (2022). Environmental emissions influencing solar photovoltaic waste management in Australia: An optimised system network of waste collection facilities. Journal of Environmental Management, 314, 115007.
[44] Outlook, I. E. (2021). IEA, Paris, 2021. Accessed: Mar. 28, 2021. https://www.iea.org/reports/india-energy-outlook-2021
[45] Palaniappan, S. K., Chinnasamy, M., Rathanasamy, R., & Pal, S. K. (2021). Recycling of Solar Panels. Materials for Solar Energy Conversion: Materials, Methods and Applications, 47–86.
[46] Priyadarshini, P., & Abhilash, P. C. (2020). Circular economy practices within energy and waste management sectors of India: A meta-analysis. Bioresource Technology, 304, 123018.
[47] Psomopoulos, C. S., Papamichael, I., Kalkanis, K., Chatzistamou, E. D., Kiskira, K., Ioannidis, G. C., & Kaminaris, S. D. (2022, August). End of life treatment of photovoltaic panels. Expected volumes up to 2045 in EU. In AIP Conference Proceedings (Vol. 2437, No. 1, p. 020084). AIP Publishing LLC.

[48] Rathore, N., & Panwar, N. L. (2022). Strategic overview of management of future solar photovoltaic panel waste generation in the Indian context. Waste Management & Research, 40(5), 504-518.
[49] Sheoran, M., Kumar, P., Sharma, S., & Bukya, M. (2022). Current situation analysis of solar PV waste management in India. Materials Today: Proceedings, 58, 773–782.
[50] Sindhu, S., Nehra, V., & Luthra, S. (2017). Solar energy deployment for sustainable future of India: Hybrid SWOC-AHP analysis. Renewable and Sustainable Energy Reviews, 72, 1138–1151.
[51] Suresh, S., Singhvi, S., & Rustagi, V. (2019). Managing India'sSPV module waste. BRIDGE TO INDIA Energy Private Limited Authors.
[52] Tan, V., Dias, P. R., Chang, N., & Deng, R. (2022). Estimating the Lifetime of Solar Photovoltaic Modules in Australia. Sustainability, 14(9), 5336.
[53] TERI, 2015. Technical Manual for Banks & FIs on Grid-Connected Rooftop Solar Power.
[54] Trypolska, G., Kurbatova, T., Prokopenko, O., Howaniec, H., & Klapkiv, Y. (2022). Wind and Solar Power Plant End-of-Life Equipment: Prospects for Management in Ukraine. Energies, 15(5), 1662.
[55] Veleva, V., & Bodkin, G. (2018). Corporate-entrepreneur collaborations to advance a circular economy. Journal of Cleaner Production, 188, 20–37.
[56] Wang, C., Feng, K., Liu, X., Wang, P., Chen, W. Q., & Li, J. (2022). Looming challenge of photovoltaic waste under China's solar ambition: A spatial–temporal assessment. Applied Energy, 307, 118186.
[57] Wang, X., Tian, X., Chen, X., Ren, L., & Geng, C. (2022). A review of end-of-life crystalline silicon solar photovoltaic panel recycling technology. Solar Energy Materials and Solar Cells, 248, 111976.
[58] Weckend, S., Wade, A., & Heath, G. A. (2016). End of life management: solar photovoltaic panels (No. NREL/TP-6A20-73852). National Renewable Energy Lab.(NREL), Golden, CO (United States).
[59] Xu, Y., Li, J., Tan, Q., Peters, A. L., & Yang, C. (2018). Global status of recycling waste solar panels: A review. Waste Management, 75, 450-458.
[60] Yang, D., Wang, W., Gueymard, C. A., Hong, T., Kleissl, J., Huang, J., ... & Peters, I. M. (2022). A review of solar forecasting, its dependence on atmospheric sciences and implications for grid integration: Towards carbon neutrality. Renewable and Sustainable Energy Reviews, 161, 112348.

[61] Yang, Q., Huang, T., Chen, F., Uche, J., Wang, Y., Yuan, P., ... & Li, J. (2022). Water saving potential for large-scale photovoltaic power generation in China: Based on life cycle assessment. Renewable and Sustainable Energy Reviews, 167, 112681

[62] Yu, H. F., Hasanuzzaman, M., Rahim, N. A., Amin, N., & Nor Adzman, N. (2022). Global challenges and prospects of photovoltaic materials disposal and recycling: a comprehensive review. Sustainability, 14(14), 8567.

5

Resilience and Renewables: Intelligent Techniques for Integration and Strengthening Power Systems

Amit Kumar, Shikha Thakur, and Yog Raj Sood

Electrical Engineering Department, NIT Hamirpur, India
E-mail: amit.ee200@gmail.com; shikhathakur274@gmail.com; yrsood@nith.ac.in

Abstract

As global energy systems advance toward sustainability and stability, incorporating renewable energy sources into power grids has become more common. However, the intermittent nature of renewable resources challenges power system resilience, needing novel methods to ensure grid stability and reliability. This chapter addresses the role of intelligent strategies in improving power system resilience through renewable energy sources. Intelligent approaches use modern algorithms, data analytics, and machine learning to provide proactive maintenance, dynamic forecasting, fault detection, and optimal resource allocation in power systems. These strategies help to strengthen and adapt the grid by forecasting equipment failures, maximizing renewable resource consumption, and detecting anomalies. Furthermore, intelligent techniques improve cybersecurity by detecting and mitigating possible threats, thereby protecting the integrity and dependability of power systems. This study gives insights into the synergistic relationship between intelligent approaches and renewable energy integration in enhancing power system resilience by conducting a thorough review of case studies and technology breakthroughs. Finally, the integration of intelligent techniques paves the way for sustainable and resilient energy infrastructure, ensuring power grid reliability and stability in the face of rising renewable energy penetration and growing grid issues.

DOI: 10.1201/9788743809852-5

Keywords: Resilience, renewable, reliability and grid stability.

5.1 Introduction

In the goal of a sustainable energy future, incorporating renewable energy sources into power systems has emerged as a critical strategy. However, this transformation creates a unique set of obstacles, the most notable of which are the intermittency and unpredictability inherent in renewable energy sources. To address these issues and assure the stability and resilience of power systems, the use of intelligent techniques has become increasingly important. This chapter investigates the critical role of intelligent strategies in improving power system resilience by seamlessly integrating renewable energy sources. Intelligent approaches provide a wide range of solutions to improve the reliability and flexibility of modern power grids, including predictive maintenance, dynamic control strategies, fault detection, and adaptive resource allocation. In the face of a fast-changing energy landscape, knowing how intelligent strategies can be used to improve renewable energy integration is critical. By employing sophisticated algorithms, machine learning, data analytics, and advanced control systems, utilities and grid operators can successfully negotiate the complexity of renewable energy fluctuation while also improving grid resilience.

In response to these issues, the incorporation of intelligent approaches has emerged as a vital strategy for improving power system resilience while accommodating the increasing use of renewable energy sources. Intelligent approaches cover a wide spectrum of modern technologies, such as artificial intelligence, machine learning, data analytics, and optimization algorithms. These methods let power system operators evaluate massive volumes of data, forecast future trends, and optimize system operations with unparalleled precision and efficiency. Intelligent techniques are vital in cybersecurity, as they safeguard power systems from cyber threats and ensure the integrity of key infrastructure. Intelligent cybersecurity solutions protect power grids from hostile assaults by detecting and mitigating potential vulnerabilities, assuring uninterrupted service and reducing the risk of disruption.

The integration of renewable energy sources, such as solar, wind, and hydroelectric power, into existing power systems represents a pivotal step toward a sustainable energy future. However, the inherent variability and intermittency of renewable energy generation pose significant challenges to the stability and resilience of power grids. Addressing these challenges requires innovative approaches that not only accommodate the fluctuations in renewable energy output but also enhance the overall resilience of the power system.

The use of intelligent approaches emerges as a promising solution for improving power system resilience and successfully integrating renewable energy sources. Intelligent approaches involve a wide range of modern technologies, such as artificial intelligence, machine learning, data analytics, and optimization algorithms. These strategies allow power system operators to use the massive volumes of data provided by renewable energy sources and grid infrastructure to make real-time choices.

One of the most difficult issues in integrating renewable energy into power networks is energy generation variability, which can cause supply and demand imbalances. Intelligent approaches provide predictive capabilities, allowing operators to anticipate changes in renewable energy output and alter grid operations accordingly. Machine learning algorithms, for example, can evaluate past data to forecast patterns in solar irradiance or wind speed, allowing grid operators to optimize energy dispatch and storage techniques.

Intelligent approaches improve power system resilience by allowing for proactive maintenance and issue identification. These techniques can predict probable equipment breakdowns by evaluating data from sensors and monitoring devices, allowing for proactive maintenance activities. In addition, automated fault detection algorithms can quickly identify and isolate grid interruptions, reducing downtime and improving power system reliability.

5.2 Literature Review

Our modern society relies heavily on reliable access to electricity. Power systems, essential for global connectivity, face multiple issues. Blackouts pose a significant threat to the industry, disrupt daily life, and incur significant recovery expenses. Power outages can occur due to natural disasters, technological problems, or purposeful human interference. Recent trends indicate a rise in natural disasters, primarily caused by severe weather and changing climate conditions. This pattern raises concerns about the dependability and resilience of power systems.

This chapter provides a comprehensive analysis of microgrids and their potential to reduce power outages caused by extreme events. This research examines microgrids' ability to maintain power supply during bad conditions, emphasizing their usefulness in improving system resilience.

Natural calamities, like hurricanes and earthquakes, can have significant impacts on grid resilience. This chapter analysed the impact of natural disasters and derived lessons from them. The report identifies future research priorities for improving resilience in power systems to better withstand and recover from these difficulties.

Resilience research primarily focuses on developing and evaluating risk assessments for infrastructure projects. To address these motives, resilience metrics must be developed and standardized, as well as formal computational methods to evaluate baseline resilience and the effectiveness of improvement strategies. Policymakers are interested in implementing evidence-based regulations, strategies, and plans to improve the resilience of current systems and develop industry guidelines to standardize resilience thinking throughout the infrastructure lifecycle.

This research reviews current papers to assess power system resilience, including its theoretical, operational, and evaluative bases. The study aimed to identify and discuss mathematical models for each specified resilience analysis phase. The stepwise procedure was applied to a real-world scenario involving the impact of a lightning strike on Great Britain's electrical infrastructure. This study adds to the current literature by standardizing computational models throughout the five stages of resilience assessment to a common reference curve. The case study used various metrics to assess the system's resilience at the infrastructure, operational, and organizational levels.

5.3 Power System Resilience

Power system resilience is an electrical power system's ability to with stand and recover from disturbances or disruptions while maintaining critical services and decreasing downtime as shown in Figure 5.1. Natural disasters such as hurricanes, earthquakes, and extreme weather occurrences, as well as equipment failures, cyberattacks, or human mistakes, can all cause disruptions. Resilience in power systems is critical since they support many parts of modern civilization, such as communication, transportation, health care, and commerce. When electricity systems are disrupted, impacts can be vast, affecting both individuals and communities.

5.3.1 Key aspects of power system resilience

The key aspects of power system resilience as shown in Figure 5.2 are as follows:

- *Reliability*: Power systems must be dependable, ensuring that electricity is always available to fulfill demand. Reliability monitors the frequency

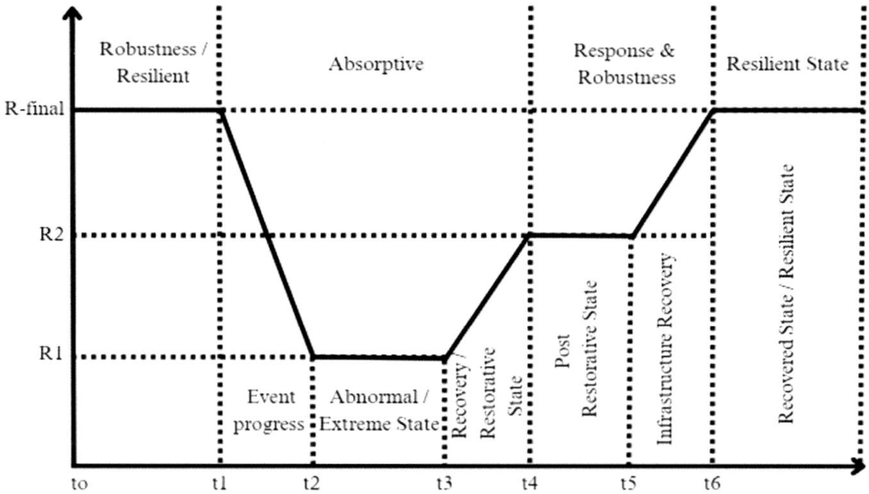

Figure 5.1 Resilience curve: disruption and recovery.

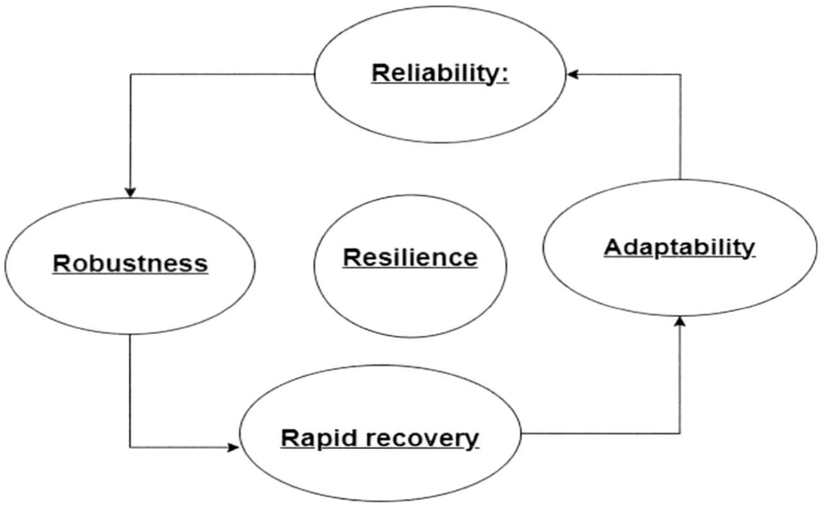

Figure 5.2 Key aspects of power system Resilience enhancement.

and duration of power outages, with the goal of minimizing customer interruption. Reliability is a critical aspect of power system performance, ensuring that electricity is available whenever and wherever it is needed to support economic activity, public safety, and quality of life. By prioritizing reliability, power system operators and policymakers can help build a more resilient and sustainable energy infrastructure capable of meeting the needs of society now and in the future.

- *Robustness*: Robustness in power systems often involves incorporating redundancy into critical components. This means having backup systems or alternative paths for electricity flow in case of failures. For example, having multiple power generation sources (such as coal, natural gas, nuclear, and renewable energy) ensures that if one source fails or is unavailable, others can pick up the slack to meet demand. Similarly, transmission lines are often designed with redundancy, so if one line goes out of service due to damage or maintenance, electricity can still be transmitted through alternate paths.
- *Rapid recovery*: Rapid recovery is a critical component of power system resilience, guaranteeing that the system can swiftly recover from disturbances and resume normal operations. Resilient power systems have efficient reaction mechanisms that allow for the rapid detection, diagnosis, and resolution of problems. This includes sophisticated monitoring and control devices that constantly scan the grid for abnormalities and anomalies. These systems identify disruptions like as equipment failures, voltage fluctuations, or line outages by analyzing real-time data from sensors, meters, and other devices.
- *Adaptability*: Resilient power systems are adaptable and can efficiently adjust to changing conditions and new threats. This could include adopting flexible generation resources, implementing smart grid technologies, or creating contingency plans for different eventualities.

Increasing power system resilience requires a combination of technological, operational, and policy interventions. This includes investing in infrastructure upgrades, deploying advanced monitoring and control systems, fortifying cybersecurity defenses, and instituting proactive maintenance practices. Additionally, collaboration among utilities, government agencies, regulators, and other stakeholders is required to properly solve resilience concerns.

5.3.2 Integrating renewable energy sources to boost grid resilience entails numerous clever ways that harness advanced technologies and strategies

Intelligent forecasting: Advanced forecasting approaches combine complex weather forecasting models with machine learning algorithms to increase the accuracy of projecting renewable energy generation. Here is an in-depth description of how this works:

Weather forecasting models: Employ past weather data, atmospheric conditions, and complicated mathematical equations to forecast future weather patterns. Temperature, wind speed, solar irradiation, and precipitation may all be forecasted for specified time periods and geographical locations using these models.

Machine learning algorithms: Machine learning algorithms are trained on historical weather data and energy-generating data in order to detect complicated patterns and correlations between meteorological factors and renewable energy output. These algorithms may learn from previous data and anticipate based on current weather forecasts, gradually improving their accuracy over time.

Data integration and fusion: Advanced forecasting systems combine data from a variety of sources, including weather stations, satellites, renewable energy installations, and grid infrastructure, to provide a complete picture of the elements impacting renewable energy generation. By combining data from multiple sources, these systems can enhance forecast accuracy and dependability.

Real-time optimization: Grid operators can continuously change forecasts based on current weather information and actual electricity generation data by employing real-time machine learning algorithms. This enables more precise forecasting and proactive grid management and resource allocation, such as scheduling energy storage discharge or dispatching flexible demand response resources to balance supply and demand.

Blockchain technology: Implementing blockchain technology for energy trading and transaction verification allows for peer-to-peer energy exchange between renewable energy producers and consumers, resulting in a more resilient and decentralized energy system.

Smart grid technologies: Using smart grid technologies such as improved sensors, communication networks, and control systems can optimize grid operations, monitor renewable energy output, and regulate energy flows in real time, improving overall dependability and resilience.

Predictive maintenance: Using data analytics and machine learning algorithms to perform predictive maintenance on grid infrastructure and renewable energy assets can help avoid equipment failures, reduce downtime, and ensure continuous functioning, ultimately enhancing overall resilience.

5.4 Renewable Energy Integration Challenges and Intelligent Techniques to Overcome It

5.4.1 Intermittency and variability

- *Intermittency*: Renewable energy sources such as solar and wind are inherently intermittent, which means that their output varies depending on weather and time of day.
- *Variability*: Renewable energy output can be unpredictable and quick, making it difficult for grid operators to maintain an equilibrium between electricity supply and demand.
- *Intelligent technique*: Advanced forecasting models and predictive analytics can estimate renewable energy generation patterns, allowing grid operators to manage grid operations accordingly. Real-time monitoring and control systems alter power flow and distribution to meet supply and demand, reducing the impact of intermittency.

5.4.2 Grid stability issues

- *Frequency and voltage changes*: The integration of fluctuating renewable energy sources can cause changes in grid frequency and voltage levels, affecting grid stability.
- *Power quality*: Variations in renewable energy generation can have an impact on the quality of power transmitted to consumers, resulting in voltage sags and harmonics.
- *Intelligent control methods*: As model predictive control (MPC) and adaptive control systems, continuously monitor grid conditions and alter settings to keep the system stable. These algorithms optimize the functioning of grid assets, such as renewable energy sources and energy storage devices, in order to maintain grid frequency and voltage.

5.4.3 Limited grid flexibility

- *Limited dispatchability*: Unlike traditional power plants, renewable energy sources often lack dispatchability, making it difficult to adapt output to meet fluctuations in demand.
- *Grid congestion*: In places with a high renewable energy penetration, grid congestion can emerge, reducing the ability to distribute electricity from renewable sources to areas in need.
- *Intelligent technique*: Smart grid technologies, such as advanced metering infrastructure (AMI) and grid automation systems, allow for

real-time monitoring and control of grid assets. Demand response programs encourage users to alter their electricity usage based on renewable energy availability, increasing system flexibility and balancing supply and demand.

5.4.4 Integration into existing infrastructure

- *Challenge*: Integrating renewable energy sources into existing grid infrastructure can be complicated and expensive.
- *Intelligent technique*: Virtual power plants (VPPs) combine and optimize the operation of dispersed energy resources, such as renewable energy sources and energy storage systems. VPPs use clever algorithms to coordinate resource dispatch, increasing grid efficiency while reducing the need for expensive infrastructure changes.

5.4.5 Traditional approaches to renewable energy integration

- *Backup generation:* Sourced from fossil fuel-based power facilities, supplements renewable energy during periods of low or high demand.
- *Limitations*: Backup generation increases carbon emissions and reliance on nonrenewable energy sources. It does not offer a long-term solution to the intermittent nature of renewable energy sources.

5.4.6 Grid expansion and reinforcement

- *Overview*: Grid expansion and strengthening entail the construction of new transmission lines and infrastructure to support the integration of renewable energy.
- *Limitations*: Grid expansion is expensive and time-consuming. It may potentially encounter pushback from local populations and environmental concerns.

5.5 Limitations of Traditional Approaches in Ensuring Grid Resilience

- *Inadequate adaptability*: Traditional techniques frequently lack the flexibility required to respond effectively to the dynamic nature of renewable energy sources.
- *Environmental impact*: Backup generation from fossil fuel-based power plants exacerbates environmental concerns while undermining attempts

to minimize greenhouse gas emissions. Curtailment and grid expansion are costly and inefficient solutions that do not address the underlying reasons for renewable energy integration issues.

- *Limited sustainability*: Traditional approaches do not promote sustainable energy practices or help to build long-term grid resilience in the face of climate change and other environmental challenges.

5.6 Future Approaches for Strengthening Power System Resilience

Improving the resilience of power systems is critical for providing a reliable energy supply, particularly in the face of evolving issues such as climate change, cybersecurity attacks, and rising demand. Here are some prospective techniques that can help increase power system resilience.

Integration of renewable energy sources: Increased use of renewable energy sources such as solar, wind, and hydroelectric power can improve power system resilience by diversifying the energy mix and lowering reliance on centralized generation facilities. Distributed renewable energy systems, such as rooftop solar panels and community-owned wind farms, can help boost local resilience.

Enhanced cybersecurity measures: Improving cybersecurity measures such as intrusion detection systems, encryption protocols, and robust authentication processes is critical for protecting power systems from cyber threats such as ransomware attacks and malicious intrusions.

Resilience planning and risk assessment: Conducting comprehensive resilience planning and risk assessments to identify vulnerabilities, assess potential impacts, and design mitigation strategies is critical for developing resilient power systems that can survive diverse hazards and disruptions.

Decentralizing energy sources: This technique is used to improve power system resiliency. On-site renewable energy generation can lessen the disturbance to a site of a grid outage, minimizing the effects of a long-term outage.

Integration of microgrids: One strategy for improving resilience in a power system is to integrate microgrids. Microgrids have demonstrated self-healing and resilient capabilities in extreme events that inflict damage beyond the typical spectrum of failures.

Connecting tie switches to recovery paths: A solution for improving system resilience has been presented by connecting tie switches to recovery paths in a radial grid design.

Climate-resilient infrastructure: Creating and adapting power infrastructure to resist the effects of climate change, such as sea-level rise, high temperatures, and increased frequency of natural disasters, is crucial for improving long-term resilience. This could entail raising substations, strengthening infrastructure against extreme weather events, and transferring susceptible assets away from flood-prone locations.

5.7 Conclusion

The incorporation of renewable energy sources (RES) into power systems creates both opportunities and problems for increasing resilience. Using sophisticated methodologies, experts can successfully solve these difficulties and maximize the benefits of RES integration, thereby enhancing power system resilience.

By utilizing the abundant and diverse renewable resources available, such as solar, wind, and hydroelectric power, power systems can diversify their energy mix, reduce reliance on fossil fuels, and lessen the effects of climate change. However, the intermittent nature of renewable power threatens grid stability and reliability.

Intelligent tools, such as improved forecasting algorithms, predictive analytics, and machine learning algorithms, are critical for maximizing renewable energy integration into power systems. These techniques allow for accurate prediction of renewable generation patterns, real-time monitoring of grid conditions, and dynamic regulation of power flows.

Furthermore, intelligent procedures make it easier to coordinate renewable power with energy storage systems, demand response programs, and grid balancing mechanisms, assuring the most efficient use of renewable resources while maintaining grid stability and resilience.

Intelligent control and optimization methodologies allow for the development of self-healing and adaptive power systems capable of autonomously responding to disturbances, optimizing resource allocation, and reducing the effects of unexpected events.

The integration of renewable energy sources through the use of intelligent approaches has enormous promise for improving power system resilience. By employing cognitive control, forecasting, and optimization skills, power systems can successfully handle the challenges of renewable integration while maximizing the benefits of clean, sustainable energy. This method lays the path for more resilient, adaptable, and sustainable power systems that can satisfy future demand.

References

[1] Gholami A, Shekari T, Amirioun MH, Aminifar F, Amini MH, Sargolzaei A. Toward a consensus on the definition and taxonomy of power system resilience. IEEE Access 2018;6:32035-53.

[2] Bompard E, Huang T, Wu Y, Cremenescu M. Classification and trend analysis of threats origins to the security of power systems. Int J Electr Power Energy Syst 2013;50:50-64.

[3] Henry D, Emmanuel Ramirez-Marquez J. Generic metrics and quantitative approaches for system resilience as a function of time. Reliab Eng Syst Saf 2012; 99:114-22

[4] A. Schweikert, L. Nield, E. Otto, and M. Deinert, Resilience and critical power system infrastructure: Lessons learned from natural disasters and future research needs. World Bank, Washington, DC, 2019.

[5] M. Mahzarnia, M. P. Moghaddam, P. T. Baboli and P. Siano, "A Review of the Measures to Enhance Power Systems Resilience," in IEEE Systems Journal, vol. 14, no. 3, pp. 4059-4070, Sept. 2020.

[6] Huang, Cheng-Liang and Wu, Yuan-Kang and Li, Yuan-Yao, 2021 7th International Conference on Applied System Innovation (ICASI), A Review on the Resilience Assessment of Power Systems under Disasters,2021.

[7] R. C. Bansal, M. Manohar, and Y. R. Sood, Eds., Electric power systems resiliency: Modelling, opportunity and challenges. San Diego, CA: Academic Press, 2022.

[8] S. K. Soonee, S. C. Saxena, K. V. S. Baba, S. R. Narasimhan, K. V. N. P. Kumar and S. Mukhopadhyay, "Grid Resilience in Indian Power System," 2018 IEEE 8th Power India International Conference (PIICON), Kurukshetra, India, 2018, pp. 1-6.

[9] E. Vugrin, A. Castillo, and C. Silva-Monroy, "Resilience metrics for the electric power system: A performance-based approach," Sandia Nat. Laboratories, Albuquerque, NM, USA, Tech. Rep. SAND2017-1493, 2017

[10] I. M. Diahovchenko, G. Kandaperumal, A. K. Srivastava, Z. I. Maslova, and S. M. Lebedka, "Resiliency-driven strategies for power distribution

system development," Electric Power Syst. Res., vol. 197, no. 107327, p. 107327, 2021.

[11] Y. Wang, C. Chen, J. Wang and R. Baldick, "Research on Resilience of Power Systems Under Natural Disasters—A Review," in IEEE Transactions on Power Systems, vol. 31, no. 2, pp. 1604-1613, March 2016

[12] S. Espinoza, M. Panteli, P. Mancarella, and H. Rudnick, "Multi-phase assessment and adaptation of power systems resilience to natural hazards," Electric Power Syst. Res., vol. 136, pp. 352-361, Jul. 2016

[13] J. Wang, W. Zuo, L. Rhode-Barbarigos, X. Lu, J. Wang, and Y. Lin, "Literature review on modeling and simulation of energy infrastructures from a resilience perspective," Rel. Eng. Syst. Saf., vol. 183, pp. 360–373, Mar. 2019

[14] Y. Liu and C. Singh, "Evaluation of hurricane impact on composite power system reliability considering common-cause failures," Int. J. Syst. Assurance Eng. Manage., vol. 1, no. 2, pp. 135-145, Aug. 2010

[15] M. R. Jamieson, G. Strbac, and K. R. W. Bell, "Quantification and visualisation of extreme wind effects on transmission network outage probability and wind generation output," IET Smart Grid, vol. 3, no. 2, pp. 112-122, Apr. 2020

[16] F. H. Jufri, J.-S. Kim, and J. Jung, "Analysis of determinants of the impact and the grid capability to evaluate and improve grid resilience from extreme weather event," Energies, vol. 10, no. 11, p. 1779, Nov. 2017

[17] A. Kumar, Y. R. Sood and A. Maheshwari, "Power Systems Resilience Enhancement through Renewable Energy Integration: Insights and Future Directions," 2024 IEEE 4th International Conference in Power Engineering Applications (ICPEA), Pulau Pinang, Malaysia, 2024, pp.140-145

6

A Comprehensive Review of IoT, Intelligent Systems, and Computing Applications in Enhancing Renewable Energy Sources

Preety Shoran[1], Ajay Nain[2], Rohit Malik[2], and Sophiya Sheikh[2]

[1]School of Sciences, CHRIST (Deemed to be University), Ghaziabad, Uttar Pradesh-201003, India
[2]School of Computer Applications, Lovely Professional University, Phagwara, Punjab-144411, India
E-mail: sunnypreety83@gmail.com; mr.ajaynain@gmail.com; rohitxmalik7@gmail.com; spohiyasheikh@gmail.com

Abstract

This chapter provides a thorough examination of the application of the Internet of Things (IoT), intelligent systems, and advanced computing in enhancing the effectiveness and sustainability of renewable energy sources such as wind, ocean, hydro, and solar energies. This study explores the incorporation of real-time monitoring, predictive maintenance, and energy forecasting facilitated by the Internet of Things (IoT) and intelligent systems. The integration of artificial intelligence (AI)-based analytics and cloud computing methodologies significantly improves the process of decision-making, grid management, and optimization of energy storage. This analysis highlights the significant impact of recent technological breakthroughs and case studies on the transformation of renewable energy generation and management, ultimately contributing to the development of a sustainable and intelligent energy landscape.

Keywords: Artificial intelligence, intelligent systems, Cloud computing, Internet of Things, energy forecasting.

DOI: 10.1201/9788743809852-6

6.1 Introduction

When society grows in all fields then the use of energy increases exponentially. Due to the enhancement of the economy and society's needs, the population has increased worldwide. With the increasing global population and the development of the economy and society, the rising demand for energy for daily life and production is an inevitable trend. Due to the high usage of fossil fuels worldwide, climate change comes rapidly with huge emissions of harmful gases that lead to global warming. These factors are

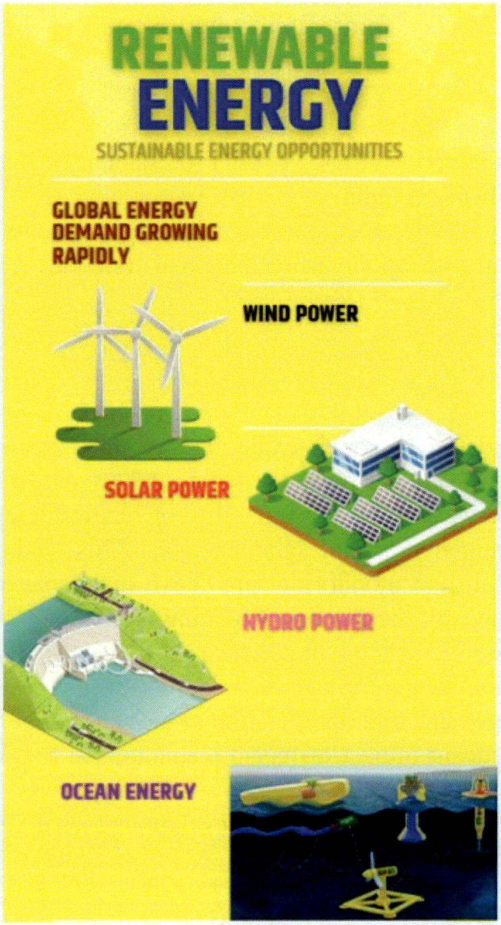

Figure 6.1 Renewable energy sources.

causes of problems such as ecologically damaging, glacier melting, and rising sea levels. All such problem leads to threats to human beings. So renewable energy is used for solving the above problems or problems caused by global warming to increase the production of energy. In past years, the whole world has come into existence for renewable energy. Different types of energy are involved like hydro, wind, solar, ocean, geothermal, biomass, etc in renewable energy. Nowadays, artificial intelligence plays an important role in all fields. It is used in renewable energy in a very efficient manner. In this chapter, we are defining how the intelligence system works in renewable energy and what are the benefits of using an intelligence system. Intelligent systems include different latest technologies like the Internet of Things, cloud computing, machine learning, etc. Figure 6.1 shows the different renewable energy opportunities. The contribution of this chapter is discussed below:

1. Deep discussion on the benefits and challenges in solar energy, wind energy, hydroelectric energy, and ocean energy using intelligent systems.
2. Enhancement of solar energy using intelligent systems like data science, machine learning, etc.
3. Role of sensors and machine learning in generation and wind farm management.

This chapter primarily covers four types of energy and the role of intelligent systems in it and a structural pictorial organization of this is shown in Figure 6.2.

6.2 Solar Energy

6.2.1 Introduction

In times dominated by increasing energy requirements and urgent environmental considerations, solar energy has emerged as a promising solution for the generation of sustainable power. The core of this disruptive transition centers around the integration of solar energy with intelligent systems and advanced computing technologies, notably encompassing the Internet of Things (IoT). Solar energy, which originates from the sun's abundant and eco-friendly resources, offers an exceptional prospect for fulfilling our energy requirements while mitigating the release of greenhouse gases. Nevertheless, the efficient capture and usage of solar energy require not just expertise in photovoltaic technologies but also the incorporation of intelligent, flexible, and reactive systems.

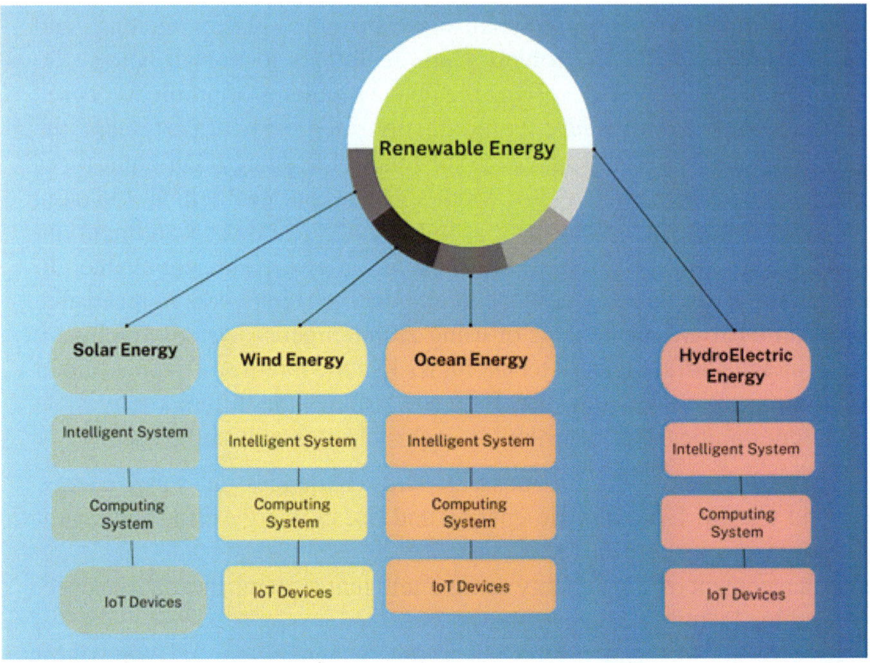

Figure 6.2 Structural chart of organization.

Intelligent systems, supported by advanced artificial intelligence and data analytics, are revolutionizing the solar energy sector. These systems enable the continuous monitoring of solar installations, collecting essential data such as solar irradiance, panel efficiency, and environmental variables in real time. The data acquired by intelligent sensors is subsequently inputted into advanced algorithms that evaluate patterns, forecast performance, and improve energy production. The synergistic relationship between data-driven insights and autonomous decision-making enables solar energy systems to function with exceptional efficiency and efficacy. Figure 6.3 presents a flowchart for dynamic solar energy predictions using machine learning algorithms.

The utilization of Internet of Things (IoT) devices serves to enhance and magnify the capabilities of intelligent solar systems. The integration of solar panels, inverters, and other components into a network through the Internet of Things (IoT) facilitates the remote control, monitoring, and administration of these devices. The utilization of real-time data transmission facilitates prompt

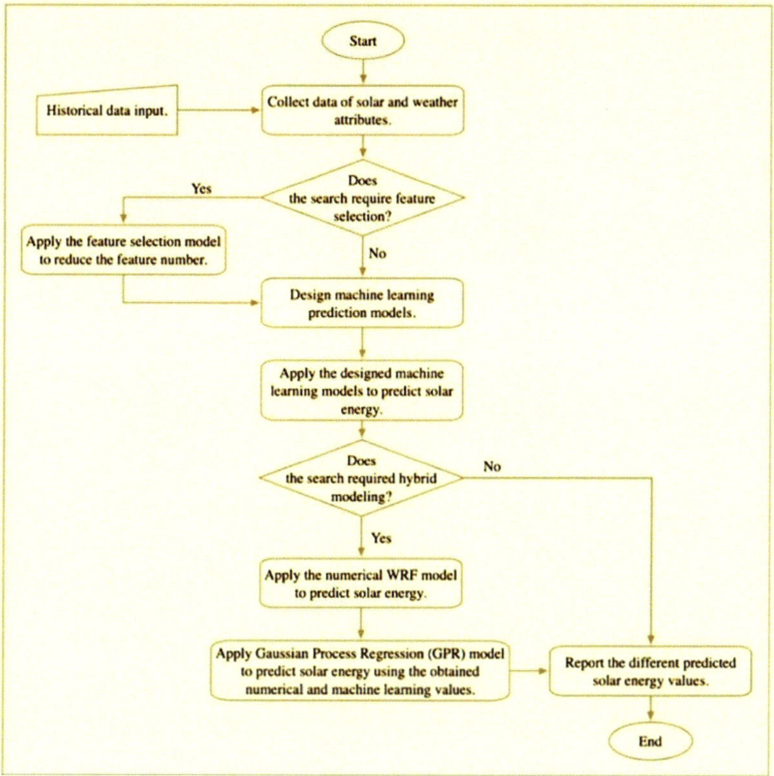

Figure 6.3 Flowchart for Dynamic Solar Energy prediction using Machine Learning based.

reactions to variations in energy supply and consumption, thus improving the stability of the grid and reducing inefficiencies.

The symbiotic association between solar energy and intelligent technologies is beyond mere enhancements in efficiency. This fosters a paradigm in which solar installations are seamlessly integrated into the larger energy ecosystem, enabling them to interact with smart grids, energy storage systems, and even electric automobiles. As global efforts progress toward achieving sustainability, the integration of solar energy with intelligent systems and computing technologies, enhanced by the connectivity provided by the Internet of Things (IoT), not only facilitates the transition to cleaner energy sources but also paves the way for a more robust, adaptable, and interconnected energy infrastructure.

6.2.2 Previous study

In the study [1], the authors employed machine learning methodologies to conduct short-term solar power forecasting. The researchers investigate the utilization of ensembles of artificial neural networks to enhance the precision of solar energy production prediction. The study demonstrates the capacity of machine learning in augmenting the management of solar energy and its integration into the power system.

The authors present a novel methodology utilizing machine learning [2] techniques to facilitate the predictive maintenance of photovoltaic panels. The research uses a range of algorithms to forecast maintenance requirements and detect potential concerns. This study emphasizes the significance of machine learning in mitigating operational downtime, optimizing maintenance expenses, and enhancing the effectiveness of solar energy production.

The authors introduced a solar energy management system that utilizes Internet of Things (IoT) technology [3]. The research demonstrates the application of the Internet of Things (IoT) in optimizing energy consumption through the integration of real-time solar data and energy demand. The study highlights the potential of the Internet of Things (IoT) in improving the efficiency of solar energy systems and its role in promoting sustainable energy management.

The authors illustrate the importance of the Internet of Things (IoT) in the surveillance and administration of solar energy installations [4]. This study presents an Internet of Things (IoT) platform that aims to facilitate real-time monitoring, remote control, and data-driven decision-making in solar installations. This study emphasizes the crucial significance of the Internet of Things (IoT) in facilitating effective energy use and its contribution to the development of smart energy systems.

The research conducted by the authors [5] centers around the acquisition of real-time solar data through the utilization of Internet of Things (IoT) technology. The research demonstrates the benefits of implementing ongoing monitoring of solar factors in order to optimize energy management. The research endeavors to enhance the efficiency and effectiveness of solar energy generation and usage by the collection and analysis of data pertaining to solar irradiance, temperature, and efficiency.

6.2.3 Intelligent systems

Intelligent systems, propelled by the integration of artificial intelligence (AI), data analytics, and Internet of Things (IoT) technologies, are instigating a profound shift within the solar energy industry. These systems utilize sophisticated computational capabilities to augment the generation, administration, and dissemination of solar energy, resulting in unparalleled levels of efficiency and sustainability. The integration of intelligent systems with solar energy infrastructure is emerging as a promising solution as the global community endeavors to migrate toward cleaner energy sources.

a. *Real-time monitoring and data insights*: Intelligent systems integrate sensors and Internet of Things (IoT) devices to consistently monitor essential variables, including solar irradiance, temperature, and panel performance. The data points undergo processing through artificial intelligence algorithms to generate real-time insights pertaining to the performance of solar panels and the prevailing environmental conditions. This facilitates proactive maintenance and optimization of performance [6].

b. *Energy forecasting and optimization*: Through the examination of historical data and the analysis of weather patterns, intelligent systems facilitate the precise prediction of energy consumption. Artificial intelligence (AI)-based algorithms have the capability to forecast solar energy generation, thereby aiding grid operators in effectively managing imbalances between energy supply and demand, as well as optimizing the distribution of energy resources [7].

c. *Predictive maintenance*: The utilization of machine learning algorithms in conjunction with gathered data has the capability to forecast maintenance requirements for solar panels and the corresponding equipment. The utilization of this predictive [8] methodology serves to mitigate periods of inactivity, diminish expenses associated with maintenance, and prolong the operational duration of solar installations.

d. *Remote monitoring and control*: The utilization of Internet of Things (IoT) technology enables the remote monitoring of solar energy systems, facilitating their centralized control and management [9]. The ability to remotely address issues enables prompt replies, leading to decreased maintenance time and expenses.

6.2.4 Computing systems

The incorporation of computing systems has become a pivotal factor in the advancement of the solar energy industry, as the global community increasingly adopts renewable energy sources. The field of computing has significantly transformed the generation, management, and optimization of solar energy through the implementation of various technologies such as artificial intelligence (AI), cloud computing, and the Internet of Things (IoT). The integration of computing systems and solar energy not only optimizes operational effectiveness but also plays a pivotal role in fostering long-term environmental sustainability.

a. *Energy forecasting and resource optimization*: Computational systems provide precise energy prediction through the examination of past data, meteorological trends, and instantaneous sun irradiation. AI-driven models play a crucial role in the prediction of solar energy production, providing valuable assistance to grid operators in effectively regulating discrepancies between energy supply and demand, as well as optimizing the allocation of resources [10].

b. *Autonomous operations and maintenance*: The integration of robotics, unmanned aerial vehicles (UAVs), and artificial intelligence (AI)-based computing systems enable the seamless execution of independent cleaning and maintenance operations for solar panels. These technologies have the potential to minimize human intervention, enhance the efficiency of energy capture, and detect panels that are not working optimally [11].

c. *Smart energy management*: The integration of Internet of Things (IoT) devices with computing systems enables the seamless monitoring of solar installations in real time. The utilization of data analytics and artificial intelligence algorithms enables the processing of information pertaining to panel efficiency, environmental conditions, and energy consumption. This facilitates the dynamic allocation of energy resources [12] to minimize wastage.

d. *Data-driven insights*: Cloud computing [13] plays an essential role in the storage and processing of enormous quantities of data acquired from solar installations. The utilization of advanced analytics and artificial intelligence algorithms enables the extraction of valuable insights from the data, hence facilitating informed decision-making and promoting ongoing enhancements in performance.

6.3 Wind Energy

6.3.1 Introduction

Wind energy is a source of renewable energy that is becoming more and more important. It helps communities all over the world to get clean, sustainable power. As the wind energy industry continues to grow, new technologies are needed to make wind energy systems more efficient, reliable, and sustainable. Intelligent systems are one of these technologies which include machine learning algorithms, decision-supportive systems, and other advanced computing technologies.

Intelligent systems are used in wind energy for a variety of tasks, such as condition monitoring, predictive maintenance, performance optimization, and wind farm management. Sensors and other sources of data are used by these systems to analyze and improve the performance of wind turbines and wind farms. For example, artificial neural networks (ANNs) and convolutional neural networks (CNNs) can be used to monitor the condition of wind turbines and predict when they will need maintenance. Fuzzy logic control systems can be used to improve wind turbine performance and lower energy costs.

Wind energy systems use more than just intelligent systems. They also use high-performance computing (HPC), cloud computing, edge computing, virtual reality (VR), and augmented reality (AR). Cloud computing is used to store, process, and analyze data, while high-performance computing (HPC) is used to run complex simulations of wind turbine performance and wind farm design. Edge computing is used to process and analyze sensor data in real time, while virtual reality (VR) [21] and augmented reality (AR) are used for training, simulation, and visualization.

The Internet of Things (IoT) is also an important technology in wind energy systems. IoT sensors are used to collect data on how well wind turbines work and how the environment is, which is then analyzed by smart systems to improve the performance of wind turbines and wind farms. Predictive maintenance is one of the most important uses of IoT in wind energy. This is because machine learning algorithms can be trained on sensor data to predict when maintenance is needed and keep expensive downtime from happening.

Overall, it is important for wind energy systems to use smart systems and other computing technologies to make them more efficient, reliable, and sustainable. As the wind energy industry continues to grow, these technologies will become even more important to make sure that wind energy stays a reliable and sustainable source of renewable energy.

6.3.2 Previous research

In this [14] paper, a FANN structure is suggested for the generation of electricity from wind energy estimation. A solution that can model the evolution of electric power considering its background and an average wind speed in an area with predominant wind energy was found and tested through the proposed system.

The author [15] proposes a comparative and thorough analysis of support vector regression, random trees, random forests, and artificial neural networks, and presents the benefits and drawbacks of using the methods. The entire thought process, including feature engineering, metrics selection, model selection, or hyperparameter tuning, is revealed using a step-by-step methodology built on the CRISP-DM data mining framework. We provide a summary highlighting the best outcomes and the trade-off between performance and the resources used to achieve these outcomes using the metrics chosen for model evaluation.

The authors [16] proposed a TCN-based model for day-ahead wind power prediction using a casual convolution architecture with residual connections in order to discover correlations between meteorological characteristics and wind power generation. The authors also proposed a scheme that effectively solves the long-distance dependency problems.

The authors [17] used IOT technologies to operate and maintain offshore wind power generation and created an economic efficiency model. Using real meteorological data from different places, the benefits of maintaining an offshore wind power system with IOT technology are compared to previous maintenance methods to match the proposed method.

6.3.3 Intelligent systems

Intelligent systems are a group of technologies that are made to act like humans in ways like reasoning, learning, and solving problems. These systems use artificial intelligence (AI) algorithms and techniques like machine learning, deep learning, and neural networks to analyze and make sense of large data sets, make predictions, and automate complex tasks. These systems can learn from their mistakes and get better over time, which makes them very flexible and able to adapt to new situations.

Intelligent systems have been used in many different fields, including the renewable energy field. Intelligent systems are used for a number of things in wind energy [18], such as monitoring the condition of wind turbines and fault diagnosis, improving performance, predicting maintenance needs, and power management. They can examine real-time data from wind turbines and sensors, find problems and outliers, and give advice on how to improve turbine performance and reduce downtime.

a. *ANN and CNN*

ANNs are computing systems that are inspired by the structure and function of biological neural networks in the human brain. They consist of many interconnected nodes or neurons, which are organized in layers and trained to recognize patterns in data. ANNs have been used in wind energy systems to predict wind speed and power output, as well as to optimize turbine performance and reduce maintenance costs.

CNNs are a type of ANN that are specifically designed for image recognition and processing. They can automatically learn and identify patterns and features in images, making them well-suited for tasks such as wind turbine blade inspection and monitoring. By analyzing high-resolution images of turbine blades, CNNs can detect and classify defects and anomalies, allowing for early detection and maintenance.

b. *Fuzzy logic control system*

Fuzzy logic control (FLC) is an intelligent control system that is based on fuzzy set theory. Unlike traditional control systems, which rely on precise mathematical models and logic rules, FLC [28] allows for uncertainty and imprecision in decision-making. This makes it well-suited for applications such as wind energy systems, where there are often complex and uncertain variables that need to be considered. It works by mapping input variables to output variables using a set of fuzzy logic rules. These rules are based on linguistic terms such as "high," "low," "medium," "hot," and "cold,", which are represented using fuzzy sets. Fuzzy sets allow for a range of values to be assigned to a particular linguistic term, rather than a precise numerical value.

The input variables to an FLC system are typically measured using sensors, such as wind speed, temperature, and power output. These input variables are then fuzzified, or converted into fuzzy sets, using membership functions that define the degree to which the variable belongs to each linguistic term. The fuzzy logic rules are then applied to the fuzzy sets, using operators such as AND, OR, and NOT, to determine the appropriate output values.

FLC has several advantages over traditional control systems. It can handle uncertainty and imprecision in data and can make decisions based on incomplete or uncertain information. It is also able to learn from experience, allowing it to adapt to changing conditions and improve performance over time. Finally, FLC can generate human-interpretable rules, making it easier to understand and debug the system.

In wind energy systems, FLC can be used to control the pitch angle and rotor speed of wind turbines, to optimize power output and reduce stress on the turbine components. It can also be used for fault detection and diagnosis, to identify and correct problems before they cause downtime or damage to the turbine.

c. *Decision supports systems*

Decision support systems are computer-based systems that are designed to help decision-makers solve complex problems by providing them with relevant information and analytical tools. In the context of intelligent systems, DSS [20] can be used to help operators and managers make better decisions by providing them with real-time data and predictive analytics.

One of the main roles of DSS in intelligent systems is to provide decision-makers with insights into the performance of the system and to help them identify areas where improvements can be made. For example, a DSS might provide real-time data on wind speed and turbine performance, along with predictive analytics that can be used to optimize the performance of the turbine.

Another role of DSS in intelligent systems is to help operators and managers respond to unexpected events or failures. For example, if a wind turbine experiences a fault or outage, a DSS might provide operators with real-time data on the status of the turbine, along with recommendations on how to respond to the issue.

DSS can also be used to support strategic decision-making in wind energy systems. For example, a DSS might provide long-term weather forecasts and market data that can be used to plan for future investments and expansion of the wind energy system.

Overall, DSS plays an important role in helping decision-makers make informed decisions in the context of intelligent systems, by providing them with real-time data, predictive analytics, and analytical tools that can be used to optimize performance and respond to unexpected events.

d. *Multiagent system*

A multiagent system (MAS) is a type of intelligent system that consists of multiple agents that interact with each other to achieve a common goal. In the context of wind energy, a multiagent system can be used to manage and optimize the performance of a wind farm.

In a wind farm, multiple wind turbines are typically installed in a specific area to generate electricity from the wind [19]. A multiagent system can be used to control the operation of each turbine in the farm and coordinate their actions to maximize the efficiency of the system.

By using a multiagent system to manage wind energy, it is possible to improve the overall efficiency and reduce the cost of producing electricity from wind. Additionally, multiagent systems can help to increase the reliability and stability of wind energy systems by providing a more flexible and adaptive approach to managing the turbines in real time.

6.3.4 IoTs

The use of Internet of Things (IoT) technology in wind energy has several benefits, including increased efficiency, improved safety, and reduced maintenance costs. IoT devices can be deployed on wind turbines to collect data on a range of parameters, such as wind speed, temperature, and vibration levels. This data can be transmitted in real time to a central monitoring system, allowing wind farm operators to monitor the performance of their turbines and identify potential issues before they become more serious. One of the key benefits of IoT technology in wind energy is that it enables condition-based maintenance, which involves performing maintenance on turbines based on their actual condition, rather than on a fixed schedule. This can help to reduce maintenance costs, as it allows for more efficient use of resources and can help to prevent unplanned downtime. IoT devices can also be used to track the performance of individual turbines and identify trends over time, allowing operators to make data-driven decisions about maintenance and repairs.

IoT technology can also help to improve safety in wind energy by providing real-time monitoring of the turbines and alerting operators to potential safety hazards. For example, sensors can be used to monitor wind speed and direction, and alert operators if conditions become unsafe for turbine operation. Overall, the use of IoT technology in wind energy has the potential to increase the efficiency and reliability of wind turbines, reduce maintenance costs, and improve safety. As such, it is becoming an increasingly important component of modern wind energy systems. A detailed description of the use of IoTs in wind energy systems is provided in Table 6.1

Predictive maintenance

Data analysis, machine learning, and other intelligent systems are used in the technique of predictive maintenance to forecast when maintenance is necessary for wind turbines. Predictive maintenance [25] seeks to spot potential problems before they develop into serious ones to reduce downtime and repair expenses.

Table 6.1 Benefits of Utilization of IoTs in Wind Energy.

Benefits of IoTs in wind energy	Description
Condition-based maintenance	IoT devices can be used to collect data on the condition of wind turbines, allowing for maintenance to be performed only, when necessary, rather than on a fixed schedule. This can help to reduce maintenance costs and prevent unplanned downtime.
Real-time monitoring	IoT devices can provide real-time monitoring of wind turbines, allowing operators to quickly identify and respond to potential issues.
Improved safety	IoT devices can be used to monitor environmental conditions, such as wind speed and direction, and alert operators if conditions become unsafe for turbine operation.
Data-driven decision making	IoT devices can collect large amounts of data on wind turbine performance, which can be analyzed to identify trends and patterns over time. This can help operators make data-driven decisions about maintenance, repairs, and other aspects of wind farm operation.
Improved efficiency	IoT devices can be used to optimize the performance of wind turbines, by adjusting their operation based on real-time data on wind speed, direction, and other parameters. This can help to increase energy production and reduce costs.
Remote monitoring	IoT devices can be monitored remotely, allowing operators to access real-time data and monitor the performance of their wind farm from a remote location.
Reduced downtime	By providing real-time monitoring and predictive maintenance capabilities, IoT devices can help reduce downtime and ensure that wind turbines operate at peak efficiency.

Blades, gearboxes, and generators in wind turbines are all subject to wear and tear over time. In order to identify anomalies and forecast when maintenance is necessary, predictive maintenance systems gather and analyze data on a variety of parameters, including temperature, vibration, and electrical current. For instance, a rise in vibration levels could signify that a gearbox is wearing out too quickly, or a shift in temperature could imply that a component is overheating.

Operators of wind farms can minimize the impact on energy production by scheduling maintenance work during times of low wind activity by anticipating maintenance needs in advance. Predictive [26] maintenance can also contribute to extending the component life of wind turbines, lowering costs over the course of the turbine's lifetime. Predictive maintenance systems help to improve the reliability and efficiency of wind energy systems by reducing unplanned downtime and maintenance costs.

Predictive maintenance can be approached in different ways depending on the equipment being monitored and the data available. Approaches such as CBM, FMEA, data-driven predictive maintenance, and prescriptive maintenance [27] can all be effective in reducing equipment downtime and maintenance costs. These all are discussed below.

Condition-based monitoring (CBM): This method employs sensors to continuously and in the moment track the health of machinery [23]. The information gathered by these sensors is then examined to look for anomalies or modifications in the way the equipment behaves that might portend a failure. Before an item of equipment fails, CBM can be used to identify early indications of wear and tear, lubrication problems, and other problems.

Failure mode and effects analysis (FMEA): FMEA is a proactive maintenance strategy that examines potential equipment failure [22] modes and their possible effects. Following this analysis, tasks for preventive maintenance that can be carried out to lessen the likelihood of equipment failure are identified.

Data-driven predictive maintenance: In this method, equipment failures [22, 24] are predicted using machine learning algorithms that analyze data from sensors and other monitoring systems. Data patterns and anomalies that might be signs of impending failures can be found by the algorithms. Equipment with complex failure modes or that operate in varying environments benefit the most from this strategy.

Prescriptive maintenance: Prescriptive maintenance goes beyond predictive maintenance [27] by advising particular actions to deal with anticipated equipment failures. To provide actionable recommendations that can stop equipment failures, this method combines predictive analytics with expertise in equipment and maintenance procedures.

6.4 Ocean Energy

6.4.1 Introduction

Ocean energy is a term for renewable energy that comes from tidal energy, wave energy, ocean current energy, and ocean thermal energy, among other things. These different kinds of energy use the power of the ocean to make electricity that is safe for the environment and will last for a long time. The moon and sun's gravity pulls on earth's oceans, making them rise and fall twice a day. This is how tidal energy is made. Tidal energy can be used by putting in place tidal turbines, which use the energy of the moving water to make electricity. This kind of energy works best in places with strong tides, like narrow channels between islands or along coastlines. The movement of ocean waves, which are caused by the wind blowing across the surface of the water, is what makes wave energy. Devices that capture the up-and-down movement of the waves and turn it into electricity can be used to get energy from waves. This type of energy has the potential to be a big source of renewable energy, especially in places like offshore locations where the waves are consistent.

Ocean current energy comes from the movement of ocean currents, which are caused by differences in wind, water temperature, and salinity. Underwater turbines that capture the movement of the water and turn it into electricity can be used to get energy from ocean currents. This type of energy works best in places where there are strong and steady currents, like the Gulf Stream off the coast of Florida. The difference in temperature between the warm water on the ocean's surface and the cold water deeper down in the ocean creates ocean thermal energy. Heat exchangers can take advantage of the difference in temperature between the ocean and the air and turn it into electricity. This type of energy works best in tropical areas where there is a big difference in temperature between the surface and deep water. One of the best things about ocean energy is that it is a clean source of electricity that can be used over and over again and does not release greenhouse gases. Unlike traditional fossil fuel sources of energy, ocean energy does not contribute to air pollution, water pollution, or climate change. Also, because the ocean is always moving, ocean energy is more reliable than some other types of renewable energy, like wind and solar power.

However, there are also some challenges associated with ocean energy. One of the biggest challenges is the cost of installing and maintaining the equipment needed to harness ocean energy. Another problem is the effect that ocean energy projects could have on marine ecosystems and coastal communities, especially in places where the projects are located. Despite these challenges, ocean energy has the potential to play an important role in the transition to a more sustainable and environmentally friendly energy system. As technology keeps getting better and prices go down, it is likely that more and more ocean energy projects will be built all over the world.

6.4.2 Previous research

The authors [29] suggested analyzing ODSIs as governance actors via Earth Systems Governance in order to comprehend their role in determining architecture and authority in governance relations, as well as the implications of integrating data and technology.

This paper [30] evaluates the current use and anticipates the future use of large and small data in ocean science. It also recommends that fully maximizing the potential of ocean data and establishing the federation of data owners and consumers can provide a solid scientific foundation for research and forecasting.

The authors [31] introduced a multi-input CNN method for predicting power generation, which demonstrated superior performance compared to conventional models. The study determined that the power output of the WEC has a positive association with wave height and that the proposed method might be utilized for condition monitoring and fault diagnostics of maritime energy converters.

The authors [32] proposed the use of ConvNets for forecasting short-term wave power. The use of these biologically inspired networks to traditional categorical problems using 2D data has been fruitful. The report also describes the context and concerns that led to the development of this particular CNN model and how more can be developed.

The authors [33] have shown that machine learning techniques can provide information about unresolved turbulent processes as a proof of concept. Also, they reduced data from a high-resolution eddy-resolving QG model with a spatial low-pass filter and trained convolutional neural networks to predict the relationship between turbulent processes and their influence on the large-scale flow.

6.4.3 Intelligent systems

Intelligent systems [34] are becoming more and more important to the development and use of technologies that use ocean energy. These systems use advanced algorithms and machine learning to improve the efficiency and reliability of ocean energy devices and make them work better. Intelligent systems are especially useful for controlling and keeping an eye on tidal turbines. These turbines work in harsh, changing environments that require precise control to get the most energy out of them while putting the least amount of wear and tear on the machines. Intelligent systems can use real-time data [35] on tide patterns, turbine performance, and environmental conditions to optimize turbine operation and make sure the turbines work at their most efficiently. Wave energy devices are also being made and designed with the help of smart systems. These devices must be able to work in the open ocean, which has unpredictable and sometimes harsh conditions, and they must also be designed to get the most energy out of the sun. Engineers can make more efficient and reliable wave energy devices by using machine learning algorithms to simulate how waves behave and improve the design of the devices. These intelligent systems have the potential to make ocean energy technologies a lot more effective, reliable, and cost-effective. As these systems continue to change and get better, ocean energy is likely to become a more important source of renewable energy in the years to come.

Also, intelligent systems like artificial neural networks (ANNs), genetic algorithms, fuzzy logic systems, reinforcement learning, and expert systems are being used more and more to develop and improve ocean energy technologies.

- Artificial neural networks [40] are used to look at large amounts of data and find patterns so that ocean energy devices can work better. ANNs are especially good at predicting ocean conditions and figuring out how to control tidal turbines and wave energy devices to get the most energy out of them.
- Genetic algorithms are used to make the best ocean energy devices by simulating [38, 40] the process of natural selection. These algorithms can be used to figure out the best way to place tidal turbines and wave energy devices. They can also be used to find the best design parameters for a certain ocean energy device.
- Fuzzy logic systems [43] are used to model complex and uncertain systems, like how waves move in the open ocean. These systems can be used to predict how well ocean energy devices will work in different situations and to improve the way these devices are controlled in real time.
- Reinforcement learning is a type of machine learning that involves teaching an algorithm to make decisions based on feedback from the environment. This method is used to improve the control of tidal turbines and wave energy devices in real time, based on how the ocean is changing.
- Expert systems are computer programs that act like a human expert when it comes to making decisions. With the help of experienced engineers and researchers, these systems are used to improve the design and control of ocean energy devices by combining their knowledge and skills.

Also, these smart systems are changing the way ocean energy technologies are designed, built, and used. By using the power of advanced algorithms and machine learning techniques, we can make ocean energy devices that are more efficient, reliable, and cost-effective. These devices can help us switch to a more environmentally friendly and sustainable energy system.

6.4.4 Computing technologies

Computing technologies play an increasingly important role in the development, deployment, and operation of ocean energy technologies. Ocean energy devices are modeled and analyzed with the help of simulation software, which helps to improve their design, performance, and control. IoT systems connect sensors and devices that use ocean energy to the internet. This lets these devices be monitored and controlled in real time. Cloud computing is a scalable and cost-effective way to store and process large amounts of data generated by ocean energy devices and sensors. Data acquisition systems collect this data and send it to a central database where it can be analyzed. Energy storage [37] systems are used to store energy made by ocean energy devices like wave energy converters or tidal turbines and release it when energy demand is high. These computer technologies are helping to speed up the switch to a more sustainable and renewable energy system [41]. They are doing this by unlocking the full potential of ocean energy and tackling important problems like climate change and energy security. Table 6.2 presents some key benefits and challenges of it.

Here are a few examples of how computers are used in the field of ocean energy.

- Simulation software: Simulation software is used to model and study how ocean energy devices work in different situations. Advanced numerical techniques [38] are used by these software tools to simulate how waves and currents move and how devices like tidal turbines and wave energy converters work. Simulation software can help improve the design, performance, and control of devices, as well as predict how these devices will affect the environment.
- IoT: The Internet of Things (IoT) [35] is being used to connect ocean energy devices and sensors to the internet. This lets these devices be monitored and controlled in real time. IoT systems can collect information about the ocean, how well devices work, and how they affect the environment. This information can then be used to improve device control and maintenance [39]. IoT systems are also used to remotely monitor and control devices that are out at sea. This lets operators make better decisions about how to run and maintain the devices [40].

Table 6.2 Benefits and Challenges of Using IoTs, Automated systems in Renewable Energy systems.

Benefits	**Challenges**
• *Renewable and sustainable:* Ocean energy is a source of energy that is both renewable and sustainable. This means that it has the potential to provide a steady source of power without using up natural resources. The energy in the ocean is always being replaced, so it is a source of energy that can be used repeatedly.	• *High capital and operational costs*: One of the biggest problems with ocean energy is that it costs a lot to start up and run. The cost of building, installing, and maintaining infrastructure for ocean energy is much higher than that of traditional energy sources. This can make it hard for ocean energy to compete in the market.
• *Low carbon emissions*: Ocean energy is a clean source of energy that does not put out much pollution or greenhouse gases. It might be able to cut carbon emissions by a lot and help fight climate change.	• *Impacts on the environment*: Ocean energy systems could have effects on the environment, such as changing the marine ecosystem and causing marine life to move around. Ocean energy should have as little of an effect on the environment as possible, so these effects need to be carefully looked at and managed.
• *Reliable*: The ocean is a predictable source of energy, with waves and tides occurring regularly and reliably. This makes it easier to predict and plan how much energy will be made and makes it a more reliable source of energy than wind and solar.	• *Technical challenges*: Ocean energy technologies are still in their early stages of development, and they have problems with reliability and storing energy, among other things. These problems can make it harder for ocean energy systems to work well.
• *Versatility*: There are many kinds of ocean energy technologies, such as systems that convert wave, tidal, and ocean thermal energy. This makes it possible to choose the best technology for a specific location, considering things like the local wave and tide patterns.	• *Legal and regulatory problems*: The legal and regulatory framework for ocean energy is still transforming, which can make investors and developers anxious. There are also problems with licenses and permits because projects that use ocean energy must follow different rules at the national and international levels.
• *Jobs creation*: The development and use of ocean energy technologies can create jobs and help the economy grow. When ocean energy infrastructure is built, installed, run, and maintained, it can create a wide range of jobs in engineering, manufacturing, and skilled trades, which helps local communities and economies.	• *Access to funding*: Accessing funding for research, development, and commercialization of ocean energy is hard, just like it is for any new technology. If there is not enough money, progress in ocean energy can slow down and it can be hard for companies to bring their technology to market.

- Cloud computing: Ocean energy devices and sensors use cloud computing to store and process large amounts of data. Cloud computing [34] is a scalable and cost-effective way to store and analyze this data. It can also be used to build machine learning models and predictive analytics tools to improve device performance and control.
- Data acquisition: Ocean energy devices and sensors send data to a central database [36] through data acquisition systems. Most of the time, these systems have sensors that measure things like wave height and current speed [42], as well as sensors that measure how well the device works and how it affects the environment.
- Energy storage: Ocean energy devices like wave energy converters and tidal turbines use energy storage [37] systems to store the energy they make. Most of the time, these systems use batteries or other ways to store energy when energy production is high and release it when energy demand is high.

These computing technologies are making it possible to make ocean energy devices that are more efficient, reliable, and cost-effective. They are also helping the transition to a more sustainable and renewable energy system move forward. By using the power of advanced computing technologies, we can use ocean energy to its fullest potential and help solve important problems like climate change and energy security. Here are some key benefits and challenges of it.

6.5 Hydroelectric Energy

6.5.1 Introduction

Hydroelectric power is a type of renewable energy that uses the flow of water to make electricity. With the development of smart systems and computer technologies, hydroelectric power plants can now be better managed, monitored, and optimized. This chapter looks at how intelligent systems and computing technologies can help improve the performance, safety, and long-term viability of hydroelectric power generation. Figure 6.4 shows the working of a hydroelectric energy generator.

Hydroelectric power plants can be monitored and controlled with the help of smart systems and computer technology. For example, sensors can be used to keep an eye on water levels, flow rates, and other important factors in real time. Using advanced algorithms, the data from these sensors can be used to predict potential problems and suggest ways to fix them to avoid

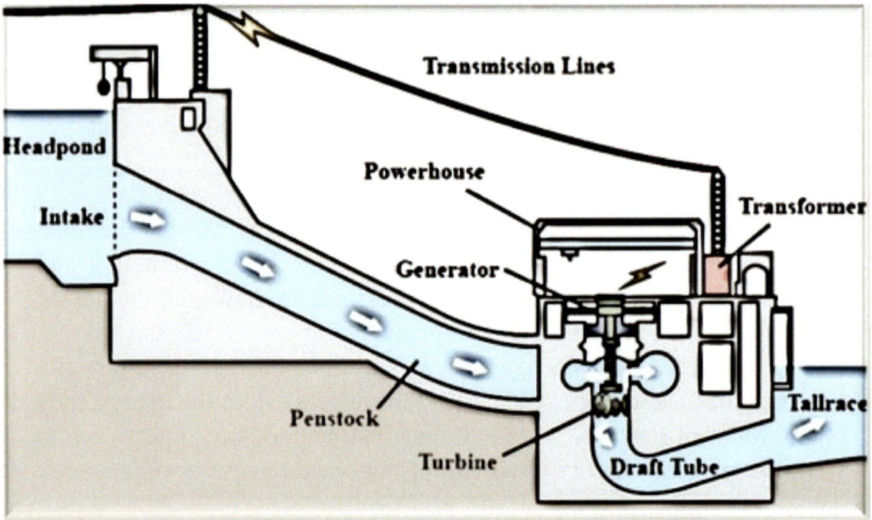

Figure 6.4 Working of a hydroelectric energy generator.

equipment downtime or damage. Intelligent systems can also improve the way power is made by looking at data from the past and suggesting the best way to maximize output while keeping maintenance costs to a minimum.

Intelligent systems and computing technologies can also be used to make hydroelectric turbines work better. Machine learning algorithms can be used to figure out the best angle for the turbine blades so that they produce the most power. Intelligent systems and computing technology can also help make hydroelectric power plants safer. Machine learning algorithms can be used to find potential safety risks and suggest steps to take to avoid accidents. These algorithms can look at data from sensors and other sources to find outliers that could mean there is a problem with safety. Also, the data can be used to predict when maintenance will be needed, which lowers risks for operators and shortens the amount of time the machine is down.

Hydroelectric energy can also be made more sustainable with the help of intelligent systems and computing technology. Advanced analytics can be used to make the best use of water resources, balancing the need to make power and save water. This helps to protect the environment and ensure a steady, long-term supply of energy. Machine learning algorithms can also be used to predict how climate change will affect the way hydroelectric power is Machine learning algorithms can also be used to predict how climate change will affect the way hydroelectric power is made.

Intelligent systems and computing technologies are changing how hydroelectric energy is made, monitored, and managed. Operators can improve power production, safety, and sustainability by using advanced analytics, machine learning algorithms, and other tools. As the need for clean and sustainable energy keeps going up, intelligent systems and computing technologies will become even more important in hydroelectric power generation.

6.5.2 Previous research

This study [44] seeks to develop and apply artificial neural network (ANN) models to forecast short- and medium-term hydroelectric production in Ecuador, using historical data such as hydropower production and precipitation. For this purpose, both one-step and multistep forecasted problems were considered, based on the prediction horizon. Using multilayer perceptron (MLP), long short-term memory (LSTM), and sequence-to-sequence (seq2seq) LSTM, 16 ANN structures were designed. More than 3000 models were configured, trained, and validated using a hyperparameter-based grid search algorithm. In both scenarios, the MLP univariate and differentiated model of a single-step scenario outperforms the other architectures analyzed. The obtained model can be a useful tool for sustainable hydropower production planning and decision-making.

The primary objective of this research [45] is to develop an Internet of Things-based water flow monitoring (WAFLOW-MT) device (IoT). This device may assist the technicians at the pico hydroelectric power plant in monitoring the rivers' water flow rate so that the water deficit is continuously recorded. The WAFLOW-MT device design and development process consisted of four stages: (1) need analysis; (2) system design; (3) system development; and (4) system testing. During the six-day testing period, data was collected from the water flow sensor and transmitted via IoT to the web server thingspeak.com. Based on the test results, it can be concluded that the WAFLOW-MT device monitored the rivers' water flow effectively.

The purpose of this research [46] is to develop a practical and rapid method for predicting and classifying the future schedules of hydropower plants in order to improve the overall energy efficiency by utilizing the water of cascaded hydropower plants. This research is important to water resource planners for the planning and management of energy-generating reservoirs. The proposed method consists of techniques and approaches for data mining. Using multiple linear regression, the energy production relationship between upstream and downstream hydropower plants is first determined. Then, with the aid of historical data, a cluster analysis is performed to identify typical generation curves. To rapidly predict generation schedules, the decision tree algorithm C4.5, IterativeDichotomiser 3-IV, improved C4.5, and chi-squared automatic interaction detection are utilized, and a comprehensive comparison of various algorithms is conducted. Using SIPINA, the decision tree algorithms are solved. The results demonstrate that the C4.5 algorithm is more practical for rapidly generating schedules for cascaded hydropower plants.

The purpose of this research [47] is to determine which of the numerous information models can provide accurate projections of renewable energy output, such as solar, wind, or pumped storage. Several machine learning techniques, including multilayer perceptions MLP, RNN CNN, and LSTM designs, are frequently employed in the field of sustainable energy forecasting. This type of modeling uses historical data to predict potential values and can predict short-term solar and wind generation patterns. The energy management system can be applied to all industries for the management of network devices by constructing a forecasting application that allows all users to check the number of available resources at a given time interval. Even the proposed AI-based method can be used to make timely decisions in emergency situations regarding the various energy sources that must be supplied.

Using high-fidelity models, surrogate modeling techniques, and optimization, this study [48] describes a significant advancement in computing hourly power generation schemes for a hydropower reservoir. The predictive power of the high-fidelity hydrodynamic and water quality model CE-QUAL-W2 is successfully emulated by an artificial neural network, which is then incorporated into a genetic algorithm optimization approach to maximize hydro-power generation subject to dam operations and water quality constraints. This methodology is applied to a multipurpose reservoir in the vicinity of Nashville, Tennessee, United States. The model successfully reproduced high-fidelity reservoir data and enabled 6.8% and 6.6% increases in hydropower production value relative to actual operations for dissolved oxygen (DO) limits of 5 and 6 mg/L, respectively, while observing a predicted decrease in power generation at more stringent DO constraints. Exploration of simultaneous temperature and DO constraints uncovered the capability to address multiple water quality constraints at particular locations. The reduced computational demands of the new modeling approach demonstrated its capacity to provide decision support for reservoir operations scheduling while maintaining high-fidelity hydrodynamic and water quality data as part of the optimization decision support routines.

This paper [49] describes an alternative electrical energy source. The designed pico hydropower generation system is ideally suited for remote areas where power transmission is impractical. To maximize the designed concept, the Pelton turbine is utilized because it is suitable for both high and low water flow rates in the Philippines. In addition, an IoT-based system that can easily monitor any undesirable issues. This pico hydropower system is more cost-effective than other electricity sources.

This paper [50] explores the current and future role of machine learning (ML) in the hydropower industry. A summary of the principal applications of machine learning (ML) in the field of hydropower operations is provided to illustrate the most prevalent topics covered in the scientific literature in recent years. The goal is to provide recommendations for novel research directions that can be pursued in the near future to cover previously unexplored areas. This paper's primary contribution is a critical examination of the current state of ML applications in hydropower scheduling. This study identifies and discusses new roles that can be covered by machine learning (ML) coupled with cyber-physical systems (CPS), with a particular focus on short-term hydropower scheduling (STHS) challenges, in light of the established literature available in recent years.

6.5.3 Intelligent systems

Intelligent systems are being used in hydroelectric energy to improve the efficiency and sustainability of power generation, optimize power plant operations, and cut down on downtime. These systems use real-time data, analytics, and other kinds of information to help power plant managers make smart decisions about power production, maintenance schedules, and other important things. Hydroelectric power plants can meet the growing need for clean, sustainable energy by using smart systems to cut costs and make the power grid more reliable while meeting the growing demand for clean, sustainable energy.

Intelligent systems like artificial neural networks (ANNs) and convolutional neural networks (CNNs), fuzzy logic systems, decision support systems, and expert systems are being used in hydroelectric energy to improve the performance of the power plant and lower expenses.

1. *ANN and CNN*: ANNs and CNNs are two types of advanced machine learning [50] algorithms that can learn to find patterns in data and make predictions about how things will work in the future. ANNs and CNNs can be used in hydroelectric energy to predict the power plant's output

based on real-time data from sensors and other sources. This can help operators get the most power out of their generators and cut down on downtime.

2. *Fuzzy systems*: Hydroelectric energy also uses fuzzy logic systems, which are another type of intelligent system. Fuzzy logic [53] systems use a set of rules to make decisions when the information they have is not complete or is not clear. Fuzzy logic [51] systems can be used to control the operation of turbines and other equipment in hydroelectric energy. This means that their performance can be changed in real time based on changes in the environment.
3. *Decision support systems*: Decision support systems, or DSS, are computer programs that help people make decisions. DSS can be used in hydroelectric energy to give operators real-time data, analytics, and other information that can help them make smart decisions about power production, maintenance schedules, and other important parts of running a power plant.
4. *Expert systems*: Expert systems are computer [52] programs that act like experts when it comes to making decisions. Expert systems can be used in hydroelectric energy to give operators advice based on their past experiences and what they know. This can help operators make smart decisions about how to run power plants, when to do maintenance, and other important parts of managing power plants.

In hydroelectric energy, the use of intelligent systems like ANN and CNN, fuzzy logic systems, DSS, and expert systems is important. These systems can help operators get the most power out of their generators, reduce downtime, and save money. By using these smart systems, hydroelectric power plants can make power generation more efficient, safe, and sustainable, and meet the growing demand for clean and sustainable energy.

6.5.4 Computing technologies

The incorporation of computing technologies in hydroelectric power plants has greatly improved their efficiency, safety, and ability to keep running for a long time. In this section, we will talk about some of the most important computing technologies that have been used in hydroelectric power generation. These include simulation software, the Internet of Things (IoT), cloud computing, data acquisition, and energy storage.

1. *Simulation softwares*: Hydroelectric power plants have been built, tested, and run with the help of simulation software [50]. Computational fluid

dynamics (CFD) software can simulate how water flows through turbines, predict how well turbines will work, and improve the design of turbines. Other software, like finite element analysis (FEA), can simulate how turbines work mechanically, predict how stress will be dispersed, and optimize the design of the turbines' structures. This makes the power plant safer and more efficient and reduces the chance that something will break.

2. *Internet of Things (IoT)*: The IoTs have also been used to monitor and control hydroelectric power plants from afar. Smart meters and sensors that are part of the Internet of Things [45] can be used to keep an eye on important things like water levels, flow rates, temperature, and humidity. This information can be gathered and sent to a central control system so it can be analyzed and used to make decisions. By using IoT devices, operators of power plants can quickly find potential problems, do preventive maintenance, and improve the plant's performance.
3. *Cloud computing*: Cloud computing has also been used to store and process data remotely in hydroelectric power generation [47]. Using cloud-based services, operators of power plants can get real-time data, analyze data, and run simulations from anywhere in the world. This can make it easier to make decisions, reduce downtime, and improve the power plant's performance.
4. *Data acquisition systems*: Getting data is another important part of using computers to make hydroelectric power. Data acquisition [46, 54] systems can get data from sensors, Internet of Things (IoT) devices, and other sources in real time. This information can be used to improve maintenance schedules, boost power production, and find potential problems.

Computing technologies have made hydroelectric power generation much safer, more efficient, and more environmentally friendly. Optimization of hydroelectric power plants has been greatly helped by simulation software, IoT devices, cloud computing, data acquisition systems, and energy storage systems. As the need for clean, renewable energy sources grows, so will the importance of computing technologies in hydroelectric power generation.

Hydroelectric energy is a renewable energy source that provides dependable electricity generation and water management advantages. However, it also has environmental and initial economic issues, is limited by regional availability and weather dependency, and requires ongoing maintenance and repair. Here are some key benefits and challenges discussed in Table 6.3.

Table 6.3 Benefits and Challenges of Using IoTs and Automated Systems for Renewable Hydroelectric systems.

Benefits	Challenges
Clean and renewable: Hydroelectric energy is a clean, renewable source of energy that does not pollute the air or release harmful greenhouse gases. This makes it a reliable source of energy that can help cut down on the amount of carbon dioxide that comes from making electricity.	*Environmental impact*: Hydroelectric energy is a renewable source of energy but building and running hydroelectric power plants can have big effects on the environment. For example, they can change the way rivers flow naturally, destroy fish habitats, and hurt fish populations.
Reliable: Hydroelectric power plants can run all the time, so they are a steady source of energy that is not affected by changes in wind or sun energy.	*High up-front costs*: Building hydroelectric power plants can be expensive because they need a lot of infrastructure, like dams, reservoirs, and powerhouses, which can be expensive to build.
Cost-effective: Once the initial infrastructure is set up, it does not cost much to make and keep hydroelectric energy running. This makes it a good source of energy for the long term.	*Limited geographic availability*: Hydroelectric energy works best in places with a lot of water flowing through them. This makes it hard to use in places where there are not enough natural resources to make it work.
Water management: Hydroelectric power plants can help manage water resources by controlling the flow of water, lowering the risk of flooding, and providing water for irrigation and drinking.	*Weather dependence*: Even though hydroelectric energy is a reliable source of power, it is still affected by weather patterns like droughts, floods, and changes in the amount of rain or snow that falls.
Energy storage: Hydroelectric power plants can also store energy by pumping water uphill into a reservoir when they have extra energy. This water can be released later to make more energy when demand is high.	*Maintenance and repairs*: Hydroelectric power plants need regular maintenance and repairs to make sure they work as well as possible. This can be hard and expensive because the infrastructure is so big and complicated.

6.6 Conclusion

This in-depth evaluation highlights the utmost importance of the Internet of Things (IoT), intelligent systems, and sophisticated computing as crucial facilitators in the transformation of renewable energy sources. Through the integration of real-time data collecting, predictive maintenance, and precise energy forecasting, these technologies have facilitated remarkable progress in enhancing the efficiency and sustainability of wind, ocean, hydro, and solar energies. The incorporation of artificial intelligence (AI)-enabled analytics and cloud computing has significantly enhanced decision-making procedures, grid integration, and the management of energy storage. The increasing prominence of renewable energy on a worldwide scale has positioned the amalgamation of the Internet of Things (IoT), intelligent systems, and computers as a pivotal force in crafting a future that necessitates clean, intelligent, and interconnected energy solutions. This study aims to present a comprehensive analysis of recent advancements in renewable energy technologies and their corresponding ramifications. By examining these breakthroughs, the review underscores the significant potential of these technologies in driving the renewable energy sector toward a more environmentally sustainable and technologically advanced era.

References

[1] Fentis, A., Bahatti, L., Mestari, M., & Chouri, B. (2017, June). Short-term solar power forecasting using Support Vector Regression and feed-forward NN. In 2017 15th IEEE International New Circuits and Systems Conference (NEWCAS) (pp. 405-408). IEEE.

[2] Huuhtanen, T., & Jung, A. (2018, June). Predictive maintenance of photovoltaic panels via deep learning. In 2018 ieee data science workshop (dsw) (pp. 66-70). IEEE.

[3] Bagdadee, A. H., Zhang, L., & Saddam Hossain Remus, M. (2020). A brief review of the IoT-based energy management system in the smart industry. Artificial Intelligence and Evolutionary Computations in Engineering Systems, 443-459.

[4] Spanias, A. S. (2017, August). Solar energy management as an Internet of Things (IoT) application. In 2017 8th International Conference on Information, Intelligence, Systems & Applications (IISA) (pp. 1-4). IEEE.

[5] Clarke, J. A., Cockroft, J., Conner, S., Hand, J. W., Kelly, N. J., Moore, R., ... & Strachan, P. (2002). Simulation-assisted control in building energy management systems. Energy and buildings, 34(9), 933-940.

[6] Baidya, S., Potdar, V., Ray, P. P., & Nandi, C. (2021). Reviewing the opportunities, challenges, and future directions for the digitalization of energy. Energy Research & Social Science, 81, 102243.

[7] Khan, M., Silva, B. N., & Han, K. (2016). Internet of things based energy aware smart home control system. Ieee Access, 4, 7556-7566.

[8] Huuhtanen, T., & Jung, A. (2018, June). Predictive maintenance of photovoltaic panels via deep learning. In 2018 ieee data science workshop (dsw) (pp. 66-70). IEEE.

[9] Iqbal, J., Khan, M., Talha, M., Farman, H., Jan, B., Muhammad, A., & Khattak, H. A. (2018). A generic internet of things architecture for controlling electrical energy consumption in smart homes. Sustainable cities and society, 43, 443-450.

[10] Cervone, G., Clemente-Harding, L., Alessandrini, S., & Delle Monache, L. (2017). Short-term photovoltaic power forecasting using Artificial Neural Networks and an Analog Ensemble. Renewable Energy, 108, 274-286.

[11] Khadka, N., Bista, A., Adhikari, B., Shrestha, A., & Bista, D. (2020, March). Smart solar photovoltaic panel cleaning system. In IOP Conference Series: Earth and Environmental Science (Vol. 463, No. 1, p. 012121). IOP Publishing.

[12] Siano, P. (2014). Demand response and smart grids—A survey. Renewable and sustainable energy reviews, 30, 461-478.

[13] Nayak, S. K., Panda, S. K., & Das, S. (2020). Renewable energy-based resource management in cloud computing: a review. Advances in Distributed Computing and Machine Learning: Proceedings of ICADCML 2020, 45-56.

[14] Dumitru, Cristian-Dragos, and Adrian Gligor. "Daily average wind energy forecasting using artificial neural networks." Procedia Engineering 181 (2017): 829-836.

[15] Buturache, Adrian-Nicolae, and Stelian Stancu. "Wind energy prediction using machine learning." *Low Carbon Economy* 12.01 (2021): 1.

[16] Lin W-H, Wang P, Chao K-M, Lin H-C, Yang Z-Y, Lai Y-H. Wind Power Forecasting with Deep Learning Networks: Time-Series Forecasting. Applied Sciences. 2021; 11(21):10335. https://doi.org/10.3390/app112110335

[17] Fen Zhou, Xuping Tu, Qingdong Wang, Research on offshore wind power system based on Internet of Things technology, International Journal of Low-Carbon Technologies, Volume 17, 2022, Pages 645–650, https://doi.org/10.1093/ijlct/ctac049

[18] Wang, J., Yu, X., Shao, S., Zhang, J., & Xu, X. (2020). A comprehensive review of high-performance computing in wind energy. Renewable and Sustainable Energy Reviews, 117, 109489.

[19] Sun, C., Cai, M., Wang, X., & Wang, Z. (2019). A review on the applications of edge computing in wind energy. Renewable and Sustainable Energy Reviews, 112, 544-553.

[20] Meng, Q., Qiu, S., Zeng, X., & Wang, W. (2020). Cloud computing-based big data platform for wind farm monitoring and operation management. Renewable Energy, 146, 1886-1897.

[21] Garg, S., & Verma, R. (2019). A review of virtual reality and augmented reality applications in the renewable energy sector. Renewable and Sustainable Energy Reviews, 101, 59-66.

[22] Mahmud, M. A., & Islam, M. R. (2021). IoT-based wind turbine condition monitoring and fault diagnosis: A review. Renewable Energy, 170, 319-331.

[23] L. T. Minh, M. M. Rahman, and H. R. Pota, "Multi-Agent System in Wind Farm Control and Management: A Review," Energies, vol. 13, no. 8, p. 1863, 2020. https://doi.org/10.3390/en13081863

[24] J. Li, Y. Zhang, Y. Li, and J. Zhang, "A Multi-Agent System for Intelligent Wind Power Management," Energies, vol. 12, no. 22, p. 4311, 2019. https://doi.org/10.3390/en12224311

[25] "Predictive Maintenance: How Big Data and IoT Can Save Your Assets" by IoT For All: https://www.iotforall.com/predictive-maintenance-big-data-iot-save-assets/

[26] "A Review of Predictive Maintenance Techniques" by O. Bendahmane and S. S. Wang: https://www.mdpi.com/1424-8220/20/10/2841/htm

[27] "Predictive Maintenance: The Future of Asset Management" by J. M. Schmitt and M. W. Crain: https://www.pumpsandsystems.com/predictive-maintenance-future-asset-management

[28] Han, B., Zhou, L., Yang, F., & Xiang, Z. (2016). Individual pitch controller based on fuzzy logic control for wind turbine load mitigation. IET Renewable power generation, 10(5), 687-693.

[29] Drakopulos, Lauren & Havice, Elizabeth & Campbell, Lisa. (2022). Architecture, agency and ocean data science initiatives: Data-driven transformation of oceans governance. Earth System Governance. 12. 100140. 10.1016/j.esg.2022.100140.

[30] Chengcheng Qian, Baoxiang Huang, Xueqing Yang & Ge Chen (2022) Data science for oceanography: from small data to big data, Big Earth Data, 6:2, 236-250, DOI:10.1080/20964471.2021.1902080

[31] Ni C, Ma X. Prediction of Wave Power Generation Using a Convolutional Neural Network with Multiple Inputs. Energies. 2018; 11(8):2097. https://doi.org/10.3390/en11082097

[32] Bento, P., Pombo, J., Calado, M. D. R., & Mariano, S. (2021). Ocean wave power forecasting using convolutional neural networks. *IET Renewable Power Generation, 15*(14), 3341-3353.

[33] Bolton, T., & Zanna, L. (2019). Applications of deep learning to ocean data inference and subgrid parameterization. *Journal of Advances in Modeling Earth Systems, 11*(1), 376-399.

[34] Peng, J., Chen, H., & Chen, Q. (2019). Smart ocean energy management system based on cloud computing and big data analytics. Journal of Cleaner Production, 236, 117687.

[35] Delahaye, A., Wang, Q., & Kurniawan, A. (2019). IoT-enabled real-time monitoring and control for ocean energy systems: A review. Renewable and Sustainable Energy Reviews, 112, 253-263.

[36] Sun, C., Sheng, W., & Yu, M. (2020). Research on data acquisition and processing in ocean energy exploitation. Journal of Marine Science and Engineering, 8(1), 28.

[37] Asif, M., Jeong, S., & Kim, Y. H. (2020). Energy storage systems for renewable energy sources: Future trends and current status. Energy Conversion and Management, 210, 112654.

[38] Hou, G., Wang, X., Gao, Y., Zhang, H., & Zhang, Y. (2020). Numerical simulation of a vertical axis tidal current turbine with a novel rudder. Renewable Energy, 155, 1169-1183.

[39] Li, Y., Li, Y., Li, C., & Li, Y. (2021). Intelligent monitoring and control system for ocean energy conversion: A review. Renewable and Sustainable Energy Reviews, 147, 111220.

[40] Nara, S., & Das, S. (2021). A review on artificial intelligence and machine learning techniques for the performance optimization of wave energy converter. Renewable and Sustainable Energy Reviews, 148, 111385.

[41] Zhang, Y., Liu, H., Liu, Y., Yang, W., & Cheng, H. (2021). Optimal power dispatching for wave energy conversion system based on deep reinforcement learning algorithm. Applied Energy, 292, 116938.

[42] Meng, J., Wu, W., Wang, X., Zhang, J., & Liu, X. (2021). An intelligent energy management system for a hybrid wind-wave-solar energy platform. Applied Energy, 303, 117609.

[43] Zhang, X., Hu, Y., & Lu, Y. (2021). A fuzzy logic-based energy management system for a wave energy converter with a hydraulic accumulator. Applied Energy, 279, 115818.

[44] Barzola-Monteses, J., Gomez-Romero, J., Espinoza-Andaluz, M., & Fajardo, W. (2022).Hydropower production prediction using

artificial neural networks: an Ecuadorian application case. Neural Computing and Applications, 1-14.

[45] Yuniarti, N., Hariyanto, D., Yatmono, S., & Abdillah, M. (2021). Design and Development of IoT Based Water Flow Monitoring for Pico Hydro Power Plant. Int. J. Interact. Mob. Technol., 15(7), 69-80.

[46] Parvez, I., Shen, J., Hassan, I., & Zhang, N. (2021). Generation of hydro energy by using data mining algorithm for cascaded hydropower plant. Energies, 14(2), 298.

[47] Alqahtani, A. S., Kshirsagar, P. R., Manoharan, H., Balachandran, P. K., Yogesh, C. K., & Selvarajan, S. (2022). Prophetic Energy Assessment with Smart Implements in Hydroelectricity Entities Using Artificial Intelligence Algorithm. International Transactions on Electrical Energy Systems, 2022.

[48] Shaw, A. R., Smith Sawyer, H., LeBoeuf, E. J., McDonald, M. P., & Hadjerioua, B. (2017).Hydropower Optimization Using Artificial Neural Network Surrogate Models of a High-Fidelity Hydrodynamics and Water Quality Model. Water Resources Research, 53(11), 9444-9461.

[49] Fortaleza, B. N., Juan, R. O. S., & Tolentino, L. K. S. (2018). IoT-based pico-hydro power generation system using Pelton turbine. Journal of Telecommunication, Electronic and Computer Engineering (JTEC), 10(1-4), 189-192.

[50] Bordin, C., Skjelbred, H. I., Kong, J., & Yang, Z. (2020). Machine learning for hydropower scheduling: State of the art and future research directions. Procedia Computer Science, 176, 1659-1668.

[51] Conteh F, Tobaru S, Lotfy ME, Yona A, Senjyu T. An effective Load shedding technique for micro-grids using artificial neural network and adaptive neuro-fuzzy inference system. AIMS Energy 2017;5:814–37. https://doi.org/10.3934/energy.2017.5.814.

[52] Zhang W, Zhang R, Wu C, Goh ATC, Lacasse S, Liu Z, et al. State-of-the-art review of soft computing applications in underground excavations. Geosci Front 2020;11:1095–106. https://doi.org/10.1016/j.gsf.2019.12.003.

[53] Kurt H, Aslan Y. Optimization of Power Output of a MicroHydro Power Station Using Fuzzy Logic Algorithm. Int J Tech Phys ProblEng2013;5:138–43.

[54] Senthil Kumar AR, Goyal MK, Ojha CSP, Singh RD, Swamee PK. Application of artificial neural network, fuzzy logic and decision tree algorithms for modelling of streamflow at Kasol in India. Water Sci Technology 2013;68:2521–6. https://doi.org/10.2166/wst.2013.491.

7

Intelligent Techniques Demand Side Management and Flexibility

Younes Noorollahi and Shiva Ansaripour

Energy Modelling and Sustainable Energy System (METSAP) Research Lab, Faculty of New Sciences and Technologies, University of Tehran, Iran
E-mail: noorollahi@ut.ac.ir; fh-aachen.de

Abstract

Comparing current power systems with the past time, more complexity can be observed, like large integration of renewable energy sources with fluctuated production curve, distributed generation, and also higher demand. Intelligent techniques are active movements of controlling the grid in a secure, reliable, and efficient way. An optimized energy network can be defined as the positive interaction of three main layers: physical power system, energy management, and socioeconomic operation. The management layer accounts for ensuring the reliability and security of the energy supply and is highly interconnected with the last layer to handle the optimized power flow in the system. Dramatic changes in the universal energy landscape have made demand side management (DSM) a challenging field that may face several limits. On the other hand, intelligent DSM aims to modify consumption time profile and grid connection parameters to achieve savings in electricity charges despite encountering instability on the grid and consequent disruptions to services. The initial step of this purpose is to carry out an analysis of the demand curve and consumer behavior that is simplified by artificial intelligence (AI) algorithms which can overcome some unexpected behavior. Implementation of different policies in DSM provokes consumers for peak demand reduction and energy-saving; however, it is worth mentioning that in smart grids, loads can also be produced and are known as active loads. The basics of DSM will be discussed in this chapter, and ultimately, the application of an AI-based DSM in performing faster response on a real-time basis to trace consumption

DOI: 10.1201/9788743809852-7

patterns and forecasting it, demand response strategies, and managing power outages will be explained and clarified.

Keywords: Demand side management (DSM), smart grids, machine learning, artificial intelligence.

7.1 Introduction

Large penetration of renewable energy sources (RES), and electro-mobility has made power systems' flexibility a challenging topic. On the other side, flexibility can be achieved by customers participating in demand response programs. To tackle these challenges, a variety of sophisticated computational techniques are considered a key solution to shift the flexibility as a duty of the power system to the customer side. Customer behavior learning, optimal pricing, power flow, and device scheduling are the optimization algorithms to handle the reliability of the system [1].

DSM can range from altering high energy-consuming devices at home to complicated dynamic load management systems. Nevertheless, it necessarily does not mean demand reduction but also may aim to shift the loads by influencing consumption patterns.

With regards to an overview in [2], DSM in the United Kingdom can be defined in three main topics as shown in Figure 7.1. Energy efficiency mostly concentrates on overall consumption reduction, while demand response mostly aims to make a balanced system by peak-shaving programs. On-site backup can also be called a physical demand response and acts in peak hours by activating a diesel generator and using storage. In terms of DSM implementers, it can be arranged by the government and applied by utilities, to be more exact, distribution network operators (DNOs). Briefly, the DNO is responsible for safeguarding the system against imbalances by using different pricing systems and issuing regulatory signals [3]. Also, the policies branch is sorted in Figure 7.1. The aggregator's role is to link the customers and manage the programs by following actions: registration and communication with users, making real-time measurements, implementing the standards the measurements for consumption verification, and calculating user engagement based on the input data [4].

The main customers in the DR programs may include residential, industrial, and also commercial users. Additionally, DR services can manifest in the form of power companies and act as virtual generators in the next-day market [5].

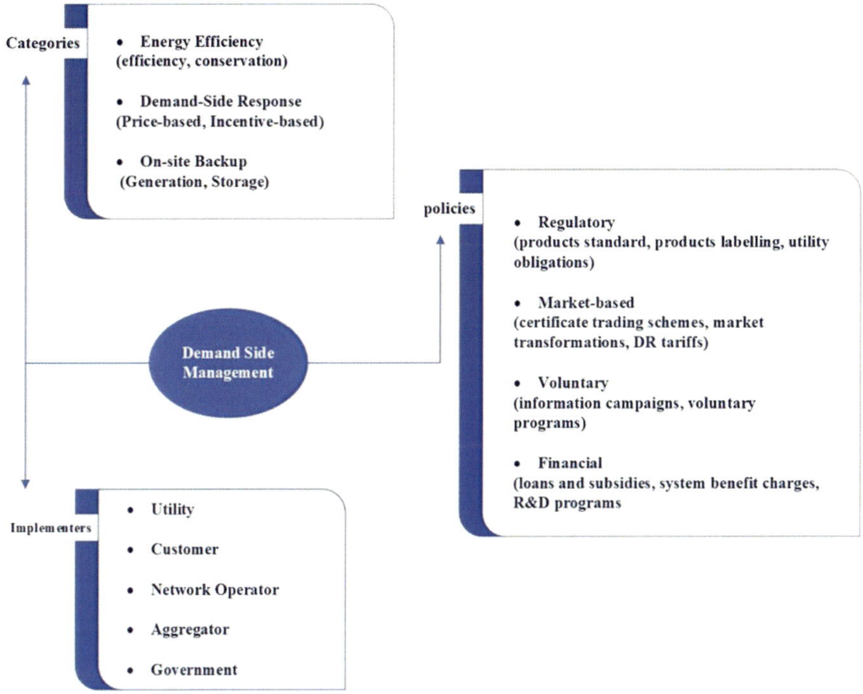

Figure 7.1 Demand side management (DSM) [2].

7.2 Literature Review

Along with electrical grid development, several DSM approaches have been discussed; in the 1920s, direct load management was started in Germany. In 1956, in France, demand responses were applied for peak shaving, and an initial term of DSM was presented in the United States around 1984 by Clark Gellings [6]. The wide use of demand response programs was developed around the globe from 1990 to 2010. Some methods like real-time trading, storage system, and flattening of the load curve have been intensely expanded since 2010 [4].

The US Department of Energy has defined demand response as "a tariff orprogram established to motivate changes in electric use by end-use customers in response to changes in the price of electricity over time, or to give incentive payments designed to induce lower electricity use at times of high market prices or when grid reliability is jeopardized" [7].

The main objective of demand response (DR) is to make consumers active in participating in the electric grid stability in coordination with systems conditions and economic signals. Based on DR programs, two main categories can be found:

1. *Price-based*: When different prices are defined for energy, consumers can reduce or shift their loads to be less in charge of paying bills.
2. *Incentive-based*: Mechanisms that rely on consumption modifications induced by nonprice signals, for which the involved users receive compensations.

There are also other sights to classify the DR [8]:

- Market DR: That contains price signals and incentives, real-time pricing. This DR relies on certain markets which are not expected to be quick in changes but in real-time pricing, the figures of the network will be forwarded to the end-users without delay. As a matter of fact, due to the limited elasticity of consumers and other physical conditions, the price alone, cannot be efficient for load shedding in a grid. That is the reason for implementing the physical DR.
- Physical DR: Defined by grid management and emergency signals. This demand response can be known as a binding request in case the infrastructure of the grid is not efficient for demand response due to a failure. An optimal system is defined as a good combination of both market and physical responses.

7.3 Optimal Power Flow and DSM

The optimal power flow problem aims to obtain an operational condition of power systems with minimized appropriate cost function and voltage variables. Since 1962 by now, several solutions like nonlinear and quadratic programming, interior-point methods, Newton–Raphson, and Lagrange coefficients have been implemented to calculate the best fitting outcome [9]. Recently, with regard to new computational development, more accurate results can be achieved by learning-based approaches like genetic algorithm (GA) [2].

The optimization problem is the load-shifting system that might be devised as below [10]

$$\text{Minimize} = \sum_{t=1}^{N} (\text{Load}\,(t) - \text{Objective}(t))^2. \tag{7.1}$$

$$\text{Load}\,(t) = \text{Forecast}\,(t) - \text{Connect}\,(t) - Disconect\,(t)\,, \tag{7.2}$$

where Objective(t) is the value of the consumption and Load(t) is calculated based on the formula.

7.4 DSM Instruments

Energy savings is a way to reduce overall consumption and promote long-term stability. Higher energy efficiency is the way that helped most states in the EU to achieve objectives like reduction in greenhouse gas emissions and energy security [11]. Consequently, DR is a tool for energy efficiency and these two areas should be assessed in parallel [12]. On the other hand, in the energy market, marginal and factual electricity consumption would be variable during the day and end-user tariffs may encourage the consumers to control their loads.

As mentioned in Figure 7.2, a wide range of optimization deployment approaches can be particular to demand side management systems. These methods can be recognized by the type of user interaction, the time scale, and the method of optimization.

Demand response programs can be grouped into three sorts, based on the main goals.

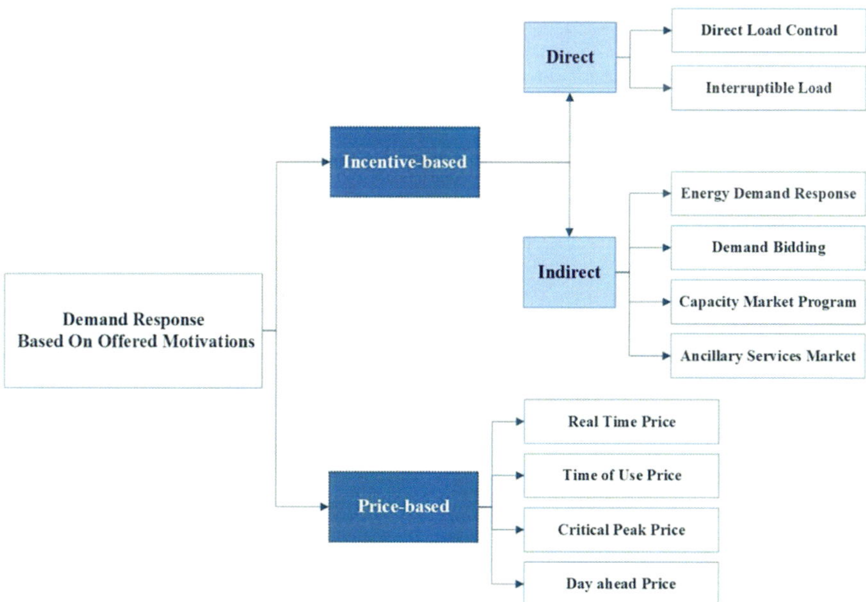

Figure 7.2 DR classification based on offered motivation.

7.4.1 DR based on the control mechanism

This kind of DR is divided into centralized and distributed mechanisms. In the first system, consumers can contact the energy company directly. Meanwhile, users in the distributed program interact with themselves to get to know about the total consumption [13].

7.4.2 DR based on offered motivations

This category comprises systems with price-based and incentive motivations or reward programs in subdivisions shown in Figure 7.2 [14]. A price-based approach can rely on different principles, like time of use (TOU), day-ahead pricing (DAP), and real-time pricing (RTP) as shown in Figure 7.3. In TOU, a static price table is applied to the electricity, and the price in peak hours is expected to be the highest rate during one day [15]. DAP is a short consumption prediction of 24 hours to highlight price based on that demand

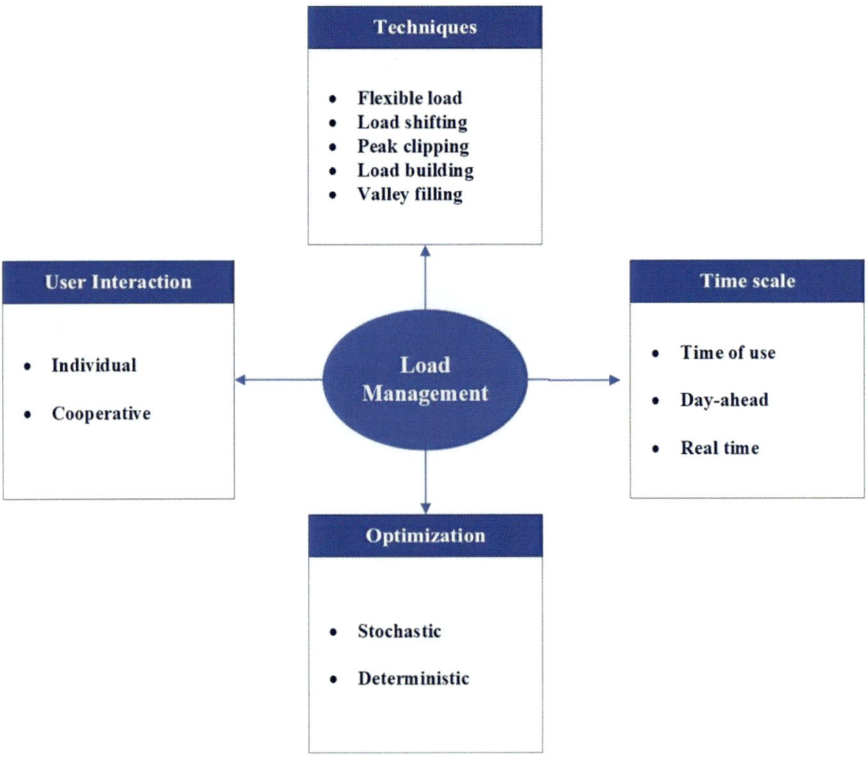

Figure 7.3 Load management techniques and aspects.

profile. However, the RTP pricing method is a dynamic scheme with up to 1-hour prediction [16].

7.4.3 DR based on the decision variable

This category distinguishes the methods by variables to determine task scheduling and power control DR systems [4]. Intense development focused on end-user behavior projects as it is the most challenging part of DSM and is relatively complex and considered a stochastic variable that directly impacts the DR programs [17].

7.5 Machine Learning-based DSM

Along with this, artificial intelligence (AI) response strategies offer smart and autonomous controllers and advanced data management software. The adequate output of consumption patterns, energy demand forecasting, and energy market participation can result in accurate real-time simulation [18].

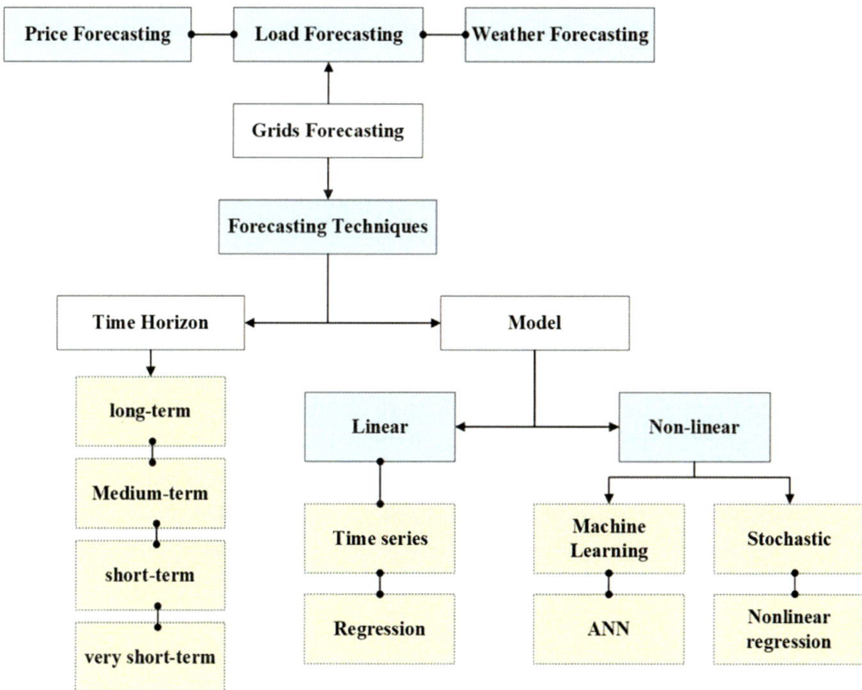

Figure 7.4 Categories of forecasting and methods in an energy system.

Machine learning includes a set of algorithms following the input data to draw information for different purposes, the main usage of machine learning in DSM can be summarized in optimal power flow, optimal device scheduling, optimal pricing, and customer behavior learning [1, 19]. Artificial neural network (ANN)-based algorithms are applied in [20] to make short-term forecasting in a price-sensitive environment by the implementation of fuzzy logic. Furthermore, deep learning algorithms were used for DSM and customer behavior learning in [21] and [22] in the residential sector. Particle swarm optimization (PSO) is used in [23] and [24] for optimal device scheduling and in [25] for OPF. Reinforcement learning ([26], [27]), robust optimization, and Bayesian models also are used in DSM programs. Figure 7.4 disputes the forecasting role in an energy system and machine learning implementation in this field [28].

7.6 Energy Demand Forecast using Machine Learning

Customer behavior learning in different time horizons is a significant branch of DSM and in the following, a review of forecasting methods is done to investigate the machine learning algorithms and also the criteria to be considered in multivariate energy demand forecasting.

A comprehensive study of 50 countries worldwide, including Iran, has investigated the impact of GDP, population growth, and renewables installation on air pollution [29]. It has been concluded that the two first criteria have a positive effect on CO_2 emission. According to Ding's study [30], a novel self-adapting intelligent Grey model is a better approach than competing natural gas forecasting methods. Using this model, Spain's primary energy consumption, combined with input-output models, resulted in the best combination for the Grey model, as done by Liu et al. [31]. The logistic model has been applied in [32] to long-term natural gas consumption forecasting in China. For getting the parameters of the logistic model Levenberg–Marquardt algorithm is adopted. In[33], the autoregressive-moving-average model with exogenous inputs (ARMAX) model has been developed for residential and commercial energy demand forecasting in Iran. Alcaraz and Villalvazo [34] presented the natural gas estimation shortage with an econometric analysis based on panel data and analyzed the natural gas interconnection shortage and GDP. In 2017, Scarpa and Bianco investigated long-term natural gas consumption, considering heating degree days, natural gas prices, and GDP per capita. To obtain this goal, they used the regression algorithm and the Kalman filter method [35]. The relationship between price and income with natural gas consumption has been shown in [31] by Liu et al. The generalized least square method was used in this study. Wang et al. [36] studied the natural gas consumption model with high accuracy by a MAPE value of 2.32% with a hybrid model based on the particle swarm optimization-wavelet neural network (PSO-WNN). In [37], artificial neural networks (ANN), multiple linear regression (MLR), and support vector regression (SVR) were presented for forecasting natural gas consumption in Istanbul with these criteria: seasonal index, temperature, price of natural gas, population and 12 years history of natural gas consumption. In [11, 12], electricity consumption forecasting in a building is done by using the GPR method. Mahdi Sharifzadeh et al. [40] did a comparative study of ANN, SVR, and GPR for forecasting residential electricity demand. Further, the Gaussian process quantile regression is used in [41] to predict Power load probability density. In some cases, the GPR method is preferred in forecasting wind forecasting [13, 15] because it is flexible to provide uncertainty representations [43].

Researchers have proposed some methods to decrease CO_2 emissions [44]. The share of renewable energy sources by 2030 in six European countries (France, Germany, Italy, Spain, United Kingdom, and Turkey) was forecasted by Utkucan Şahin, calculating the highest amount for Germany with 53.8 Mtoe renewable energy production. Dominkovic et al. studied southeast Europe to make a 100% renewable energy system for 2050 to

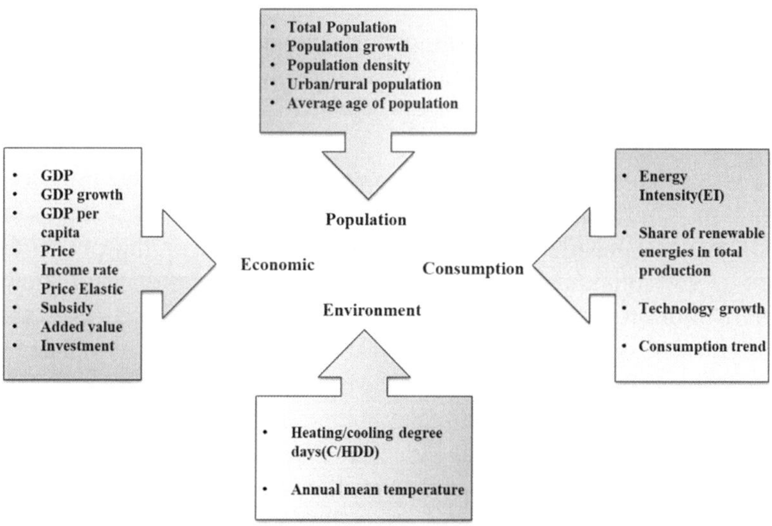

Figure 7.5 Criteria and variables in demand forecasting.

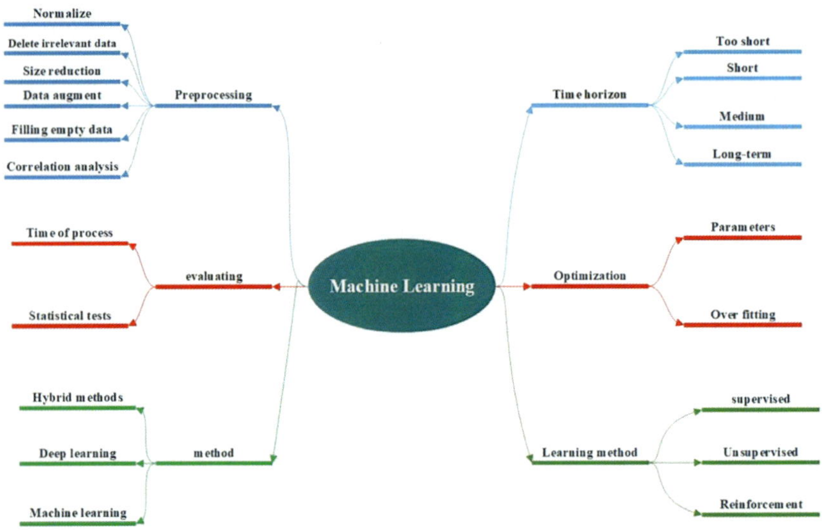

Figure 7.6 Machine learning principals.

achieve a zero-carbon energy community. For getting this object, biomass and other renewable energies have been used. Also, improving energy efficiency is considered a reformation strategy for decreasing CO_2 [45]. Daví-Arderius et al. expounded on the impact of electricity losses on CO_2 decrement. Construction of DGs near the consumers and covering different generations with renewable energy are the two most essential policies suggested in [46]. Technical progress, energy structure, and economic level are the variables that are considered in [47] to analyze their impact on CO_2 emission in China. Also, it is concluded that technological advances have about a 1% effect on China's CO_2 emission. Supporting, 14 years (2000–2014) research in China indicates that technological developments had a remarkable impact on CO_2 emission [48]. The South Asian Association for Regional Cooperation (SAARC) countries have been weighted and ranked based on the CO_2 emission issue by grey relational analysis (GRA) in [49]. The results have shown substantial pollution problems in India with the first rank; hence, renewable energy installation and adoption of ISO14001 certification were introduced to solve this problem. Toward reduction of carbon emission, [50] offered three assortments that contain adding a clean energy supply, developing energy conservation, and negative emission strategies like using CCS technology.

Figures 7.5 and 7.6, respectively, show the most important criteria for demand forecasting and the different stages that should be considered in a machine learning algorithm.

7.6.1 Long-term prediction using Gaussian process regression (GPR)

In this study, the consumption curve of natural gas in a province in Iran is forecasted by Gaussian process regression (GPR). GPR is a nonparametric probabilistic kernel method that can model arbitrary complex systems [40]. This method combines arbitrary variables with a number describing the joint Gaussian distribution [41]. GPR models a probability distribution by functions and can be parameterized with statistical functions like mean *m(x)* which is the expected value of *f(x)* and covariance $k(\mathbf{x}\text{-}\mathbf{x}')$ which defines the similarity between data points. It can be shown as

$$y = f(x) \sim GP(m(x), \kappa(x - x')), \tag{7.3}$$

where *x* and *y* are the input and output in the training data set, respectively, and *f(x)* is called the latent variable. For simplification, mostly *m(x)* is considered to be 0.

A variety of covariance functions can be used. Some of the most common ones are squared exponential (SE) eqn (7.4), Matern (MA) eqn (7.5), and rational quadratic (RQ) eqn (7.6).

$$\kappa_{\text{SE}}(x - x') = \theta_{\text{f}}^2 exp(\frac{\|x - x'\|^2}{\theta_{\text{l}}^2}). \tag{7.4}$$

$$\kappa_{\text{Ma}}(x - x') = \sigma^2 \frac{2^{1-v}}{\Gamma(v)}(\sqrt{2v}\frac{x - x'}{l})^v \kappa_{\text{v}}(\sqrt{2v}\frac{x - x'}{l}),\ v,\ l > 0. \tag{7.5}$$

$$\kappa_{\text{RQ}}(x - x') = \sigma^2(1 + \frac{x - x'}{2\alpha l^2})^{-\alpha},\ \alpha, l > 0. \tag{7.6}$$

The SE function is infinitely differentiable, so the GPR is so smooth using it, and it is too strict for physical action [42]. θ_f and θ_l are parameters that control the length scale.

K_v in Matern covariance was modified Bessel function. The function becomes simple when *v* is half floating-point *v* = p +1/2, where *p* is an integer. *V* is mostly considered to be *v* = 5/2 and *v* = 3/2.

Along with this, two effective methods were proposed to evaluate machine learning algorithms' performance: Holdout-test, which mostly applies to large data sets, and cross-validation (CV), which is called K-fold validation. This approach divides the data set into parts and estimates each fold's accuracy to prevent overfitting the output. Also, by increasing the number of folds, more reputable outputs can be derived.

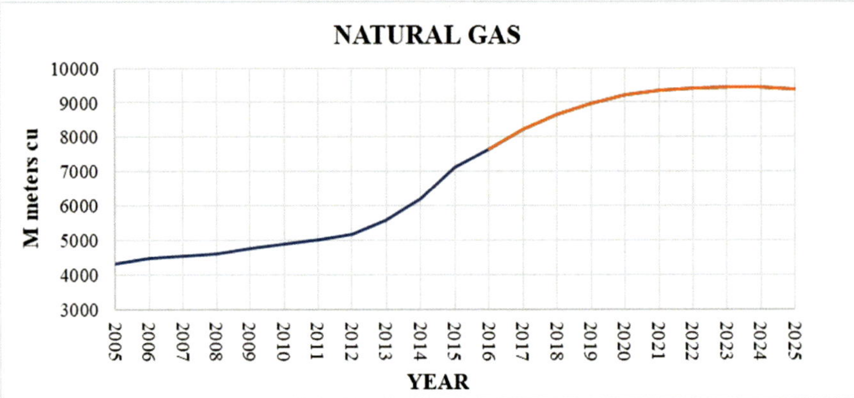

Figure 7.7 The results of natural gas demand prediction with long-term prediction.

7.7 Result and Discussion

Despite several intelligence DSM methods, there are some fundamental challenges to selecting an appropriate system: (1) continuous input data for real-time decisions; (2) the limited time of process; (3) encountering various periods rather than just a daily slot. Heuristic algorithms can be called nature-inspired techniques which enable them to tackle optimization challenges. Particle swarm optimization (PSO), evolutionary algorithm, and ant colony optimization may be distinguished among other approaches [4]. Gaussian process regression (GPR) can also be known as a practical algorithm for long-term prediction. Figure 7.7 shows the output of 5 years ahead of prediction with GPR and considers the population and gross domestic production (GDP), as variables. The best-fitted covariance function is selected by comparing R-squared, "Matern" with R-squared equal to 0.96.

7.8 Conclusion

An energy system with high penetration of Renewable Energy Sources (RES), needs a well-integrated and bidirectional demand model to compensate for the fluctuations of RES and keep the grid flexible.

In this context, some control strategies for traditional power systems are going to be changed. As evident, it is less expensive to intelligently control the load than the installation of new storage systems or power plants.

Demand side management (DSM) was developed for Demand and supply balance by adding a certain portion of "smart" to the loads and participating consumers as active users. In this system, customers would rearrange their habits and schedule to minimize their bills.

Additionally, demand response (DR), is widely known as a critical element not only as a control strategy that regulates the system but also as a significant plan for the next generation of virtual power plants (VPPs).

References

[1] T. A. Nakabi, K. Haataja, and Pekka Toivanen, "Computational Intelligence for Demand Side Management and Demand Response Programs in Smart Grids," 8th Int. Conf. Bioinspired Optim. methods their Appl. Paris, 2018., 2018, [Online]. Available: https://www.researchgate.net/publication/331382569_Computational_Intelligence_for_Demand_Side_Management_and_Demand_Response_Programs_in_Smart_Grids.

[2] P. Warren, "A review of demand-side management policy in the UK," Renewable and Sustainable Energy Reviews, vol. 29. pp. 941–951, 2014, doi:10.1016/j.rser.2013.09.009.

[3] L. Cioara, T.; Anghel, I.; Pop, C.; Bertoncini, M.; Croce, V.; Ioannidis, D.; D'Oriano, "Enabling New Technologies for Demand Response Decentralized Validation Using Blockchain.," in IEEE International Conference on Environment and Electrical Engineering and 2018 IEEE Industrial and Commercial Power Systems Europe (EEEIC/I&CPS Europe), Palermo, Italy, 2018, pp. 1–4.

[4] V. Stanelyte, D. Radziukyniene, N. Radziukynas, "Overview of Demand-Response Services: A Review," Energies, vol. 15, no. 5, 2022.

[5] S. Radenkovi'c, M.; Bogdanovi'c, Z.; Despotovi'c-Zraki'c, M.; Labus, A.; Lazarevi'c, "Assessing consumer readiness for participation in IoT-based demand response business models.," Technol. Forecast. Soc. Chang., vol. 150, p. 119715, 2020.

[6] C. W. Gellings, "The Concept of Demand-Side Management for Electric Utilities," Proc. IEEE, vol. 73, no. 10, pp. 1468–1470, 1985, doi:10.1109/PROC.1985.13318.

[7] J. S. Vardakas, N. Zorba, and C. V. Verikoukis, "A Survey on Demand Response Programs in Smart Grids: Pricing Methods and Optimization Algorithms," IEEE Commun. Surv. Tutorials, vol. 17, no. 1, pp. 152–178, 2015, doi:10.1109/COMST.2014.2341586.

[8] P. Palensky and D. Dietrich, "Demand side management: Demand response, intelligent energy systems, and smart loads," IEEE Trans. Ind.

Informatics, vol. 7, no. 3, pp. 381–388, 2011, doi:10.1109/TII.2011.2158841.

[9] A. R. Kumar, R. S., Raghav, L. P., Raju, D. K., & Singh, "Intelligent demand side management for optimal energy scheduling of grid connected microgrids.," Appl. Energy, vol. 285, no. 116435, 2021.

[10] B. . Pavithra, N.; Esther, "Residential demand response using genetic algorithm," Proc. 2017 Innov. Power Adv. Comput. Technol., pp. 1–4, 2017.

[11] S. Olkkonen, V.; Rinne, S.; Hast, A.; Syri, "Benefits of DSM measures in the future Finnish energy system," Energy, vol. 137, pp. 729–738, 2017.

[12] W. Wohlfarth, K.; Worrell, E.; Eichhammer, "Energy efficiency and demand response—Two sides of the same coin?," Energy Policy, vol. 137, no. 11070, 2020.

[13] J. Hui, H.; Ding, Y.; Shi, Q.; Li, F.; Song, Y.; Yan, "5G network-based Internet of Things for demand response in smart grid: A survey on application potential," Appl. Energy, vol. 257, no. 113972, 2020.

[14] K. T. Wu, Y.K.; Tang, "Frequency Support by Demand Response—Review and Analysis.," Energy Procedia, vol. 156, pp. 327–331, 2019.

[15] F. M. T. Voulis, N.; Van Etten, M.J.J.; Chappin, É.J.L.; Warnier, M.; Brazier, "Rethinking European energy taxation to incentivise consumer demand response participation.," Energy Policy, vol. 124, pp. 156–168, 2019.

[16] H. Bergaentzlé, C.; Clastres, C.; Khalfallah, "Demand-side management and European environmental and energy goals: An optimal complementary approach.," Energy Policy, vol. 67, pp. 858–869, 2014.

[17] G. T. Chatzigeorgiou, I.M.; Manolas, D.; Gkaragkouni, T.; Andreou, "Demand Response in Greece: An Introductory Mobile Application," in In Proceedings of the 2018 IEEE International Conference on Environment and Electrical Engineering and 2018 IEEE Industrial and Commercial Power Systems Europe (EEEIC/I&CPS Europe), Palermo, Italy, 2018, pp. 1–5.

[18] H. R. K. Kiran Chaurasia, "New Approach using Artificial Intelligence-Machine Learning in Demand Side Management of Renewable Energy integrated Smart Grid for Smart City," 2021.

[19] O. Hirotaka, T., Naoto, T., Shou, K. and Atsumi, "A design method for incentive-based demand response programs based on a framework of social welfare maximization.," IFAC-PapersOnLine, vol. 51, no. 28, pp. 374–379, 2018.

[20] A. Khotanzad, E. Zhou, and H. Elragal, "A neuro-fuzzy approach to short-term load forecasting in a price-sensitive environment," IEEE Trans. Power Syst., vol. 17, no. 4, pp. 1273–1282, 2002, doi:10.110 9/TPWRS.2002.804999.

[21] R. L. Heng Shi, Minghao Xu, "Deep Learning for Household Load Forecasting – A Novel Pooling Deep RNN," " IEEE Trans. Smart Grid, vol. 1, no. 99, 2017.

[22] D. Zhang, S. Li, M. Sun, and Z. O'Neill, "An Optimal and Learning-Based Demand Response and Home Energy Management System," IEEE Trans. Smart Grid, vol. 7, no. 4, pp. 1790–1801, 2016, doi: 10.1109/TSG.2016.2552169.

[23] H. A. Ehsan Dehnavi, "'Optimal pricing in time of use demand response by integrating with dynamic economic dispatch problem'," Energy, vol. 109, pp. 1086–1094, 2016.

[24] R. V Sofana Reka S., "Demand Side Response Modeling with Controller Design Using Aggregate Air Conditioning Loads and Particle Swarm in Advanced Computing and Communication Technologies Optimization. In:," springer, singapoore, pp. 573–581, 2016.

[25] M. F. Abdi H, Dehnavi E, "Dynamic economic dispatch problem integrated with demand response (DEDDR) considering non-linear responsive load models."," IEEE Trans. Smart Grid, vol. 7, no. 6, pp. 2586–2590, 2015.

[26] J. W. S. K. Dehghanpour, M. Hashem Nehrir and N. C. Kelly, "Agent-Based Modeling of Retail Electrical Energy Markets With Demand Response," IEEE Trans.on smart grid, 2016.

[27] Y. Liang, L. He, X. Cao, and Z. J. Shen, "Stochastic control for smart grid users with flexible demand," IEEE Trans. Smart Grid, vol. 4, no. 4, pp. 2296–2308, 2013, doi:10.1109/TSG.2013.2263201.

[28] S. R. Battula, A. R., Vuddanti, S., & Salkuti, "Review of Energy Management System Approaches in Microgrids.," Energies, vol. 14, no. 17, pp. 1–32, 2021.

[29] L. F. Anny Key de Souza Mendonça, Gabriel de Andrade Conradi Barni, Matheus Fernando Moro, Antonio Cezar Bornia, Emil Kupek, "Hierarchical modeling of the 50 largest economies to verify the impact of GDP, population and renewable energy generation in CO2 emissions," Sustain. Prod. Consum., vol. 22, pp. 58–67, 2020, doi:10.1016/j.spc.20 20.02.001.

[30] S. Ding, "A novel self-adapting intelligent grey model for forecasting China's natural-gas demand," Energy, vol. 162, pp. 393–407, Nov. 2018, doi:10.1016/j.energy.2018.08.040.

[31] J. Liu, G., Dong, X., Jiang, Q., Dong, C., Li, "Natural gas consumption of urban households in China and corresponding influencing factors," Energy Policy, vol. 122, pp. 18–26, 2018.

[32] Q. Shaikh, F., Ji, "Forecasting natural gas demand in China: logistic modelling analysis," Int. J. Electr. Power Energy Syst, vol. 77, pp. 25–32, 2016.

[33] A. Shakouri, G.H., Kazemi, "Selection of the best ARMAX model for forecasting energy demand: case study of the residential and commercial sectors in Iran," Energ. Effic, vol. 9, pp. 339–352, 2016.

[34] S. Alcaraz, C., Villalvazo, "The effect of natural gas shortages on the Mexican economy," Energy Econ, vol. 66, pp. 147–153, 2017.

[35] V. Scarpa, F., Bianco, "Assessing the quality of natural gas consumption forecasting: an application to the Italian residential sector," Energies, pp. 10, 1879, 2017.

[36] H. Wang, D., Liu, Y., Wu, Z., Fu, H., Shi, Y., Guo, "Scenario analysis of natural gas consumption in China based on wavelet neural network optimized by particle swarm optimization algorithm," Energies, vol. 11, p. 825, 2018.

[37] O. F. Beyca, B. C. Ervural, E. Tatoglu, P. G. Ozuyar, and S. Zaim, "Using machine learning tools for forecasting natural gas consumption in the province of Istanbul," Energy Econ., vol. 80, pp. 937–949, May 2019, doi:10.1016/j.eneco.2019.03.006.

[38] D. W. Van Der Meer, M. Shepero, A. Svensson, J. Widén, and J. Munkhammar, "Probabilistic forecasting of electricity consumption , photovoltaic power generation and net demand of an individual building using Gaussian Processes," Appl. Energy, vol. 213, no. January, pp. 195–207, 2018, doi:10.1016/j.apenergy.2017.12.104.

[39] M. Shepero, D. Van Der Meer, J. Munkhammar, and J. Widén, "Residential probabilistic load forecastingÂă: A method using Gaussian process designed for electric load data," Appl. Energy, vol. 218, no. March, pp. 159–172, 2018, doi:10.1016/j.apenergy.2018.02.165.

[40] M. Sharifzadeh, A. Sikinioti-lock, and N. Shah, "Machine-learning methods for integrated renewable power generation : A comparative study of arti fi cial neural networks , support vector regression , and Gaussian Process Regression," Renew. Sustain. Energy Rev., vol. 108, no. February, pp. 513–538, 2019, doi:10.1016/j.rser.2019.03.040.

[41] Y. Yang, S. Li, W. Li, and M. Qu, "Power load probability density forecasting using Gaussian process quantile regression," Appl. Energy,

vol. 213, no. August 2017, pp. 499–509, 2018, doi:10.1016/j.apenergy.2017.11.035.

[42] X. Ma, F. Xu, and B. Chen, "Journal of Wind Engineering & Industrial Aerodynamics Interpolation of wind pressures using Gaussian process regression," J. Wind Eng. Ind. Aerodyn., vol. 188, no. February, pp. 30–42, 2019, doi:10.1016/j.jweia.2019.02.002.

[43] H. Cai, X. Jia, J. Feng, W. Li, Y. Hsu, and J. Lee, "Gaussian Process Regression for numerical wind speed prediction enhancement," Renew. Energy, vol. 146, pp. 2112–2123, 2020, doi:10.1016/j.renene.2019.08.018.

[44] Ahmadi Esmaeil; Mohammadi-Ivatloo Behnam; Anvari-Moghaddam Amjad; Noorollahi Younes, "Stochastic Operation of a Solar-Powered Smart Home: Capturing Thermal Load Uncertainties," Sustainability, vol. 12, no. 5089, pp. 1–18, 2020.

[45] D. F. Dominković et al., "Zero carbon energy system of South East Europe in 2050," Appl. Energy, vol. 184, pp. 1517–1528, 2016, doi: https://doi.org/10.1016/j.apenergy.2016.03.046.

[46] E. T.-B. Daniel Daví-Arderius, María-Eugenia Sanin, "CO2 content of electricity losses," Energy Policy, vol. 104, pp. 439–445, 2017, doi:10.1016/j.enpol.2017.01.011.

[47] L. X. Zheng, You Fei Li, Hai Tao Wu, Rong Jun Wang, "Impact of technology advances on China's CO2 emission reduction," Chinese Sci. Bull., vol. 55, no. 19, pp. 1983–1992, 2010, doi:10.1007/s11434-010-3241-1.

[48] Z. W. Bo Wang, Yefei Sun, "Agglomeration effect of CO2 emissions and emissions reduction effect of technology: A spatial econometric perspective based on China's province-level data," J. Clean. Prod., vol. 204, pp. 96–106, 2018, doi:10.1016/j.jclepro.2018.08.243.

[49] S. Z. A. S. Muhammad Ikram, Qingyu Zhang, Robert Sroufe, "Towards a sustainable environment: The nexus between ISO 14001, renewable energy consumption, access to electricity, agriculture and CO2 emissions in SAARC countries," Sustain. Prod. Consum., vol. 22, pp. 218–230, 2020, doi:10.1016/j.spc.2020.03.011.

[50] J. Jiang, B. Ye, and J. Liu, "Research on the peak of CO2 emissions in the developing world: Current progress and future prospect," Appl. Energy, vol. 235, no. October 2018, pp. 186–203, 2019, doi:10.1016/j.apenergy.2018.10.089.

8

How to Sustainably Balance Wind Energy

Jan Emblemsvåg

NTNU, Ålesund, Norway
E-mail: jan.emblemsvag@ntnu.no

Abstract

Wind power plants are commonly proposed as technology for reducing climate gas emissions, but several studies show that the reductions of climate gases emissions are low due to the required fossil fueled balancing power. In fact, wind power plants using fossil fuel balancing can never result in a low carbon electricity grid. Thus, it seems impossible with today's commercially available technologies except through Battery Energy Storage Systems (BESS) and/or Power Purchasing Agreements with hydroelectric power plants. The BESS or the hydroelectric power plants will serve as balancing power to the wind energy. The purpose of the chapter is to analyze the environmental viability of these two alternatives using Norway as a case. Norway is an ideal case with its very high hydroelectric share of the grid, which implies that if it does not work there, it cannot work anywhere else where the hydroelectric share of the grid is much lower. The conclusion is clear; wind power plants can only contribute towards a low-carbon electricity grid through formal Purchasing Power Agreements with hydroelectric power plants. Unfortunately, this is currently not the current practice in Norway. The BESS has far too high environmental burden to constitute a low carbon grid solution. Today's BESS can therefore best be viewed as peak shaving technologies if they are combined with wind power plants.

Keywords: Battery Energy Storage System (BESS), Externalities, Levelized Cost of Energy (LCOE), Life-Cycle Assessment (LCA), Monte Carlo simulation, Norway, Purchasing Power Agreement (PPA).

DOI: 10.1201/9788743809852-8

8.1 Introduction

Fossil energy has historically been most common balancing power from 1990–2013 in 26 OECD countries [129]. Indeed, wind and gas are unlikely allies in Ireland [28] because gas turbines must balance wind. Thus, establishing low-carbon electricity grids by using wind power, or 'wind' in the remainder of this chapter, is impossible with today's technologies when the backup must come from fossil sources [40]. In fact, the main conclusion of [119] is that balancing wind power at high wind penetration is far more costly than generally believed.

Similar results are found on large grids also. According to a gas industry representative [137], Germany spent 31 billion euros at subsidizing 80 million tonnes of climate gas emission cuts in 2020, for which the same amount of money could have bought 1.24 billion tonnes of CO_2 certificates in the European emissions trading. Indeed, the Federal Audit Office of Germany claims that the cost of the Energiewende will reach 525 billion euros by 2025, and they state that "The Federal Audit Office sees the danger that the energy transition in this form will endanger Germany as a business location and overwhelm the financial strength of electricity-consuming companies and private households" [21], and translated by [132].

Note that the term 'low-carbon electricity grid' is to equal an electricity grid where 80-95% of the life-cycle GHG emissions, are eliminated compared to a grid predominantly driven by fossil fuels [40], which is derived from the target of the European Commission [48] of reducing GreenHouse Gas (GHG) emissions to 80–95% below 1990 levels by 2050. Depending on the grid situation in 1990, this definition will give different results. The direct emissions (not life-cycle emissions) of the EU-28 were 5,669 million tonnes CO_2-equivalents in 2020 [34] while the electricity production in 1990 was 3,274 TWh [61]. Thus, the suggested targets imply between 87–346 g CO_2-eq./kWh, which for convenience is rounded off to 100 g CO_2-eq./kWh, or a grid with 100 kg CO_2-eq./MWh.

The UN's Sustainable Development Goal (SDG) 7, "Ensure access to affordable, reliable, sustainable and modern energy for all", [124] is the background for the low carbon grid concept. Importantly, "achieving sustainable development in its three dimensions—economic, social and environmental—in a balanced and integrated manner" is the ultimate objective [125]. One of the most commonly applied tools for assessing the environmental performance of energy sources is the Global Warming Potential (GWP) [111].

Modern is difficult to measure, but affordability is not. However, 'affordable' is not just a question of price but also of utility. That discussion is outside the scope of this chapter due to its context specificity. For example, relatively poorer villagers in Mali and Uganda are willing to pay about ten times higher price than the typical prevailing price in developed countries [13]. Thus, the net economic benefits can be very different for the same costs [69].

'Reliable' is perhaps best defined and measured. Power system planners in the US typically aim at maintaining a LOLE value of 0.1 days/year, or 2.4 hours per year based on the target of one outage-day every 10 years [71] where LOLP (Loss of Load Probability) is defined as the probability of a loss of load event in which the system load is greater than available generating capacity during a given time period and LOLE (Loss of Load Expectation) is the sum of LOLPs for a given planning horizon and equals the expected number of time periods in which a loss of load event occurs [78].

This begets the question what technologies can today, and in the near future, balance wind to achieve a low-carbon electricity grid? Indeed, wind cannot realistically be considered an energy source on its own in a modern grid due to their significant balancing needs [40]. We must always consider wind together with something else – the total system. Therefore, we cannot discuss wind without discussing the balancing. Given the discussion before, there are three alternatives.

The first alternative is pumped hydroelectric power, or 'pumped hydro' in short, which accounts for 96% of worldwide storage capacity today [130]. Unfortunately, it has very specific location requirements [53] along with other concerns [115]. Also, pumped hydro cannot handle seasonal variations due to the duration and the sheer amount of energy that must be stored. Furthermore, efficient markets, such as NordPool, and a roundtrip efficiency of around 65-80% [59] make it difficult to compete because the margin is too low as shown in Section 8.4.

Furthermore, the El Hierro island in the Canary Island archipelago is an interesting case. The island has advantageous topography with a natural (upper) reservoir located at 700 meters above sea level that can take 380,000 m^3 of water and the island guarantees high wind levels of 7.24–8.42 m/s[1]. Indeed, the El Hierro Island "embodies the first major Megawatt-level energy project by linking energy storage systems with wind power generation using

[1] See Gorona del Viento at www.goronadelviento.es for details about the island (accessed on 20 November 2017).

water storage delivered by the pumping system between two artificial lakes" [55], but "Studies have demonstrated that in the majority of locations, the achievement of the full energy self-sufficiency is an unrealistic purpose; the average annual percentage of renewable energy sources in the energy balance is estimated at the level between 30 and 80%" [52]. Renewable energy is sometimes reaching up to 75% share [55] with an average share in the first half of 2018 of 59.67% [52]. The fundamental problem is the same as in Ireland – wind requires balancing/backup – and incapable of reaching the low carbon grid target.

The second alternative is to use hydro to balance the wind so that there is a secured total output. This is essentially an integrated hydro-wind portfolio approach where the two are strictly linked so that 1 kWh of wind implies 1 kWh less hydro produced at that time, but the sum of the two is defined at a guaranteed level to the market over a year [43]. Norway, for example, had in 2019 5.5 TWh wind power whereas hydro constituted 127.7 TWh, or more than 95%[2]. The installed capacity of hydro has doubled globally over the last 25 years, and 14 countries generated more than 50 TWh in 2019 [64]. In fact, according to IEA[3], hydro generated 4,325 TWh in 2018 whereas wind and solar PV generated 554 TWh and 1,273 TWh, respectively. Hence, the research presented here is applicable for a long list of countries.

The third alternative is to use Battery Energy Storage Systems (BESS). Currently, the commercial viability of co-located projects is limited, 10.0 GW per yearend 2020 [62], and most of the existing installations are pilot projects or demonstrators [133]. With significant improvements in BESS technology over the years, however, the BESS alternative warrants exploration.

Lithium-ion (Li-ion) batteries is currently dominating BESS investments [60], which together with the strong anticipated investment cost reductions mean that by 2030 vanadium redox flow and Li-ion are likely to be the most cost efficient for the application discussed here despite an operating life of only 8 to 13 years, respectively [104]. Thus, this work finds the Li-ion technology a good reference technology for discussing BESS.

We sometimes read about electric vehicles serving as BESS for the grid, but the usefulness is highly questionable. For example, Shaner *et al.* [105] estimate that the 150 million light duty vehicles in the U.S., each with a BESS of 100 kWh, would provide 1.5 TWh if 10% were discharging, or about

[2] Calculated using data from https://www.statnett.no/for-aktorer-i-kraftbransjen/tall-og-data-fra-kraftsystemet/last-ned-grunndata/

[3] Accessed 2021-04-21 at https://www.iea.org/subscribe-to-data-services/renewables-statistics.

3 hours of mean power demand. In other words, storage by using electric vehicles in grid context is insignificant.

Now, using BESS as peak shavers to smooth out grid load may bring advantages for all stakeholders in the energy system, see [121], including lower costs [23], less emissions [102] and better quality [117]. These advantages, however, are is irrespective of the overall grid configuration with or without wind power plants, and therefore uninteresting for our discussion in this chapter.

Finally, note that nuclear energy is currently incapable of following the fluctuations of wind power. However, load-following nuclear reactors are expected to enter commercial operations within a few years, see [90], including molten-salt reactors [41].

Then, we have biogas, which in Germany constituted 5.0% of the electricity production in 2019 [16]. Hence, biogas is relatively modest for the total, but there are debates concerning how the biological material is obtained and the impact on the environment, see [97], and the American South is complaining about the collection of biomass [72] triggering a letter from almost 796 scientists to stop the destruction [1]. Obviously, if the biological material can be obtained in an environmentally benign way using leftovers and the like, and without being transported around world, biogas will most likely be a positive contribution albeit negligible one for the larger question addressed here. Hence, biogas is excluded from this study.

Other approaches such as demand management are not discussed because such technology makes sense regardless simply because nothing is better than the energy we save. Therefore, this chapter discusses the combination of wind power with hydro and/or BESS to secure a low carbon electricity grid by testing whether or not

Wind power can be combined with hydro and/or BESS to achieve a low carbon electricity grid.

The two alternatives are therefore 1) 'Wind + Hydro' and 2) 'Wind + BESS'. Norway will serve as case due to its data availability very large share of hydro. Essentially, this work is an extension of the work of [40, 43].

With the discussion so far, we understand that the analysis will use Life-cycle emissions for the emissions and reliability metrics must be met. The life-cycle emissions must be analyzed using Life-Cycle Assessment (LCA) focusing on GWP. Such integrated environmental- and economic analyses are relatively rare, but an early approach is developed by [37]. The approach in this chapter borrows ideas from that.

In terms of building the models for the research, we should be aware of a number of serious caveats concerning estimating life-cycle emissions, see [40] for a lengthy discussion. Also [92] review several LCA studies and document a staggering spread of results. Unfortunately, studies often ignore important externalities for wind as discussed in Section 8.3. Some shortcomings can be handled analytically by incorporating uncertainty in the model by using Monte Carlo simulations, see [38] for detailed presentation of such analyses. The case is presented in Section 8.4, after which research is discussed and future work identified. Concluding remarks are found in Section 8.6.

8.2 A Brief Literature Review of Balancing Wind in Grids with High Penetration

There is considerable uncertainty and debate over the effect of wind on the displacement of emissions [118], fueled by practices such as using the average emissions of the whole system when wind will actually displace only generators operating on the margin. A key issue is that integration of Variable Renewable Energy (VRE) becomes unsatisfactory because of intermittency, non-dispatchable- and variable/uncertain energy output [135]. This impacts both costs and emissions.

This review is therefore limited to literature that demonstrates a credible handling of fluctuations and balancing, or integration, concerning system reliability. For example, literature suggesting 100% non-synchronous energy sources are excluded because it has never been demonstrated for larger grids technically and let alone economically. Thus, the review builds on [40] who focused on fossil balancing of wind.

The two keywords ‘balancing’ and ‘integration’ are used when reviewing the literature that has a credible discussion on system reliability. Interestingly, hydro combined with wind is relatively unexplored; “very little scientific literature deals with how hydropower can be used together with solar- and wind power, despite the fact that they share common driving meteorological and climatological phenomena, and they are supposed to be the three master pieces of a 100% renewable energy scenario” [45]. There are, however, numerous LCAs of wind, see Table 8.4, that this study utilizes.

Interestingly, the cost literature is richer, but the proper measurement of integration costs is a hotly debated subject in academic and policymaking circles [2]. Indeed, a generally accepted definition of “integration costs” does not exist and calculations are subject to large uncertainties and controversies, including predictions about future development of power system.

In fact, the "value factor" of VREs – the average market value of a MWh of VRE energy divided by the average price of electricity in USD/MWh—tends to decline with increasing VRE share [20, 134]. Thus, when the integration cost of VRE is included, the optimal shares of VRE in the grid will decline significantly [9, 139]. There are substantial indirect costs of VRE integration and highlights the complementarity of investments in different generation technologies for a successful decarbonization process [129].

Interestingly, a similar debate is virtually non-existent concerning climate gas emissions [40] and completely missing when considering hydro as balancing power. However, Thomson *et al.* [118] offer some insight albeit without the LCA perspective and uncertainty. Both [119] and [122] include emissions but only [119] take the life-cycle perspective, and none of the papers offer the level of detail provided here.

When it comes to BESS, there are many papers. However, the fluctuations of the wind lead to a BESS discharge duration of a minimum of 3 weeks [105], and only [56] have relevant perspective. With 6 hours of data resolution, significant inaccuracies are introduced in their analysis, as shown later. Furthermore, the meager 13.6% inter-trade in Europe [12], is ignored. Thus, huge investments in BESS and new wind capacity would be required. The analysis of [18] even question the viability of their own analysis, and [10, 136] only focus on the day-trading market.

Solar PV combined with BESS, however, differs substantially from wind combined with BESS because the BESS can be dimensioned far more reliably due to the higher regularity of the sun compared to the wind [42]. For example, [98] find that Solar PV and BESS would be effective at curbing California's domestic electricity grid mix carbon emissions by 50%, and reducing demand for non-renewable primary energy by 66%. Unfortunately, both the cost perspective is unclear, and the climate gas reductions are insufficient to constitute a low-carbon grid.

Furthermore, most BESS analyses only consider investment costs and ignore operational costs despite the importance of the number of charging cycles [104]. Thus, BESS design must be closely defined by the specific applications, and given the geophysical constraints of wind [105], large-scale applications seem prohibitively costly.

Importantly, the true environmental benefits of new storage capacity are challenging to discern due to the overall dynamic interactions between power plants and storage inherent to the operations of an electric grid [68], and to their knowledge, there has yet to be a comprehensive evaluation of the life cycle emissions associated with different configurations of renewable capacity additions, storage capacity additions, and carbon pricing options at the scale of grids (i.e., rather than individual technologies). In this study, Wind + BESS is nonetheless analyzed, similar to what Emblemsvåg [40] did concerning wind power and gas power, to evaluate whether there is any realism behind that alternative for an environmental point of view. This is clearly a novel contribution to the LCA literature but also to the VRE literature.

8.3 Externalities

The discussion of externalities is important because externalities can become internalized later through new legislations and the likes. Also, they differ from technology to technology. With the research question in this chapter, the externalities of hydro, wind and BESS are important to later aid interpretation.

8.3.1 Hydro externalities

Hydro is mostly concrete (99.5% by mass), a relatively small amount of steel (0.5%) and negligible amounts of other materials [126]. The impacts of both concrete and steel are well captured in LCAs. Hence, the externalities are not so much related to material flow as for BESS and Wind discussed shortly. In fact, in a literature review covering 29 publications of the externalities of hydro [82] point to mainly deteriorations of landscape, vegetation and wildlife. The latter two concerns biodiversity, which in hydro context is location specific because it relates to the construction of dams and the like, and their impact local flora and fauna.

In some cases, such as the Three Gorges Dam, it also impacts human life. Key problems in the public debate include the danger of dam collapse, the displacement of some 1.3 million people (critics insist the figure was actually 1.9 million) living in more than 1,500 cities, towns, and villages along the river, and the destruction of magnificent scenery and countless rare architectural and archaeological sites and at the same time increase the risks for landslides and even earthquakes [44].

8.3.2 Wind externalities

Wind is not too different from hydro in that there is a large amount of concrete (78% by mass) and steel (17.5%), but in addition there are glass (0.9%) and 'other materials' (3.6%) [126]. It is these 'other materials' that are of interest in many contexts including externalities because they involve large amounts of Rare Earth Elements (REE), as discussed in the review of LCA shortcomings [40].

Interestingly, in neither [81] nor [138] these issues are not discussed in their literature review of externalities of wind. Both end up with a shortlist consisting of 1) visual effects of the wind turbines and 2) local effects on fauna and birdlife that are killed by the rotors, and [138] also included noise. These literature reviews, covering almost 500 articles, take an economic view and focus on the operational phase for wind. Such approaches are too narrow.

For example, when it comes to the killing of birds – it is probably true that domestic cats kill many more birds than wind power plants do, as claimed by [47], but wind power plants are located where the wind resources are good and such places often correspond to specific birdlife. In the Irish sea, where there are major offshore wind power plants, seabird populations have plummeted [57].

The 'polluter pays' principle is ignored in all analysis this author has seen concerning wind power plants, particularly economic analyses. This is a significant omission because the turbine blades represent a major waste challenge at their end of life, see [77] for an excellent review. They estimate that the cumulative blade waste in 2050 will lie between 21.4 and 69.4 million tonnes with the most probable waste level being 43.4 million tonnes. This waste is generated by the fact that composites account for more than 90% of the weight of wind turbine blades [76], and this problem will increase when the blade length is doubled, since the blade volume is increased 8-fold for the same material and same design [77]. Unfortunately, there are currently no established industrial recycling routes for these composites [67], and this is still the case [80] almost ten years later, but it is likely that a solution can be found that will reduce the problem [95].

Finally, we have the area usage of wind power plants. Wind has grown as the wind turbines and -rotors have grown in size, but interestingly the area productivity remains more or less constant. In their review of 19 studies, van Zalk and Behrens [127] finds that median wind power density gives a lower value than the mean with a median of 1.84 m^2/W(e) compared to $\mu = 2.73 \pm 0.5$ m^2/W(e), which may be due to overrepresentation of a few exceptional sites in the literature. The implication is that large-scale expansion of wind will, for the first time in history, lead to a reversal of the trend of increasingly higher power densities [107]. Furthermore, as wind expands, the Power Capacity Factors (PCF), i.e. the ratio between the actual production and the theoretical, continuous electricity production, tend to decrease as the best resources are developed first [22]. The implications of this reversal of energy densities are yet to be seen, but taken together with VREs in general, [79] suggest that we will run out of land.

Another issue with the area requirements of wind power plants is that they lead to conflict over land usage. After studying the Isthmus of Tehuantepec region of Oaxaca, Mexico, Dunlap [32] concludes that while wind energy development is presented as a 'solution' to climate change not only distracts from its dependence on fossil fuels and mining, but renews and continues a slow industrial genocide, assimilating and targeting (indigenous) people who continue to value their land, sea and cultural relationships. This is not only the case for indigenous people in a less developed region in Mexico, but also taking place in one of the richest countries in the world where the indigenous people finally won a supreme court case supporting their rights [116].

8.3.3 BESS externalities

The number of studies reviewing the externalities of utility-scale BESS is very limited. This author has not found any useful except the important realization offered by Cerdas [25] that "...the material supply chain of traction batteries has been treated as a big black-box [in the context of LCA], with little consideration given to the complexity of these production networks". However, we can get some insights given some key issues.

First, electronic equipment of all kinds generates large amounts of waste despite recycling efforts. On average, the global consumption (excluding photovoltaic panels) of electrical and electronic equipment (EEE) increases annually by 2.5 million tonnes and the amount of e-waste is expected to reach 74.7 million tonnes by 2030 [51] from the record of 53.6 million tonnes per year globally in 2019. Around 75–80% of these products are sent to developing countries mainly in Asia and Africa for manual disassembly, material recovery and disposal [93] using rudimentary recycling techniques. Clearly, recycling does not imply per se an environmental benefit as its impact is affected by different issues such as the quality of the material recovered, the energy and material consumption by the process itself and the efforts caused by the required logistics [24].

REE needed for wind and electronics in general will grow 7 times in volume [63], and furthermore:

> *In climate-driven scenarios, mineral demand for use in EVs and battery storage is a major force, growing at least thirty times to 2040. Lithium sees the fastest growth, with demand growing by over*

40 times in the SDS [Sustainable Development Scenarios] by 2040, followed by graphite, cobalt and nickel (around 20-25 times). The expansion of electricity networks means that copper demand for power lines more than doubles over the same period.

Obviously, the environmental burden will be very large including the impact on biodiversity. Habitat loss and degradation currently threaten over 80% of endangered species, while climate change directly affects 20% [83], and the mining required for REE is a major culprit [110].

If we go directly to the BESS and lithium, we find that in Chile's Salar de Atacama 65% of the water in the region are used by mining and extraction – for every tonne of lithium extracted 1,900 tonnes of water is used [73]. Worse, existing mines and projects under construction is estimated to meet only half of projected lithium requirements by 2030 [63]. Recycling is not expected to contribute significantly in foreseeable future because more than half of the lithium produced worldwide goes to dispersive uses from where it cannot be recovered [26]. The current recycling rates of REE and lithium is less than 1%, cobalt at approximately 30% and copper at roughly 45% [63].

Hence, there are significant externalities related to BESS that suggest caution in our interpretation of the results. On the positive note, however, it should be mentioned that the life-span of utility-scale BESS is longer than the one used here because it can in principle have secondary usage after the end of life as a utility-scale BESS. However, the adoption of second-life BESS has been slow for various economic reasons [114]. Thus, from a modeling perspective, this positive effect cannot be included because it translates into another utility-scale BESS being required after the first has ended its useful life with a very uncertain potential usage for its second life.

Next, the hypothesis is tested using a grid with very high degree of hydro. To do this, Norway is used. Norway is an ideal case due to it large installed base of hydro, which means that there is sufficient balancing capacity for wind at almost any level. Furthermore, the data availability for the analysis is good.

8.4 The Norwegian grid case

Table 8.1 offers some descriptive statistics concerning the Norwegian grid. COVID-19 is probably the cause of the small fall in demand from 2019 to 2020, but the large increase in the trade is driven by an expanding installed base of wind and more importantly the increased production of hydro due

Table 8.1 The Norwegian grid in 2019 and 2020 using data from NVE[4] and Statnett[5] and calculated by the author.

Summary	2019 & 2020	2019	2020	Difference
Demand [GWh]	266,327	133,471	132,855	-616
Cumulative wind capacity year-end [MW]		2,920	3,977	1,057
Wind [GWh]	15,443	5,532	9,912	4,380
Hydro [GWh]	271,131	127,746	143,385	15,639
Import [GWh]	11,671	9,488	2,183	-7,304
Export [GWh]	-31,919	-9,294	-22,625	-13,331
Net import [GWh]	-20,248	194	-20,442	-20,636

to record high water reservoirs filling and a subsequent record production in 2020 [4]. The results was a record amount of export. However, such aggregated numbers can be deceptive.

The to the variability of the wind, it is important to study the variations, and data from 2019 and 2020 with 1 hour resolution are presented in Figure 8.1. Such resolution is sufficient for this purpose, but to understand intermittent energy resources 15-minute or 5-minute resolutions are better. This is easy to understand when taking into account all the variation that can take place in 15 minutes with wind. This is also why the system reserves are increased when wind is in the grid [29].

This increase in reserves is intuitively necessary since demand is relatively random and wind is random, and two random sources gives much more randomness since variance is additive. Hydro capacities improve this situation, but the challenges should not be underestimated because grid complexities also add, and many studies assume perfect grids and no losses. To improve the visibility of the two years, Figure 8.2 contains one year above the other. The demand curves are relatively similar, but the production profiles are more distinct. Notably, there was virtually no net export in 2019, and the connections to Europe provided large influx of energy that prevented blackouts.

There is, in other words, a real risk of too little supply of electricity. With more than 20% of households lacking alternative source of heating [88], a blackout in the winter could have dire consequences. Indeed, DSB

[4]Accessed 2021-03-09 at https://www.nve.no/energiforsyning/vindkraft/utbygde-vindkraftverk/

[5]Accessed 2021-03-09 at https://www.statnett.no/for-aktorer-i-kraftbransjen/tall-og-data-fra-kraftsystemet/last-ned-grunndata/

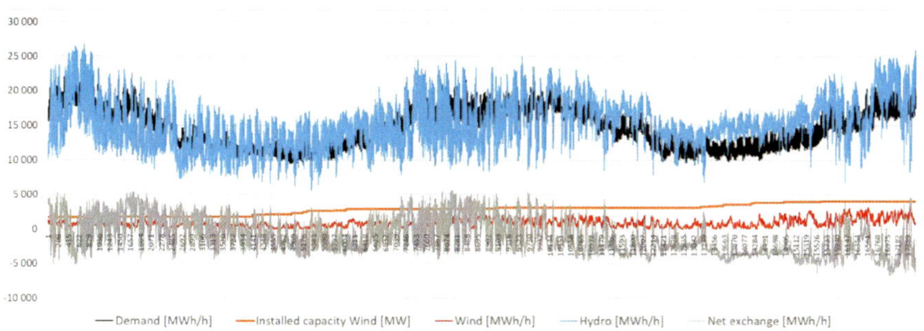

Figure 8.1 Demand and production data with 1 hour resolution from 2019 and 2020 in Norway. Data are obtained from NVE[3] and Statnett[4] and calculated by the author.

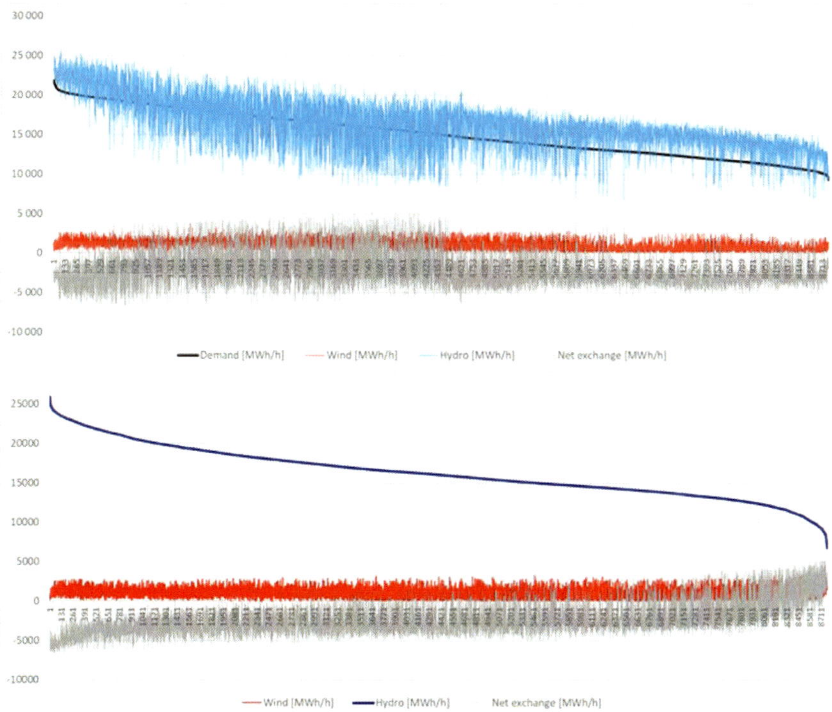

Figure 8.2 Grid profile for Norway in 2020 using 1 hour data resolution. Author calculations using data from https://www.statnett.no/for-aktorer-i-kraftbransjen/tall-og-data-fra-kraftsystemet/last-ned-grunndata/.

Table 8.2 Correlation table between various factors in 2019 and 2020.

Factors	**Correlation**
Wind vs Demand	30.2 %
Hydro vs Wind	10.0 %
Hydro vs Demand	70.2 %
Demand vs Price	33.7 %
Hydro vs Price	5.7 %
Wind vs Price	-31.9 %
Wind capacity vs Wind	47.1 %

[31] has modelled some crisis scenarios that can hit Norway, where one includes major blackout. The rationale of this scenario is that two consecutive years of poor precipitation have led to record low water reservoirs. Then, a cold winter with poor access to Swedish nuclear power and one cable to the continent malfunctioning leads to rolling blackouts. They guesstimate that the probability for such a scenario is about 10% over a 100 years period. The grid operator, Statnett, guesstimates that the probability for electricity rationing of is somewhere between 5 and 50% over 50 years [112].

Crucially, nobody considers the failure of wind, which is understandable when the wind share is low, but it will become significant if the growth in installed capacity continues as planned. The risk just mentioned, is more likely to emerge because wind and hydro do not seem to be planned as a system given the data and the correlations in Table 8.2. It is important to stress that correlations are not causations and must be interpreted with care.

For example, the fact that wind and demand correlates, albeit relatively little and on the same level as identified by Sinden [106] in the UK, could be due to and increasing installed capacity towards the end of 2020 when demand also rose due to Winter. We also observe the negative correlation between wind and price (the more wind, the lower the price). This was also seen in real life as electricity prices were record-low in Norway in 2020 (and record high in early 2021). Finally, the correlation between hydro and wind is weak, which implies that hydro is not used to balance wind. In other words, hydro does not act as a 'battery to renewables', as often portrayed in the public debate and exemplified by the discussion of Hope [58].

The same finding can also be derived from Figure 8.2. The figure on top shows the hourly demand falling and the associated production of wind and hydro. The hours with the largest demand are found to the left and the smallest to the right. Clearly, we see that hydro is what satisfies demand. Then, on the figure in the bottom of Figure 8.2, arises when we eliminate demand and reorganize the figure according to falling hydro. Then, we see that hydro also handles import / export. In both figures, wind appear as noise at the bottom with little or no relations to demand or hydro.

Indeed, the evidence suggests that wind and hydro are independent profit maximizing entities, which may become problematic the day when the water reservoirs are low, the electricity surplus we have enjoyed in Europe for decades is gone and there is little wind and the weather is cold. Thus, the current energy policy has set Europe on course for less reliable surplus capacity and greater supply risks [33].

The situation just described has received a new term in German – Dunkelflaute – first introduced by Borggrefe *et al.* [17]. Dunkelflaute is relatively common. The accurate MERRA-2 dataset provided by NASA [54] starting from 1980, is used by Ohlendorf and Schill [91] to analyze Dunkelflaute. They conclude that every year, a period of around five consecutive days with an average wind PCF below 10% occurs, and every ten years a respective period of nearly eight days.

While an hour is far better resolution than 3- or 6 hours, it is still insufficient for reliability analyses when there are large shares of VRE sources. For example, the critical event, near-blackout, in the European grid on January 8th in 2021 started at 14:05 and was finished about 15:08 [46] – in other words, an hour resolution would have missed most of the event. Nevertheless, the scenario discussed earlier is essentially the scenario outlined by [31], but brought into the reality of the development seen in the European grid.

Over the next 10 years, coal and nuclear with a total capacity of around 100 GW will be shut down in Europe, equivalent to Germany's thermal power capacity alone, according to grid operator data and quoted by [33]. With the stressed energy situation in Europe in 2021, these cuts in dispatchable power will increase the risks for blackouts further. Indeed, the world-leading expert on blackouts [103], states that within five years a pan-European blackout can take place.

Some argue that what is needed is more VRE, such as the President of the European Commission, Ursula von der Leyen, who stated on October 7th 2021 that "In these very weeks, we all see how crucial it is to reduce our dependency on fossil fuels such as gas, oil and coal. So, in the long term, the European Green Deal and renewables are the solution to the rising electricity prices. Every euro spent on renewable energies does help our planet. And it equally helps the consumers. But it is also an investment in the resilience of our economies. So, we have to speed up our work on the European Green Deal to become more energy independent" [49]. The reality

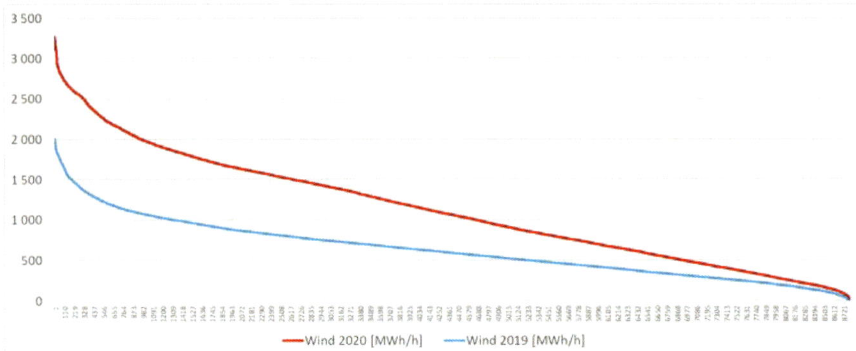

Figure 8.3 Wind profile for Norway per year using 1 hour data resolution. Author calculations using data from https://www.statnett.no/for-aktorer-i-kraftbransjen/tall-og-data-fra-kraftsystemet/last-ned-grunndata/.

is not so straightforward. From Figure 8.3 we see that despite major increases in installed capacity of wind (36%), see Table 8.1, there is about 100 days of low production irrespectively of capacity, and a similar figure exists for the entire ENTSOE [40].

Crucially, there is too strong correlation coefficient between the wind power plants in Norway, which means that when there is little wind there is little wind for most. Hence, the option of diversification through geographical dispersion does not work despite Norway being 1,752 km long[6], which can be seen from Figure 8.4. The figure shows the probability density function of the correlation coefficient between the hourly production in 2020 of all the wind power plants in Norway. The average correlation coefficient is 30%, but it ranges from -16% to 94%.

Next, more specific information necessary for calculating the emissions is presented.

8.4.1 Generic data for the analysis

An analysis must start from the premise that any energy source must be able to guarantee an output in line with the Reliability, Availability and Maintainability (RAM) requirements, see for example [100]. It is important to realize that electricity is not electricity – electricity for a modern grid has to

[6]Source: https://www.nationsencyclopedia.com/Europe/Norway-LOCATION-SIZE-AND-EXTENT.html

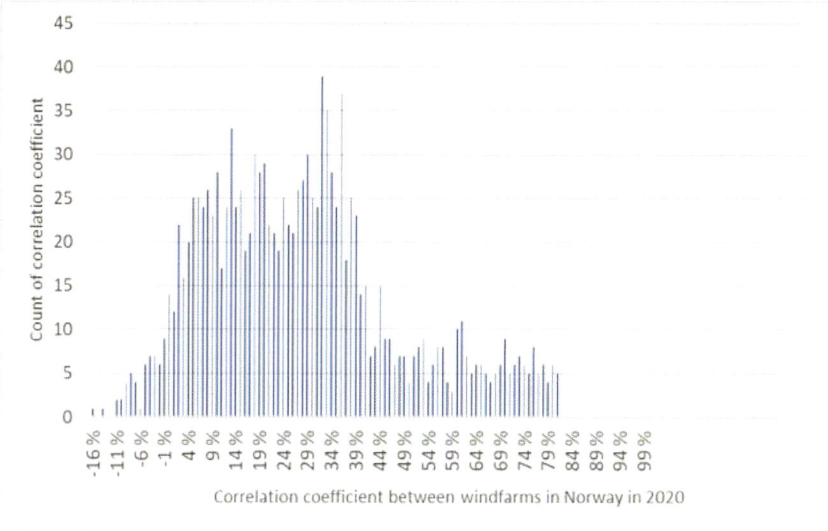

Figure 8.4 Probability distribution of correlation coefficients between all wind power plants in Norway in 2020. Calculated by the author using data from NVE[7].

be associated with the right voltage or frequency to be valuable, as discuss at length in the Irish case [40], because neither VREs nor lithium-ion BESS can provide these reliability services. Synchronous energy sources are required, which is why the total system reliability must be maintained. For Norway, this is not an issue since the ration of synchronous hydro is so large compared to wind, so output is merely a matter of the amount of electricity that must be provided by wind.

Unfortunately, wind has the fundamental problem of fluctuating- and low reliability/availability. By analyzing the frequencies of various modes for the demand and wind, the uncertainty distributions shown in Figure 8.5 are obtained. Globally, the average onshore- and offshore wind power plants PCF is about 42%–43% with large variations depending on location [66] while in 2020 it was 32.7% in Norway.

Each graph is based on a data set of 8,784 data points since 2020 had 366 days, but to make a more generic model the uncertainty is modelled from month to month as shown in Table 8.3. The purpose is to approximate the

---

[7]See https://www.nve.no/media/11713/vindproduksjon_2002-2020_faktisk_produksjonutc-1.xlsx

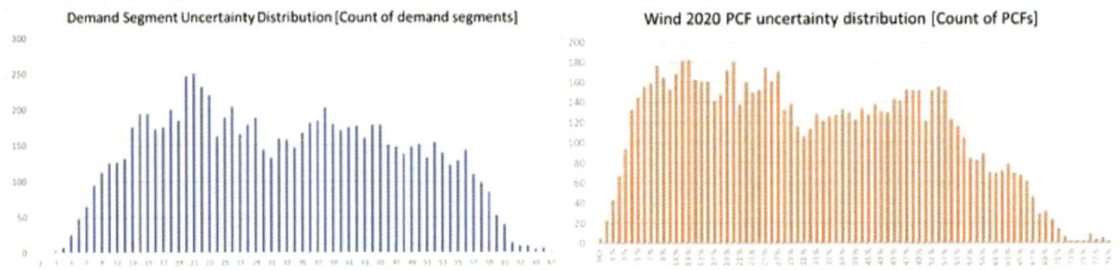

Figure 8.5 Demand- and Wind uncertainty distributions derived from 2020 data. Calculated by the author using data from NVE[8].

Table 8.3 Modeling demand and wind.

	Demand (MWh/h)			Wind (MWh/h)		
Month	**Average**	**Standard deviation**		**Median**	**Min**	**Max**
1	17,953	1,458	8%	1,441	431	2,030
2	18,372	1,542	8%	1,249	138	2,212
3	17,626	1,202	7%	1,350	302	2,118
4	15,266	1,261	8%	871	104	2,153
5	13,740	1,438	10%	751	8	1,942
6	11,859	1,061	9%	432	34	1,591
7	11,793	847	7%	619	23	1,957
8	12,108	1,023	8%	484	44	2,276
9	13,162	1,066	8%	1,185	78	2,475
10	14,983	1,582	11%	1,020	169	2,893
11	16,522	1,792	11%	1,990	456	3,108
12	18,204	1,535	8%	1,868	401	3,276

real-life situation as closely as possible without being too focused on 2020. This is also the reason why the deterministic charging/discharging profile is based on months, as shown in Figure 8.8. Seasonal variations are clearly visible, which is handled in the model by using monthly averages, standard deviations, medians and min/max values.

Note that due to the numerical nature of Monte Carlo simulations, it is the shape of the distributions that count and not any particular statistics properties. Hydro is balancing wind by filling in the difference between demand and wind. Hence, we ignore import/export to handle for simplicity, but those energy flows also add little to the research question.

[8]See https://www.nve.no/media/11713/vindproduksjon_2002-2020_faktisk_produksjonutc-1.xlsx

8.4.2 LCA data

Koffi *et al.* [74] provide the best compilation of data this author has found, see Table 8.4. According to NVE [89], the average emissions of electricity is merely 8 grams/CO_2 equivalent, but this number is not a life-cycle emissions number and irrelevant for this study.

Note that the data highlighted in grey, were used to estimate the minimum-, maximum- and expected values for the other data points to complete the data set. In total, 537 cases form the basis for the LCA data, which enable the modeling of uncertainties as triangular distributions with minimum-/maximum values together with the average values as expected values.

Norway has similar problems as those Emblemsvåg [40] describe from puncturing peatland in Ireland. The fact is that wind built on undegraded peatland is unlikely to ever reduce carbon emissions [108]. However, this situation is ascribed to uninformed development and not the technology *per se* and therefore excluded from the analysis.

For the Wind+BESS case, LCA data for the BESS must be found. As for the other data presented earlier, there are major variations – probably due to the highly problematic task of tracking down all effects including rare metals as vividly described by [96]. Therefore, a review of LCAs is therefore a better approach than to try to come up with something entirely new. The reason is that the highly complex supply chain of BESS prevents finding data when large parts of the supply chain involved countries where data is scarce. By explicitly modeling the uncertainty, more realistic estimates are found.

As of 2019, there were over 100 research articles that cover the environmental impacts from lithium-ion batteries dating back to as early as 1999 [84], and current knowledge regarding the life cycle environmental impacts of the BESS used in electric vehicles strongly varies among the studies currently available in the scientific literature due to different choices and assumptions in the modeling [25]. The results vary from 38 kg CO_2-eq/MWh up to 356 kg CO2-eq/MWh for traction BESS through the production phase [36].

Note that data for BESS in grid is nonexistent to the knowledge of this author, but for large, industrial applications, or traction BESS, there are some as shown in Table 8.5. Thus, the life-cycle emissions are in the order of 1000 times higher than the emissions from wind and hydro. Note that, to date, LCAs use only an attributional approach for EV batteries [3], which means that indirect effects are ignored [19] – including those effects is referred to as consequential LCA. A challenge with the results from Aichberger and

Table 8.4 Emission Factors (EF) for wind and hydro, reduced from [74], and [92] is added by this author.

Source	References	Case studies	Min LCA EF [gCO2/kWh]	Max LCA EF [gCO2/kWh]	Average LCA EF [gCO2/kWh]
Hydro	[5]	11	2	60	20
Hydro	[6]	6	15	40	28
Hydro	[8]	11	2	75	12
Hydro	[128]	3	4	237	120
Hydro	[50]	48	12	148	41
Hydro	[70]	9	1	609	20
Hydro	[101]	39	0	152	3
Hydro	[120]	12	1	20	11
Hydro	[131]	4	1	34	18
Hydro	[92]	1	78	109	94
Wind	[8]	20	6	46	9
Wind	[14]	34	1	185	11
Wind	[128]	10	0	124	62
Wind	[50]	48	4	84	25
Wind	[75]	72	8	124	66
Wind	[101]	63	5	55	18
Wind	[120]	22	3	41	22
Wind	[131]	8	8	30	19
Wind	[87]	39	0	364	182
Wind	[92]	1	4	12	34
Wind, offshore	[5]	5	5	24	13
Wind, offshore	[7]	13	8	33	16
Wind, onshore	[5]	14	2	81	16
Wind, onshore	[7]	44	7	56	20

Jungmeier [3], is that they concern BESS which are not strictly speaking suitable for grid applications, but they are included in Table 8.5 to provide an optimistic, lower-end estimate.

Furthermore, Ellingsen *et al.* [36] report data simply between 38 kg CO_2/kWh and 356 kg CO_2/kWh. Note the units – it is 1000 times higher than the energy sources in Table 8.2. The average is subsequently calculated and we end up with a life-cycle emission of BESS as shown in Figure 8.6. If this uncertainty is critical for the conclusion, more efforts must be made to find primary data of higher quality.

Table 8.5 Emission Factors (EF) for BESS.

Source	References	Cases	Min LCA EF [kgCO2/MWh]	Max LCA EF [kgCO2/MWh]	Average LCA EF [kgCO2/MWh]
BESS	[3]	50	70	175	120
BESS	[25]	12	39	480	260
BESS	[36]	6	38	356	197
BESS	[94]	79	49	253	110

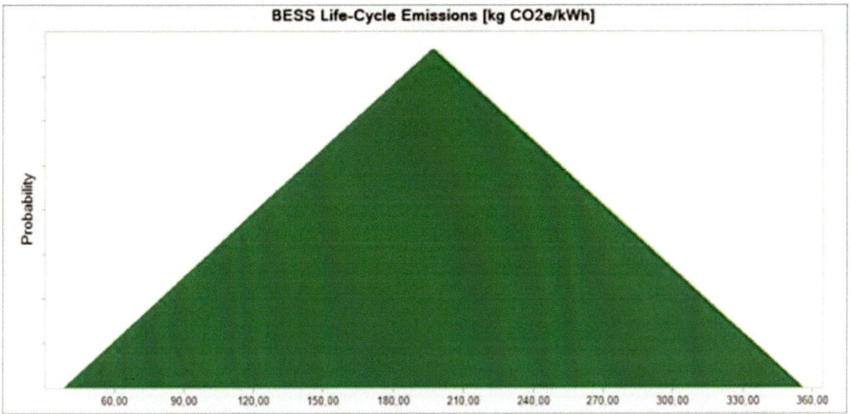

Figure 8.6 Life-cycle emission data for BESS, based on data from Ellingsen *et al.* [36].

When it comes to the life-span of the BESS, we must first define the end of life. End-Of-Life (EOL) is defined when the battery degrades to a point where only 70-80% of Beginning-Of-Life (BOL) capacity is remaining under nameplate conditions [109]. A second life can then begin, but not is a grid.

Most of the literature places the life-span in the range of 10–20 years where the majority uses the lower end and indicates the need for major maintenance and battery replacement to keep the system operational, but this is too optimistic [85]. 10-year life-span is possible provided the battery is cycled within a restricted 54% Depth Of Discharge (DOD) operating range and temperatures are actively managed [109]. It is assumed that the DOD can be improved to 80% in the model through technological development. Li-ion BESS has a typical usable life of approximately 10 years and requires major maintenance usually every 5 to 8 years to remain operational [11]. The life-span is therefore modelled as 9 years with a range from 7 to 10. In line with most LCA studies, minor maintenance and EOL issues are ignored because these issues are highly uncertain [36] although EOL issues can be substantial as discussed in Section 8.3 for both wind and BESS.

Concerning DOD, most in the literature seem to assume 100% DOD such as Ellingsen *et al.* [35]. The numbers must therefore be adjusted for a more realistic DOD. Regardless, we see that the Wind+BESS alternative in Table 8.6 has a life-cycle emission similar to natural gas, or 573 kg CO_2-eq/MWh on average [40]. When the BESS numbers on top miss several parts of the life-cycle as discussed earlier, we can already conclude that

Table 8.6 Summary of Tables 8.2 and 8.5 including actual generation in 2020.

Emission Source	Cases	Min LCA EF [kgCO2/MWh]	Max LCA EF [kgCO2/MWh]	Average LCA EF [kgCO2/MWh]	Total generation [MWh/yr]	Total emissions in [tonne/yr]
Hydro	144	11.6	148.4	36.5	143,385,412	5,227,115
Wind	393	4.3	89.9	36.6	9,911,635	363 ,55
Actuals 2020	**537**			**36.5**	**153,297,047**	**5,590,270**
Wind + Hydro						
Hydro	144	11.6	148.4	36.5	123,157,984	4,489,724
Wind	393	4.3	89.9	36.6	9,697,045	355,293
SUM	**537**			**36,5**	**132,855,029**	**4,845,017**
Wind + BESS						
Hydro	144	11.6	148.4	36.5	123,157,984	4,489,724
Wind	393	4.3	89.9	36.6	9,697,045	355,293
BESS	158	49,000	315,938	171,625	1,799,166	42,886,363
SUM	**684**			**359.3**	**132,855,029**	**47,733,589**

Wind+BESS cannot approach a 100 kg CO_2-eq/MWh grid technology since the emissions are at 359 kg CO_2-eq/MWh deterministically.

Then, we must account for the losses in the charge/discharge cycle. Losses in a Lithium-Ion BESS is 2% - 15% range [27], and the modeling here used the midpoint as expected value. Those LCA studies that consider losses, they typically assume 10% losses, or 90% charge-discharge efficiency [94].

The utility-scale duration varies from about 0.5 to 4 hours between the 10th and 90th percentiles [53]. Thus, the BESS required to fulfill the requirements presented in this study is beyond any BESS ever built with a huge margin. Its technical feasibility is therefore an issue, but it is assumed that there will be no problems with building such a large BESS.

Computing the numbers, the model indicates that the BESS must have a net capacity of around 1.6 TWh. Corrected for the DOD, the BESS must be dimensioned for 2.0 TWh. Such a BESS, if constructed, will provide wind power plants in Norway the capability of guaranteeing an annual nominal net output of 1,104 MWh/h excluding losses.

With only two notable energy sources in Norway, the summary in Table 8.6 becomes simple. The total life-cycle emissions in Norway associated with wind and hydro is 5.6 million tonnes CO_2 equivalents.

8.4.3 Alternative 1 – Wind + Hydro

We start by modeling the capacity side of the grid using the historical data presented earlier. Due to the variability of the wind, it is unwise to demand 100% guaranteed capacity due to all the balancing such an ambitious target would demand. Hence, we reduce the ambitions to 80% of installed capacity. Practically speaking, the wind power plants must establish Purchasing Power Agreements (PPA) with the hydro plants to secure balancing. Today, this is rarely done. In 2020 the overall correlation was a meager -3% indicating that mostly wind and hydro operate independently of each other.

The next step is to increase the grid reserve for two reasons. First, to reduce the problems of block failures in unlined hydropower tunnels due to frequent and fast stops of water columns, see [86]. The long-term implications of block failures are unknown, but they are obviously nonzero. Second, hydropower reservoirs can lead to biodiversity impacts and requires careful selection, construction and operations [30].

Third, the variability of the wind will in itself demand a greater grid reserve. For example, using 1,500 MW of installed wind power capacity in Ireland, Doherty and O'Malley [29] found that the best-case scenario is a 12% increase in the amount of reserve needed above the case with no wind, while the worst-case scenario is a 44% increase. They also found that 'For a reliability criterion of three load shedding incidents (LSI) per year and a forecast horizon of 3 hours, the reserve needed on the system with no wind capacity is 470MW'. Using a 3-hour forecast in the Norwegian case to reduce maintenance of hydro, the grid reserves should be increased by 465 MW compared to today with 6σ reliability. Today, however, primary, secondary and tertiary reserves are achieved through 'market solutions' [113]. Assuming a marked-based solution, the grid reserve issue is ignored in the model. In an independent, synchronous grid, the grid reserve would have to be included as Emblemsvåg [40] did when analyzing Ireland.

Strictly speaking, with the grid reserve being achieved through trade, the European energy mix would be of interest. However, it is relatively negligible compared to the larger picture as shown later. Furthermore, the grid reserve issues are solved when using a BESS. Hence, the grid reserves are excluded in the modeling for all alternatives.

8.4.4 Alternative 2 – Wind + BESS

Initially, pumped hydro is discussed, but there are practical challenges. However, Figure 8.7 is sufficient. With a difference of less than 5 EUR/MWh, there are no margins for operating pumped hydro in a market as efficient as the NordPool market, as discussed earlier, except is very good locations.

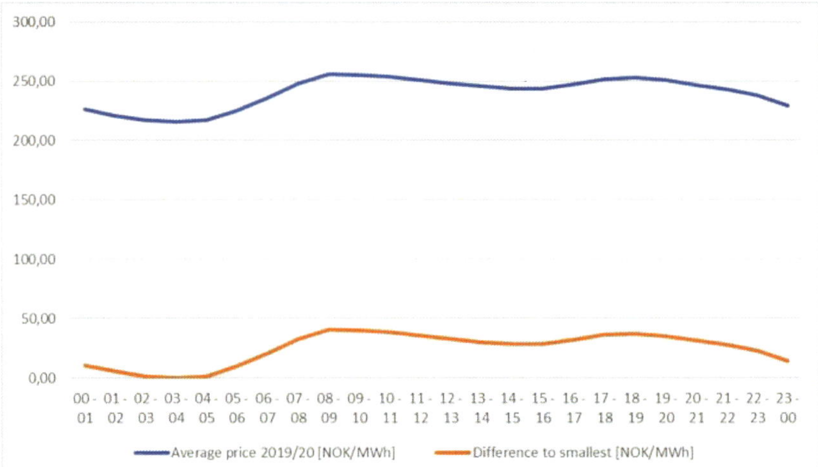

Figure 8.7 Day-profile for electricity prices on NordPool[9] in 2019/20. Note that 10 NOK is equal to 1 Euro.

To be meaningful, BESS capacity of Norwegian wind power plants must be large enough to secure an output at some level determined by the capability of wind power plants to produce enough electricity to also charge the BESS. In Figure 8.8, the deterministic charge/discharge profile is shown including the net BESS balance using an hourly resolution. The deterministic charge/discharge profiles are also calculated on a monthly basis to take into account seasonal variations. As we see, the idea is that the previous year will provide enough electricity to bridge the Summer.

In the next section, the results are presented after running a Monte Carlo simulation with 10,000 trials using Latin Hypercube sampling. The uncertainty is modelled as discussed in earlier sections.

8.4.5 Life-cycle emission factor results

In Figure 8.9, the results for the alternatives are shown. Clearly, the Wind+BESS alternative will never reach the 100 kg CO_2/MWh limit while Wind+Hydro will almost always reach that target. Taken together with the results from Emblemsvåg [40], we reach the conclusion that wind can only provide a sustainable solution when balanced by hydro.

[9]Obtained on 2021-03-03 from https://www.nordpoolgroup.com/historical-market-data/

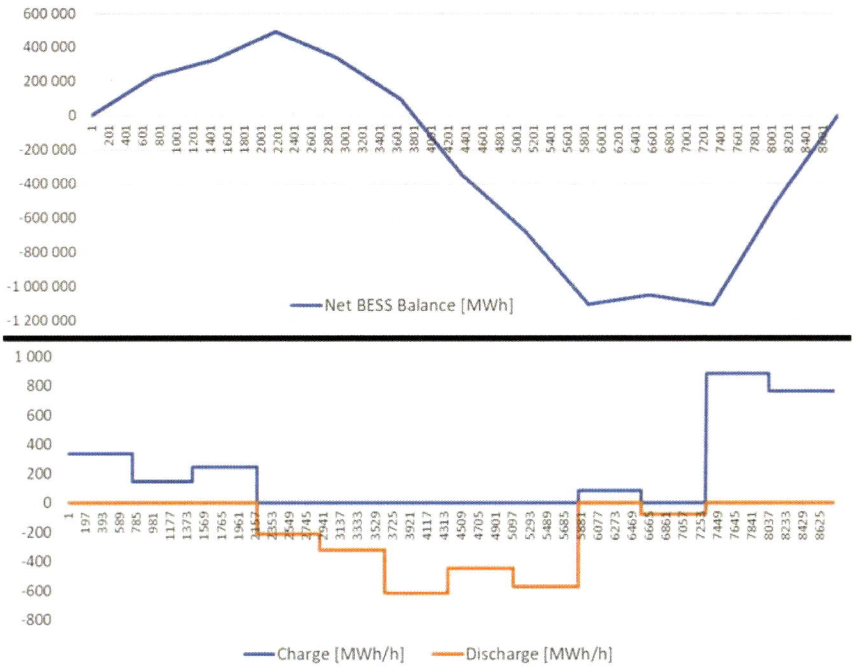

Figure 8.8 Deterministic charge and discharge profiles and net BESS balance. Author calculations.

Nuclear could conceivably work, but today's nuclear facilities are used for baseload because they are slow to ramp up and down. This could, of course, change once fourth generation nuclear technologies are available because they will provide safe, clean and proliferation resistant solutions that actually can follow the load [41]. In fact, China has put into commercial operation towards the end of 2021 a 200 MW pebble-bed modular High-Temperature Gas-cooled Reactor (HTGR), heating up helium instead of water to produce power [15]. The question is then whether or not wind is sensible at all?

If the Wind+BESS alternative is to be explored further in any future work, we can use the sensitivity analyses provided by the simulation which pinpoint the sources of uncertainty. Not surprisingly, we see from Figure 8.10 that it is the emission data from the life-cycle that is key – the design of the BESS and other factors are negligible in comparison. Clearly, there must be significant improvements for making any sustainable solution from the Wind+BESS alternative. Despite the popularity of the Li-ion BESS, it seems

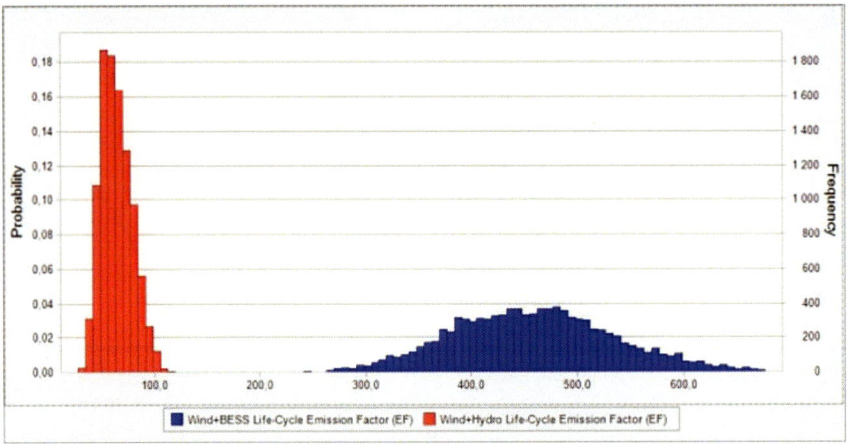

Figure 8.9 Wind+BESS versus Wind+Hydro life-cycle emission factors (EF) [kg CO_2/MWh].

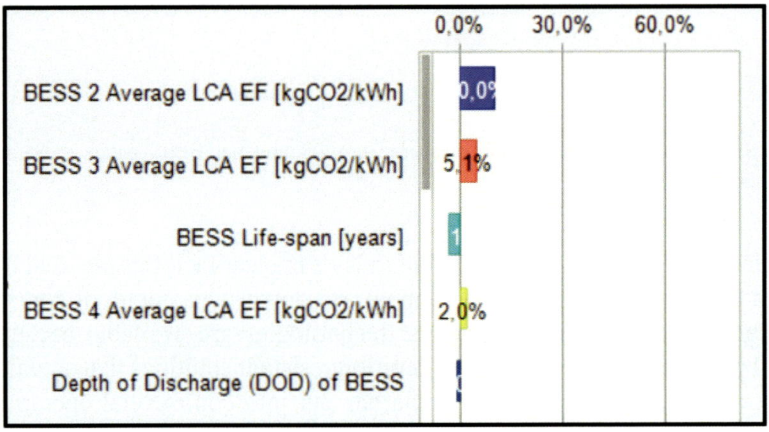

Figure 8.10 Sensitivity analysis of the Wind+BESS alternative.

to be a technological dead-end when it comes to emissions for utility scale balancing applications.

A different type of BESS must be developed for Wind+BESS to have any future from a sustainability point of view, but also from a cost point of view as discussed in the next section.

When it comes to the Wind+Hydro alternative, which is a sustainable alternative, it comes with a couple of hard requirements. First, hydro must be used for balancing wind so that there is a guaranteed total output, which means that PPA must be signed firmly linking the two together. Second, wind cannot be developed in such a way that externalities are ignored, which is also true for hydro. Third, the increased maintenance of hydro must be paid for by wind.

8.4.6 LCOE results

Given the accurate modelling of the grid data, adding a LCOE discussion is easy although it is strictly speaking not a part of the research. However, it is interesting because sustainability is also about economics. For economic evaluations of technologies, Levelized Cost of Energy (LCOE) is a commonly used decision criterion worldwide [65]. There are a number of shortcomings [39, 42], which makes it important to calculate a System LCOE [99, 123]. Including the entire power system, the System LCOE becomes higher than the narrowly defined LCOE used by the wind industry. For example, Emblemsvåg [39] finds that the LCOE for wind, in the ideal case of random weather systems, is at least 3 times higher.

Since the hydro has a low LCOE the Wind+Hydro alternative also yields a very low LCOE, provided a PPA is signed tying the two together to secure delivery. Wind+BESS, however, is very different. Even when an 80% further reduction in BESS costs is assumed, the Wind+BESS alternative becomes extremely costly due to the size of the BESS as shown in Figure 8.11.

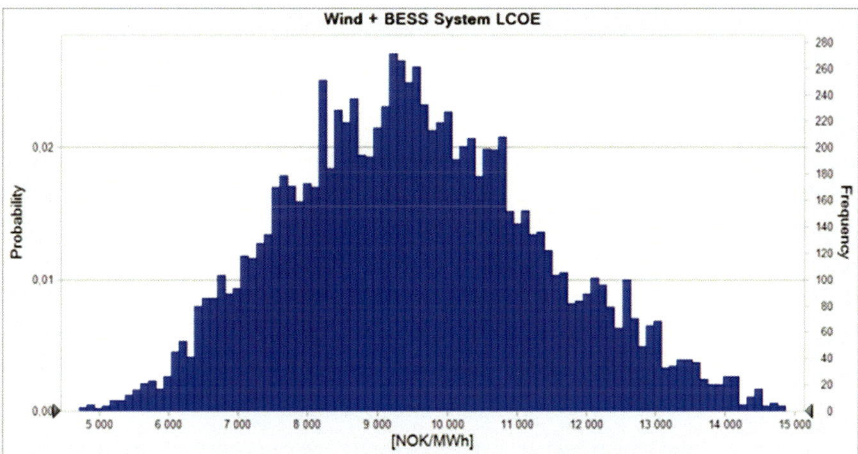

Figure 8.11 Wind+BESS System LCOE.

With 10 NOK equaling 1 euro, it costs is approximately 1 EUR/kWh ± 0.50 EUR/kWh. Basically, dimensioning a BESS for the irradicate behavior of wind becomes too costly.

In short, Wind+BESS is neither affordable nor sustainable whereas Wind+Hydro is both affordable and sustainable provided PPAs are signed to tie their performance to each other so that they in combination can deliver reliably to the grid.

8.5 Critical Evaluation of the Research and Potential Future Work

The results from the model are robust since we have used high resolution of data, multiple LCA studies to triangulate the input data and keeping the model overall conservative and the uncertainty is modelled using Monte Carlo simulations. Hence, there is under no circumstances any doubt about the main conclusion, i.e., that the Wind+BESS alternative is neither economically unviable with today's commercially available technologies even when assuming 80% cost reductions nor environmentally sustainable. However, if new technologies can be developed with significantly lower life-cycle emissions than Wind+BESS today, this conclusion can change.

The externalities of the model should be considered in future research, which will take significant amount of work. Similar to the US nuclear industry, which provides financial assurance for decommissioning through their operations[10], wind power plants should do the same. Alternatively, they should assume direct responsibility for the waste similar to many other industries.

Another possibility is to create a hybrid system comprising of solar photovoltaics (PV)-, wind- and hydro. Such studies are very rare [45], but adding another source of energy will obviously improve the outcome in terms of reliability, but it will not fundamentally alter the outcome because solar PV production is highly variable in many countries and even bridging solar PV overnight is expensive [42]. Furthermore, during the Dunkelflaute this option is not a realistic one because it will essentially become the same as the configuration discussed here, but more expensive due to the Solar PV costs. However, in suitable geographical locations around the world where the sun

[10] Information obtained on 2021-03-18 at https://www.nrc.gov/waste/decommissioning/finan-assur.html

resources are much more stable and abundant, it is definitively a solution to explore in future work.

8.6 Conclusion

Most grid operators rely on fossil power to balance wind power plants to secure reliable grid performance, which effectively hinders the establishment of a low-carbon electricity grid. However, there are two other alternatives – one where hydro is used and one where BESS is used.

Both alternatives are analyzed in this chapter, and the conclusion is beyond doubt that wind power plants together with BESS at utility-scale cannot be the foundation of a realistic, reliable, cost-efficient low carbon grid. However, wind power plants and hydro power plants can together provide the basis for a low-carbon electricity grid if formal PPAs are signed that in effect tie these power plants together to ensure that a reliable output is generated. Essentially, wind power plants must purchase capacities with hydro to deliver a guaranteed output taking variations into account, and BESS can serve as a peak shaver to handle peak loads and the like to reduces the grid reserves required.

This conclusion comes with two important caveats. First, the hydro share of the grid must be more than 50% of total, otherwise balancing a large amount of wind will be impossible. Second, the balancing must be executed in such a way that hydro power plants do not run into major maintenance issues or negative bio-diversity impact. The consequences of these two caveats are that the solution presented in this chapter is really applicable only to a small number of countries.

Acknowledgement

The author is thankful for the critical review of parts of the initial manuscript by Mads Løkeland-Stai, Sverre Sivertsen, Kjell Traa and Per Ivar Wethe. Furthermore, I would like to thank the editor of the book for both guidance and the opportunity to contribute.

References

[1] 800 scientists, *Letter from Scientists to the EU Parliament regarding forest biomass.* 2018: https://sites.tufts.edu/gdae/files/2019/10/LetterFromScientistsToEuParliament_ForestBiomass_Jan2018.pdf.

[2] Agora Energiewende, *The Integration Costs of Wind and Solar Power: An Overview of the Debate on the Effects of Adding Wind and Solar Photovoltaic into Power Systems.* 2015, Agora Energiewende: Berlin.

[3] Aichberger, C. and G. Jungmeier, *Environmental Life Cycle Impacts of Automotive Batteries Based on a Literature Review.* Energies, 2020. **13**: doi:10.3390/en13236345.

[4] Amelang, S., *EU indecision over gas as green investment imperils supply security – German utilities, in Clean Energy Wire.* 2021, Clean Energy Wire CLEW: Berlin. https://www.cleanenergywire.org/news/eu--indecision--over--gas--green--investment--imperils--supply--security--german--utilities.

[5] Amponsah, N.Y., M. Troldborg, B. Kington, I. Aalders, and R.L. Hough, *Greenhouse gas emissions from renewable energy sources: A review of lifecycle considerations.* Renewable and Sustainable Energy Reviews, 2014. **39**: p. 461–475.

[6] Ardente, F., M. Beccali, M. Cellura, and V.L. Brano, *Energy performances and life cycle assessment of an Italian wind farm.* Renewable and Sustainable Energy Reviews, 2008. **12**(1): p. 200–217.

[7] Arvesen, A. and E.G. Hertwich, *Assessing the life cycle environmental impacts of wind power - a review of present knowledge and research needs.* Renewable and Sustainable Energy Reviews, 2012. **16**: p. 5994–6006.

[8] Asdrubali, F., G. Baldinelli, F. D'Alessandro, and F. Scrucca, *Life cycle assessment of electricity production from renewable energies: Review and results harmonization.* Renewable and Sustainable Energy Reviews, 2015. **42**: p. 1113–1122.

[9] Bachner, G., K.W. Steininger, K. Williges, and A. Tuerk, *The economy-wide effects of large-scale renewable electricity expansion in Europe: The role of integration costs.* Renewable Energy, 2019. **134**: p. 1369–1380.

[10] Badings, T.S., A. Hartmanns, N. Jansen, and M. Suilen, *Balancing Wind and Batteries: Towards Predictive Verification of Smart Grids, in NASA Formal Methods (NFM 2021), Lecture Notes in Computer Science, vol 12673*, A. Dutle, M.M. Moscato, L. Titolo, C.A. Muñoz, and I. Perez, Editors. 2021, Springer, Cham. p. doi.org/10.1007/978--3--030--76384--8_1.

[11] Balducci, P., M.J.E. Alam, V. Viswanathan, D. Wu, A.J. Crawford, et al., *The Salem Smart Power Center: An Assessment of*

Battery Performance and Economic Potential. 2017, Pacific Northwest National Laboratory (PNNL), US Department of Energy.

[12] Batalla-Bejerano, J., J. Paniagua, and E. Trujillo-Baute, *Energy Market Integration and Electricity Trade.* Economics of Energy & Environmental Policy, 2019. **8**(2): p. 53–67.

[13] Bazilian, M.D., *Power to the Poor: Provide Energy to Fight Powerty.* Foreign Affairs, 2015. **94**(2): p. 133–138.

[14] Bertasiene, A., R.P. Borg, and B. Azzopardi, *A review of Life Cycle Impact Analysis of Wind Turbines, in Trends and Challenges for Wind Energy Harvesting,* C.C. Baniotopoulos, C. Borri, B. Blocken, H. Hemida, M. Veljkovic, T. Morbiato, R.P. Borg, S. Huber, E. Efthymiou, and C. Rebelo, Editors. 2015: Coimbra. p. 205–214.

[15] Bloomberg News, *China is Home to World's First Small Modular Nuclear Reactor.* 2021: https://www.bloomberg.com/news/articles/2021-12-21/new-reactor-spotlights-china-s-push-to-lead-way-in-nuclear-power.

[16] BMWi, *Erneuerbare Energien in Zahlen: Nationale und internationale Entwicklung im Jahr 2019.* 2020, Bundesministerium für Wirtschaft und Energie (BMWi): Berlin.

[17] Borggrefe, F., T. Pregger, H.C. Gils, K.-K. Cao, M. Deissenroth, et al., *Kurzstudie zur Kapazitätsentwicklung in Süddeutschland bis 2025 unter Berücksichtigung der Situation in Deutschland und den europäischen Nachbarstaaten.* 2014, Deutsches Zentrum für Luft- und Raumfahrt (DLR), Institut für Technische Thermodynamik, Abt. Systemanalyse und Technikbewertung and Institut für Energiewirtschaft und Rationelle Energieanwendung (IER), Universität Stuttgart: Stuttgart.

[18] Bouman, E.A., M.M. Øberg, and E.G. Hertwich, *Environmental impacts of balancing offshore wind power with compressed air energy storage (CAES).* Energy, 2016. 95: p. 91–98.

[19] Brander, M., R. Tipper, C. Hutchison, and G. Davis, *Consequential and Attributional Approaches to LCA: A Guide to Policy Makers with Specific Reference to Greenhouse Gas LCA of Biofuels.* Ecometrica Technical Paper, 2009. TP090403A(January).

[20] Brown, P.R. and F.M. O'Sullivan, *Shaping photovoltaic array output to align with changing wholesale electricity price profile.* Applied Energy, 2019. 256(December): doi.org/10.1016/j.apenergy.2019.113734.

[21] Bundesrechnungshof, *Bericht nach § 99 BHO zur Umsetzung der Energiewende im Hinblick auf die Versorgungssicherheit und*

Bezahlbarkeit bei Elektrizität (Report according to § 99 BHO on the implementation of the energy transition with regard to security of supply and affordability of electricity). 2021, Bundesrechnungshof: Bonn.

[22] Caralis, G., K. Rados, and A. Zervos, *The effect of spatial dispersion of wind power plants on the curtailment of wind power in the Greek power supply system.* Wind Energy, 2010. 13(4): p. 339–355.

[23] Cartes, D.A., J. Ordonez, J. Harrington, D. Cox, and R. Meeker, *Novel integrated energy systems and control methods with economic analysis for integrated community based energy systems, in 2007 IEEE Power Engineering Society General Meeting.* 2007, IEEE: Tampa, FL. p. 1–6.

[24] Cerdas, F., S. Andrew, S. Thiede, and C. Herrmann, *Environmental Aspects of the Recycling of Lithium-Ion Traction Batteries, in Recycling of Lithium-Ion Batteries. Sustainable Production, Life Cycle Engineering and Management*, A. Kwade and J. Diekmann, Editors. 2018, Springer: Cham. p. 267–288.

[25] Cerdas, F., *Integrated Computational Life Cycle Engineering for Traction Batteries.* Sustainable Production, Life Cycle Engineering and Management, ed. C. Herrmann and S. Kara. 2021, Cham: Springer.

[26] Christmann, P., E. Gloaguen, J.-F. Labbé, J. Melleton, and P. Piantone, *Chapter 1 - Global Lithium Resources and Sustainability Issues, in Lithium Process Chemistry: Resources, Extraction, Batteries and Recycling,* A. Chagnes and J. Światowska, Editors. 2015, Elsevier: Amsterdam. p. 1–40.

[27] Deloitte, *Energy storage: Tracking the technologies that will transform the power sector.* 2015, Deloitte: Frankfurt.

[28] Devlin, J., K. Li, P. Higgins, and A. Foley, *Gas generation and wind power: A review of unlikely allies in the United Kingdom and Ireland. Renewable and Sustainable Energy Reviews,* 2017. **70**: p. 757–768.

[29] Doherty, R. and M. O'Malley, *A New Approach to Quantify Reserve Demand in Systems With Significant Installed Wind Capacity.* IEEE Transactions on Power Systems, 2005. **20**(2): p. 587–595.

[30] Dorber, M., A. Arvesen, D. Gernaat, and F. Verones, *Controlling biodiversity impacts of future global hydropower reservoirs by strategic site selection.* Scientific Reports, 2020. 10: doi.org/10.1038/s41598--020--78444--6.

[31] DSB, *Analyser av krisescenarioer* 2019. 2019, Direktoratet for samfunnssikkerhet og beredskap (DSB): Oslo.

[32] Dunlap, A., *The 'solution' is now the 'problem:' wind energy, colonisation and the 'genocide-ecocide nexus' in the Isthmus of Tehuantepec, Oaxaca.* The International Journal of Human Rights, 2018. **22**(4): p. 550–573.

[33] Eckert, V., *Does renewables pioneer Germany risk running out of power?,* in *Reuters.* 2019: Frankfurt.

[34] EEA, *Annual European Union greenhouse gas inventory 1990–2019 and inventory report 2021.* 2021, European Environmental Agency (EEA): Brussels.

[35] Ellingsen, L.A.-W., G. Majeau-Bettez, B. Singh, A.K. Srivastava, L.O. Valøen, et al., *Life Cycle Assessment of a Lithium-Ion Battery Vehicle Pack.* Journal of Industrial Ecology, 2014. **18**(1): p. 113–124.

[36] Ellingsen, L.A.-W., C.R. Hung, and A.H. Strømman, *Identifying key assumptions and differences in life cycle assessment studies of lithium-ion traction batteries with focus on greenhouse gas emissions.* Transportation Research Part D: Transport and Environment, 2017. **55**: p. 82–90.

[37] Emblemsvåg, J. and B. Bras, *Activity-Based Cost and Environmental Management: A Different Approach to the ISO 14000 Compliance.* 2000, Boston: Kluwer Academic Publishers..

[38] Emblemsvåg, J., Life-Cycle Costing: Using *Activity-Based Costing and Monte Carlo Methods to Manage Future Costs and Risks.* 2003, Hoboken, NJ: John Wiley & Sons..

[39] Emblemsvåg, J., *On the levelized cost of energy of windfarms.* International Journal of Sustainable Energy, 2020. **39**(7): p. 700–718.

[40] Emblemsvåg, J., *Wind is not sustainable when balanced by fossil energy.* Applied Energy, 2021. 305: doi.org/10.1016/j.apenergy.2021.117748.

[41] Emblemsvåg, J., *Safe, Clean, Proliferation Resistant and Cost-Effective Thorium-based Molten Salt Reactors for Sustainable Development.* International Journal of Sustainable Energy, 2021. 40: doi.org/10.1080/14786451.2021.1928130.

[42] Emblemsvåg, J., *On the levelized cost of energy of Solar Photovoltaics.* International Journal of Sustainable Energy, 2021. **40**: doi.org/10.1080/14786451.2020.1867139.

[43] Emblemsvåg, J. *How to sustainably integrate wind. in Proc. of the International Conference on Electrical, Computer, Communications and Mechatronics Engineering (ICECCME), 16-18 November 2022.* 2022. Maldives: IEEE.

[44] Encyclopaedia Britannica, *Three Gorges Dam.* 2021, Encyclopaedia Britannica, Inc.: https://www.britannica.com/topic/Three-Gorges-Dam.

[45] Engeland, K., M. Borga, J.-D. Creutin, B. François, M.-H. Ramos, *et al., Space-time variability of climate variables and intermittent renewable electricity production – A review.* Renewable and Sustainable Energy Reviews, 2017. 79: p. 600–617.

[46] ENTSOE, *Continental Europe Synchronous Area Separation on 8 January 2021 Interim Report.* 2021, European Network of Transmission System Operators for Electricity (ENTSOE): Brussels.

[47] Erickson, W.P., G.D. Johnson, and D.P. Young Jr., *A Summary and Comparison of Bird Mortality from Anthropogenic Causes with an Emphasis on Collisions. , in Bird Conservation Implementation and Integration in the Americas. Proceedings of the Third International Partners in Flight Conference. 2002 March 20-24, Volume 2,* C.J. Ralph and T.D. Rich, Editors. 2005, General Techical Reports. PSW-GTR-191. Albany, CA: U.S. Dept. of Agriculture, Forest Service, Pacific Southwest Research Station: Asilomar, California. p. 1029–1042.

[48] European Commission, *A Clean Planet for all - A European strategic long-term vision for a prosperous, modern, competitive and climate neutral economy.* 2018, The European Commission, The European Council, the European Economic and Social Committee of the Regions and European Investment Bank: Brussels.

[49] European Commission, *Daily News - 07/10/2021.* 2021, The European Commission: Brussels.

[50] Evans, A., V. Strezov, and T.J. Evans, *Assessment of sustainability indicators for renewable energy technologies.* Renewable and Sustainable Energy Reviews, 2009. **13**(5): p. 1082–1088.

[51] Forti, V., C.P. Baldé, R. Kuehr, and G. Bel, *The Global E-waste Monitor 2020: Quantities, flows, and the circular economy potential.* 2021, United Nations University (UNU)/United Nations Institute for Training and Research (UNITAR) – co-hosted SCYCLE Programme, International Telecommunication Union (ITU) & International Solid Waste Association (ISWA): Bonn/Geneva/Rotterdam.

[52] Frydrychowicz-Jastrzebska, G., *El Hierro Renewable Energy Hybrid System: A Tough Compromise.* Energies, 2018. **11**: doi:10.3390/en11102812.

[53] Fu, R., T. Remo, and R. Margolis, *2018 U.S. Utility-Scale Photovoltaics-Plus-Energy Storage System Costs Benchmark.* 2018, National Renewable Energy Laboratory (NREL): Golden, CO.

[54] Gelaro, R., W. McCarty, M.J. Suárez, R. Todling, A. Molod, *et al.*, *The Modern-Era Retrospective Analysis for Research and Applications, Version 2 (MERRA-2).* Journal of Climate, 2017. **30**(14): p. 5419–5454.

[55] Godina, R., E.M.G. Rodrigues, J.C.O. Matias, and J.P.S. Catalão, *Sustainable Energy System of El Hierro Island, in International Conference on Renewable Energies and Power Quality (ICREPQ'15).* 2015: La Coruña.

[56] Grams, C.M., R. Beerli, S. Pfenninger, I. Stafell, and H. Wernli, *Balancing Europe's wind-power output through spatial deployment informed by weather regimes.* Nature Climate Change, 2017: doi: 10.1038/NCLIMATE3338.

[57] Hill, R.W., N.G. Morris, K.A. Bowman, and D. Wright, *The Isle of Man Seabird Census Report on the census of breeding seabirds in the Isle of Man* 2017-18 2019, Manx BirdLife: Isle of Man.

[58] Hope, E., *Vindkraft og vannkraft- Norge som kraftbatteri for Europa*? MAGMA, 2011(2): p. 57–65.

[59] Ibrahim, H., A. Ilincaa, and J. Perron, *Energy storage systems—Characteristics and comparisons.* Renewable and Sustainable Energy Reviews, 2008. **12**: p. 1221–1250.

[60] IEA, *Tracking Energy Integration. 2019*, International Energy Agency: Paris. https://www.iea.org/reports/tracking--energy--integration.

[61] IEA, *IEA Electricity Information.* 2020, International Energy Agency (IEA) / Organization for Economic Co-operation and Develop (OECD): https://www.iea.org/regions/europe.

[62] IEA, *Energy Storage.* 2021, International Energy Agency (IEA): Paris. https://www.iea.org/reports/energy--storage.

[63] IEA, *The Role of Critical Minerals in Clean Energy Transitions, in World Energy Outlook Special Report.* 2021, International Energy Agency (IEA): Paris.

[64] . IHA, *2020 Hydropower Status Report: Sector trends and insights.* 2021, International Hydropower Association (IHA): London.

[65] IRENA, *Renewable Energy Technologies: Cost Analysis Series.* 2012, International Renewable Energy Agency (IRENA), IRENA Innovation and Technology Centre: Abu Dhabi, United Arab Emirates.

[66] IRENA, *Renewable Power Generation Costs in 2017.* 2018, International Renewable Energy Agency (IRENA), IRENA Innovation and Technology Centre: Abu Dhabi, United Arab Emirates.

[67] Job, S., *Recycling glass fibre reinforced composites – history and progress.* Reinforced Plastics, 2013. **57**(5): p. 19–23.

[68] Jordaan, S.M., Q. Xu, and B.F. Hobbs, *Grid-Scale Life Cycle Greenhouse Gas Implications of Renewable, Storage, and Carbon Pricing Options.* Environmental Science & Technology, 2020. **54**: p. 10435–10445.

[69] Joskow, P.L., *Comparing the Costs of Internmittent and Dispatchable Electricity Generating Technologies.* American Economic Review, 2011. **101**(3): p. 238–241.

[70] Kadiyala, A., R. Kommalapati, and Z. Huque, *Evaluation of the Life Cycle Greenhouse Gas Emissions from Hydroelectricity Generation Systems.* Sustainability, 2012. **8**(6): p. doi.org/10.3390/su8060539.

[71] Kahn, E.P., *Effective Load Carrying Capability of Wind Generation: Initial Results with Public Data.* The Electricity Journal, 2004. 17(10): p. 85–95.

[72] Kamp, M.d.P., *How marginal communities in the South are paying the price for 'green energy' in Europe.* 2021, Cable News Network (CNN): https://edition.cnn.com/interactive/2021/07/us/american-south-biomass-energy-invs/?

[73] Katwala, A., *The spiralling environmental cost of our lithium battery addiction,* in Wired. 2018. https://www.wired.co.uk/article/lithium--batteries--environment--impact.

[74] Koffi, B., A.K. Cerutti, M. Duerr, A. Iancu, A. Kona, *et al., Covenant of Mayors for Climate and Energy: Default emission factors for local emission inventories.* 2017, Publications Office of the European Union: Luxembourg.

[75] Lenzen, M. and J. Munksgaard, *Energy and CO2 life-cycle analyses of wind turbines-review and applications.* Renewable Energy, 2002. **26**(3): p. 339–362.

[76] Liu, P. and C.Y. Barlow, *The environmental impact of Wind Turbine Blades (Presentation), in 37th Risoe International Symposium on Material Science, 1–16.* 2016. p. 229–240.

[77] Liu, P. and C.Y. Barlow, *Wind turbine blade waste in 2050.* Waste Management, 2017. **62**: p. 229–240.

[78] Madaeni, S.H., R. Sioshansi, and P. Denholm, *Comparison of Capacity Value Methods for Photovoltaics in the Western United States.* 2012, National Renewable Energy Laboratory (NREL): Golden, CO.

[79] Madsen, D.N., J.P. Hansen, and J. Emblemsvåg, *Population-Density Bounds for 100% Domestic Renewable Energy Generation.* PRX Energy, 2024. 3(1): doi.org/10.1103/PRXEnergy.3.013002.

[80] Martin, C., *Wind Turbine Blades Can't Be Recycled, So They're Piling Up in Landfills, in Bloomberg.* 2020. Accessed 2023–12–27 from https://www.bloomberg.com/news/features/2020--02--05/wind--turbine--blades--can--t--be--recycled--so--they--re--piling--up--in--landfills.

[81] Mattmann, M., I. Logar, and R. Brouwer, *Wind power externalities: A meta-analysis.* Ecological Economics, 2016. **127**: p. 23–36.

[82] Mattmann, M., I. Logar, and R. Brouwer, *Hydropower externalities: A meta-analysis.* Energy Economics, 2016. **57**: p. 66–77.

[83] Maxwell, S.L., R.A. Fuller, T.M. Brooks, and J.E.M. Watson, *Biodiversity: The ravages of guns, nets and bulldozers.* Nature, 2016. 536: p. 143–145.

[84] Melin, H.E., *Analysis of the climate impact of lithium-ion batteries and how to measure it.* 2019, Circular Energy Storage Research and Consulting.

[85] Mongird, K., V. Fotedar, V. Viswanathan, V. Koritarov, P. Balducci, *et al., Energy Storage Technology and Cost Characterization Report.* 2019, Pacific Northwest National Laboratory, Argonne National Laboratory and Oak Ridge National Laboratory: Oak Ridge, TN.

[86] Neupane, B. and K.K. Panthi, *Evaluation on the Effect of Pressure Transients on Rock Joints in Unlined Hydropower Tunnels Using Numerical Simulation.* Rock Mechanics and Rock Engineering, 2021. **54**: p. 2975–2994.

[87] Nugent, D. and B.K. Sovacool, *Assessing the lifecycle greenhouse gas emissions from solar PV and wind energy: a critical meta-survey.* Energy Policy, 2014. **65**: p. 229–244.

[88] NVE, *Oppvarming i boliger: Kartlegging av oppvarmingsutstyr og effektiviseringstiltak i husholdningene. NVE Rapport 85/2014.* 2014, Noregs Vassdrags og Energidirektorat (NVE). English; Norwegian Water Resources and Energy Directorate.: Oslo. https://publikasjoner.nve.no/rapport/2014/rapport2014_85.pdf.

[89] NVE, *Hvor kommer strømmen fra?* 2021, Noregs vassdrags- og energidirektorat (NVE): Oslo. https://www.nve.no/energi/energisystem/kraftproduksjon/hvor--kommer--strommen--fra/

[90] OECD/NEA, *Technical and Economic Aspects of Load Following with Nuclear Power Plants.* 2011, Organization for Economic Cooperation and Development (OECD), Nuclear Energy Agency: Paris.

[91] Ohlendorf, N. and W.-P. Schill, *Frequency and duration of low-wind-power events in Germany.* Environmental Research Letters, 2020. **15**: doi.org/10.1088/1748--9326/ab91e9.

[92] Pehl, M., A. Arvesen, F. Humpenöder, A. Popp, E.G. Hertwich, *et al., Understanding future emissions from low-carbon power systems by integration of life-cycle assessment and integrated energy modelling.* Nature Energy, 2017. **2**(December): p. 939–945.

[93] Perkins, D.N., M.-N.B. Drisse, T. Nxele, and P.D. Sly, *E-Waste: A Global Hazard.* Annals of Global Health, 2014. **80**(4): p. 286–295.

[94] Peters, J.F., M. Baumann, B. Zimmermann, J. Braun, and M. Weil, *The environmental impact of Li-Ion batteries and the role of key parameters - A review.* Renewable and Sustainable Energy Reviews, 2017. **67**: p. 491–506.

[95] Pinna, M., *Recycling turbine blades: the Achilles heel of wind power, in euronews.* 2021: https://www.euronews.com/2021/06/25/recycling-turbine-blades-the-achilles-heel-of-wind-power-and-the-controversy-engulfing-ren.

[96] Pitron, G., *The Rare Metals War: the dark side of clean energy and digital technologies.* 2021, London: Scribe UK.

[97] Rangaraju, S., *10 years of EU fuels policy increased EU's reliance on unsustainable biofuels.* 2021, Transport & Environment: https://www.transportenvironment.org/publications/10-years-eu-fuels-policy-increased-eu%E2%80%99s-reliance-unsustainable-biofuels.

[98] Raugei, M., A. Peluso, E. Leccisi, and V. Fthenakis, *Life-Cycle Carbon Emissions and Energy Return on Investment for 80% Domestic Renewable Electricity with Battery Storage in California (U.S.A.).* energies, 2020. **13**: doi:10.3390/en13153934.

[99] Reichenberg, L., F. Hedemus, M. Odenberger, and F. Johnsson, *The marginal system LCOE of variable renewables - Evaluating high penetration levels of wind and solar in Europe.* Energy, 2018. **152**: p. 914–924.

[100] RTO/NATO, *Methods and Models for Life Cycle Costing (Méthodes et modèles d'évaluation du coût de possession), RTO Technical Report No. TR-SAS-054.* 2007, North Atlantic Treaty Organisation (NATO), Research and Technology Organisation (RTO): Neuilly-sur-Seine Cedex.

[101] Raadal, H.L., L. Gagnon, I.S. Modahl, and O.J. Hanssen, *Life cycle greenhouse gas (GHG) emissions from the generation of wind and hydro power.* Renewable and Sustainable Energy Reviews, 2011. **16**(7): p. 3417–3422.

[102] Sardi, J., N. Mithulananthan, M. Gallagher, and D.Q. Hung, *Multiple community energy storage planning in distribution networks using a cost-benefit analysis.* Applied Energy, 2017. 190(March): p. 453–463.
[103] Saurugg, H., *Europe on the road to a major disaster.* International Journal for Nuclear Power, 2021. 66(3): p. 29–34.
[104] Schmidt, O., S. Melchior, A. Hawkes, and I. Staffell, *Projecting the Future Levelized Cost of Electricity Storage Technologies.* Joule, 2019. **3**(January): p. 81–100.
[105] Shaner, M.R., S.J. Davis, N.S. Lewis, and K. Caldeira, *Geophysical constraints on the reliability of solar and wind power in the United States.* Energy & Environmental Science, 2018. **11**: p. 914–925.
[106] Sinden, G., *Characteristics of the UK wind resource: long-term patterns and relationship to electricity demand.* Energy Policy, 2007. **35**: p. 112–27.
[107] Smil, V., *Energy Transitions: History, Requirements, Prospects.* 2010, Santa Barbara, CA: Praeger. 178.
[108] Smith, J., D.R. Nayak, and P. Smith, *Wind farms on undegraded peatlands are unlikely to reduce future carbon emissions.* Energy Policy, 2014. **66**: p. 585–591.
[109] Smith, K., A. Saxon, M. Keyser, B. Lundstrom, Z. Cao, *et al., Life Prediction Model for Grid-Connected Li-ion Battery Energy Storage System.* 2017, National Renewable Energy Laboratory (NREL): Golden, CO.
[110] Sonter, L.J., M.C. Dade, J.E.M. Watson, and R.K. Valenta, *Renewable energy production will exacerbate mining threats to biodiversity.* Nature Communications, 2020. **11**(4174): doi.org/10.1038/s41467--020--17928--5.
[111] Stanek, W., B. Mendecka, L. Lombardi, and T. Simla, *Environmental assessment of wind turbine systems based on thermo-ecological cost.* Energy, 2018. **160**: p. 341–348.
[112] Statnett, *SAKS 2014: Gjennomgang av og behov for SAKS-tiltak.* 2014, Statnett: Oslo. https://www.statnett.no/contentassets/67e76050727445cbb380d6546c41ff60/saks--2014------gjennomgang--av--og--behov--for--saks--tiltak.pdf.
[113] Statnett, *Market reserves.* 2021, Statnett: https://www.statnett.no/en/for-stakeholders-in-the-power-industry/system-operation/the-power-market/reserve-markets/.
[114] Steckel, T., A. Kendall, and H. Ambrose, *Applying levelized cost of storage methodology to utility-scale second-life lithium-ion battery*

energy storage systems. Applied Energy, 2021. **300**: doi.org/10.1016/j.apenergy.2021.117309.

[115] Strang, K.D., *Feasibility of a hidden renewable energy hydro power storage battery.* Journal of Energy Storage, 2017. 13: p. 164–175.

[116] Supreme Court of Norway, *Licences for wind power development on Fosen ruled invalid as the construction violates Sami reindeer herders' right to enjoy their own culture.* 2020, Høyesterett: https://www.domstol.no/globalassets/upload/hret/decisions-in-english-translation/hr-2021-1975-s.pdf.

[117] Thompson, C.C., P.E.K. Oikonomou, A.H. Etemadi, and V.J. Sorger, *Optimization of data center battery storage investments for microgrid cost savings, emissions reduction, and reliability enhancement.* IEEE Transactions on Industry Applications, 2016. **52**(3): p. 2053–2060.

[118] Thomson, R.C., G.P. Harrison, and J.P. Chick, *Marginal greenhouse gas emissions displacement of wind power in Great Britain.* Energy Policy, 2017. **101**: p. 201–210.

[119] Tsagkaraki, M. and R. Carollo, *CO2 Emissions Variations in CCGTs Used to Balance Wind in Ireland, in Energy Matters.* 2016. http://euanmearns.com/co2--emissions--variations--in--ccgts--used--to--balance--wind--in--ireland/.

[120] Turconi, R., A. Boldrin, and T. Astrup, *Life cycle assessment (LCA) of electricity generation technologies: overview, comparability and limitations.* Renewable and Sustainable Energy Reviews, 2013. **28**: p. 555–565.

[121] Uddin, M., M.F. Romlie, M.F. Abdullah, S.A. Halim, A.H.A. Bakar, *et al., A review on peak load shaving strategies.* Renewable and Sustainable Energy Reviews, 2018. **82**: p. 3323–3332.

[122] Udo, F., *Wind energy in the Irish power system.* 2011: https://www.clepair.net/IerlandUdo.html.

[123] Ueckerdt, F., L. Hirth, G. Luderer, and O. Edenhofer, *System LCOE: What are the costs of variable renewables?* Energy, 2013. **63**: p. 61–75.

[124] UN DESA, *17 Sustainable Development Goals - 17 Partnerships.* 2015, United Nations Department of Economic and Social Affair (UNDESA), Division for Sustainable Development: New York.

[125] UN ESCAP, *Integrating the three dimensions of sustainable development: A framework and tools, in Greening of Economic Growth Series.* 2015, United Nations Economic and Social Commission for Asia and the Pacific (ESCAP), Environment and Development Division: Bangkok.

[126] US DOE, *Quadrennial Technology Review: An Assessment of Energy Technologies and Research Opportunities.* 2015, United States Department of Energy: Washington DC.

[127] van Zalk, J. and P. Behrens, *The spatial extent of renewable and non-renewable power generation: A review and meta-analysis of power densities and their application in the U.S.* Energy Policy, 2018. **123**: p. 83–91.

[128] Varun, G., I.K. Bhat, and R. Prakash, *LCA of renewable energy for electricity generation systems—a review.* Renewable and Sustainable Energy Reviews, 2009. **13**(5): p. 1067–1073.

[129] Verdolini, E., F. Vona, and D. Popp, *Bridging the gap: Do fast-reacting fossil technologies facilitate renewable energy diffusion? Energy Policy*, 2018. **116**: p. 242–256.

[130] WEC, *Energy Storage Monitor: Latest trends in energy storage.* 2019, World Energy Council, Future Energy Leaders' Programme: London.

[131] Weisser, D., *A guide to life-cycle greenhouse gas (GHG) emissions from electric supply technologies.* Energy, 2007. **32**(9): p. 1543–1559.

[132] Wetzel, D., *Jetzt wird die Energiewende zur Gefahr für ganz Deutschland (Now the energy transition is becoming a danger for all of Germany), in Die Welt.* 2021: Berlin. https://www.welt.de/wirtschaft/article229449033/Energieversorgung--Bundesrechnungshof--warnt--vor--Stromluecke.html.

[133] Wind Europe, *Wind energy and on-site energy storage systems; exploring market oppotunities 2017.* 2018, Wind Europe: Brussels.

[134] Wiser, R., A. Mills, J. Seel, T. Levin, and A. Botterud, *Impacts of Variable Renewable Energy on Bulk Power System Assets, Pricing, and Costs.* 2017, Lawrence Berkeley National Laboratory: Berkeley, CA..

[135] Yao, X., B. Yi, Y. Yu, Y. Fan, and L. Zhu, *Economic analysis of grid integration of variable solar and wind power with conventional power system.* Applied Energy, 2020. **264**: doi.org/10.1016/j.apenergy.2020.114706.

[136] Yumeng, Z., W. Ligang, W. Ningling, L. Duan, Y. Zong, *et al., Balancing wind-power fluctuation via onsite storage under uncertainty: Power-to-hydrogen-to-power versus lithium battery.* Renewable and Sustainable Energy Reviews, 2019. 116: doi.org/10.1016/j.rser.2019.109465.

[137] Zaboji, N., *Ökostromförderung erreicht Rekord (Green electricity subsidies reach record), in Frankfurter Allgemeine.* 2021: Frankfurt. https://www.faz.net/aktuell/wirtschaft/klima--energie--und--umwelt

/30--9--milliarden--euro--oekostromfoerderung--erreicht--rekord--17141915.html.

[138] Zerrahn, A., *Wind Power and Externalities.* Ecological Economics, 2015. **141**: p. 245–260.

[139] Zhou, S., Y. Wang, Y. Zhou, L.E. Clarke, and J.A. Edmonds, *Roles of wind and solar energy in China's power sector: implications of intermittency constraints.* Applied Energy, 2018. **213**: p. 22–30.

9

A Comprehensive Analysis of Integrated Fault Tolerant Mechanism Capability of a Grid-connected Doubly-fed Induction Generator-based Wind System

Raghavendran C. R.[1], Preetha Roselyn J.[2], and Devaraj D.[3]

[1]Assistant Professor, Easwari Engineering College, India
[2]Associate Professor, SRM Institute of Science and Technology, India
[3]Director-Academic and Dean-SEET, Kalasalingam Academy of Research and Education, India
E-mail: raghavendran.cr@eec.edu; preethaj@srmist.edu.in; ddevaraj@klu.ac.in

Abstract

Wind energy has emerged as a crucial source of renewable power across many countries. However, integrating wind farms into the grid introduces variations in dynamic behavior, leading to network imbalances – an essential challenge in modern smart grids. To comply with grid code regulations, wind turbine generators must remain connected and actively contribute to system stability during grid faults. One of the critical grid code requirements is the fault tolerance mechanism (FTM) capability of wind power generation farms. This chapter provides an in-depth analysis of various FTM techniques in a doubly-fed induction generator (DFIG)-based wind energy system under both symmetrical and asymmetrical grid fault conditions. The study evaluates the effectiveness of different FTM strategies, including the transient suppression resistor, DC voltage suppression unit, series fault current limiter (SFCL), and voltage optimization strategy. To enhance FTM performance, integrated approaches combining these techniques such as transient suppression resistor

DOI: 10.1201/9788743809852-9

with an RL Circuit, DC voltage suppression unit with transient suppression resistor, DC voltage suppression unit with a series dynamic resistor, and voltage optimization strategy with transient suppression resistor are explored under diverse fault scenarios. The proposed methods are simulated using MATLAB/Simulink, and their hardware implementation is validated using the hardware setup (OPAL-real time test bench). The results demonstrate that these Integrated FTM schemes significantly improve system stability and performance under various grid fault conditions in a wind energy power generation system.

Keywords: Fault tolerant mechanism scheme, grid code requirement, DC interlink voltage, power interface point (PIP), doubly-fed induction generator.

9.1 Introduction

Renewable energy sources are vital for modern power grids, ensuring sustainability and efficiency.Wind energy, in particular, has seen rapid global adoption due to its scalability and environmental benefits.Advancements in turbine technology have significantly improved energy output and reliability.Many countries are investing in wind farms to reduce dependence on fossil fuels and lower emissions.As a key component of the energy transition, wind power continues to shape the future of electricity generation. Several countries have established operational guidelines for integrating distributed generation in accordance with various grid codes [1–3].

Doubly-fed induction generators (DFIG) are extensively utilized in wind power systems for their distinct advantages. They enable flexible speed operation and allow separate regulation of real and reactive power.These features enhance efficiency and grid stability, making DFIG a preferred choice for wind energy applications [4–6]. The stability of a DFIG-based system can be maintained using transient suppression resistor protection, while the shaft side converter (RSC) and grid side converter (GSC) help regulate reactive power transfer and control DC Interlink voltage fluctuations during grid disturbances [7–9].

During fault conditions, DFIG-based wind systems are often disconnected from the grid to safeguard the RSC and DC interlink capacitor [10–13]. However, as per grid code requirements, distributed generation must remain connected to the grid to a certain extent and support system stability. Therefore, effective fault tolerance mechanism (FTM) schemes are necessary to regulate active and reactive power flow while protecting power electronic components. Common FTM techniques include transient suppression resistor protection, dynamic voltage restorers (DVRs), superconducting fault current limiters (SFCL), magnetic core damping resistors (SFCL), energy capacitor systems (ECS), and energy storage systems (ESS) [14–16].

Morren et al. [14] proposed a crowbar-based FTM scheme that engages in parallel with the shaft winding during faults. Kasem et al. [15] investigated a transient suppression resistor FTM scheme, but it disconnects before fault clearance, failing to suppress the shaft current. Other studies [16, 17] have examined varying transient suppression resistor values for reactive power support, though high shaft resistance voltages may damage shaft-side switches. Advanced transient suppression resistor designs have been introduced [18, 19] to ensure activation only during faults per grid code standards.

Jesus et al. [20] explored low voltage ride through (LVRT) schemes to inject reactive power during faults. Okedu et al. [21] proposed a DC voltage suppression unit circuit with a braking resistor in parallel with the DC Interlink capacitor to limit voltage rise. However, some studies [22] reported that after fault clearance, DC interlink voltage drops significantly, requiring braking resistor removal. SFCL circuits with coordinated switching mechanisms have been analyzed under different fault conditions [23], while DVR-based methods regulate terminal voltage at the expense of increased operational costs [24, 35].

To leverage the advantages of various FTM strategies, integrated approaches have been explored. Yang et al. [26] presented an advanced control scheme combining shaft transient suppression resistor and DC voltage suppression unit techniques to mitigate shaft overcurrent and control DC interlink voltage during faults. Anca D. Hansen et al. [27] reviewed grid code requirements for wind turbines under different load variations, evaluating system performance in cases where turbines either disconnect during faults or remain connected with FTM controllers.

Other studies have utilized linear quadratic regulator (LQR) optimization to enhance system stability under grid disturbances [28]. Various protection schemes for shaft current control and DC interlink voltage suppression have been assessed [29], with some solutions involving series-connected resistors to mitigate voltage sags. The combination of shunt and series passive networks has been explored in [30]. Okedu et al. [31] proposed a combination of DC voltage suppression unit and SFCL circuits to limit DC overvoltage and shaft current. Justo et al. [32] developed a transient suppression resistor with an RL circuit to enhance PIP voltage using reactive power regulators. LVRT behavior in brushless DFIG systems has been analyzed under three-phase voltage dips in asymmetrical fault conditions [33].

Arifujjaman et al. [34, 35] implemented hardware-in-the-loop (HIL) setups for wind energy conversion systems to validate practical control approaches. Further studies have explored maximum power point tracking (MPPT) using optimization techniques such as artificial bee colony (ABC) algorithms [36]. Kunwar et al. [37] investigated power system stability concerns for high-voltage direct current (HVDC) integration with DFIG and direct drive synchronous generator (DDSG) wind farms. Stability analysis improves overall system reliability and facilitates increased DFIG penetration in wind energy generation.

The literature review highlights that conventional and Integrated FTM strategies have not been extensively evaluated under both symmetrical and asymmetrical fault conditions for all system parameters. This chapter systematically examines traditional FTM controllers, including DC voltage suppression unit, transient suppression resistor, series fault current limiter (SFCL), and voltage control techniques, along with integrated methods such as transient suppression resistor with RL circuit, transient suppression resistor with SFCL, SFCL with DC voltage suppression unit, and voltage control with transient suppression resistor protection. The study evaluates system parameters including PIP voltage, DC bus voltage, armature current, field current, postfault dynamic behavior, and real/reactive power delivery.

The chapter is structured as follows: Section 9.2 discusses FTM parameters and grid code requirements. Section 9.3 details the proposed FTM strategies, covering both conventional and integrated schemes. Section 9.4 presents simulation results and analysis.

9.2 FTM Grid Codes

The grid codes of Germany, Spain, Denmark, and the United States for distributed generation (DG) under fault conditions are illustrated in Figure 9.1. Table 9.1 outlines the fault clearing time, nominal system voltage, and fault duration. V_{pf} and V_{f} represent different nominal system voltages during faults, indicating the required voltage levels, and fault clearing times. V_{f} specifies that 15% of the nominal system voltage must be maintained during fault conditions, while V_{pf} denotes the minimum voltage required for normal wind turbine operation. The curve illustrates that the wind turbine must remain connected to the grid during abnormal conditions.

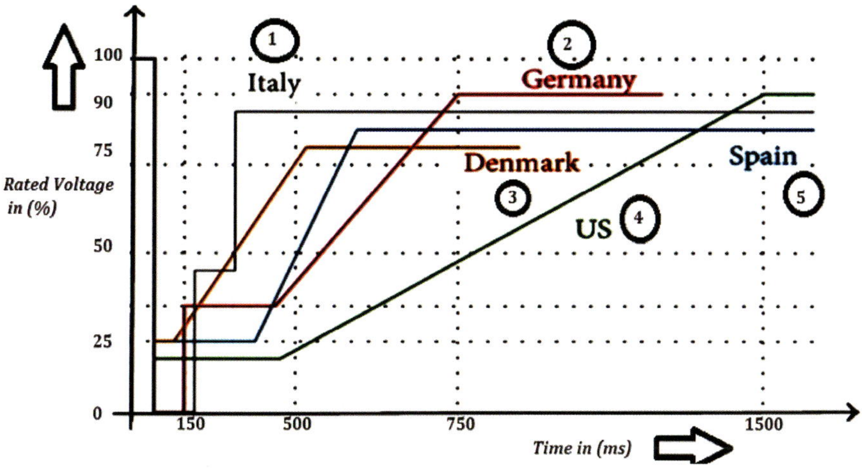

Figure 9.1 Various countries grid code FTM curve.

Table 9.1 Indian grid code requirement.

Nominal System Voltage(kV)	Fault Clearing Time (ms)	V_{pf}(kV)	V_f(kV)
400	100	360	60
220	160	200	33
132	160	120	19.8
110	160	96.25	16.5
66	300	60	9.9

Table 9.1 addresses the Indian grid code time length for which the breeze turbine should be synchronized to the framework at the various degrees of shortcoming appearance. This guarantees that the wind system capably supplies the capacity to the network quickly clearing the issue of keeping up with system stability.

9.3 Proposed FTM Methodology

9.3.1 Conventional schemes

The boundaries of the DFIG, for example, the preshortcoming wind speed and generator, yield power which has a high need. Whenever a three-phase fault occurs, the shaft current increases, causing a drop in the magnetic core voltage level. Under these conditions, the generator can deliver its rated power only at the synchronous operating speed. The FTM system lessens the shaft inrush current and the DC-connect overvoltage inside the more secure breaking point. The regular plans considered in this part are the transient suppression resistor, DC-interface chopper, series fault current limiter (SFCL), and voltage control conspiring. These plans are broken down under both even and unsymmetrical issue conditions, which are not investigated exhaustively in numerous writing papers.

9.3.1.1 FTM scheme based on transient suppression resistor

The transient suppression resistor is connected across the RSC terminals as displayed in Figure 9.2. The transient suppression resistor is designed to limit sudden surges in rotor current and control DC-link overvoltage during faults, thereby enhancing the low voltage ride-through (LVRT) capability of the system. The shaft top current regards should be kept up with inside explicit cutoff with a particular ultimate objective to update LVRT capacity. As needs be, the defensive circuit of the DFIG ought to be redesigned. Magnetic core motion damping is updated in powerful LVRT limit in the midst of voltage plunge. With the extension of one more resistive unit in the unique LVRT procedure, a high current is restricted in both the shaft side converter and structure side converter. By connecting this security unit to the network, DFIG is guaranteed to be safeguarded during the framework issue.

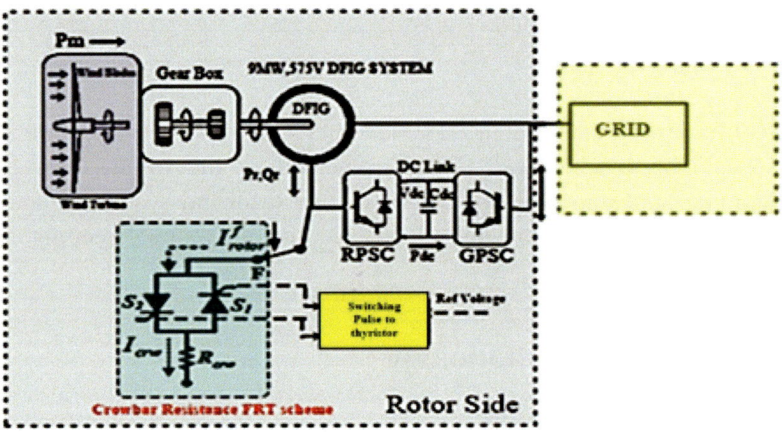

Figure 9.2 Schematic representation of transient suppression resistor FTM scheme.

9.3.1.2 FTM scheme based on DC interlink chopper

The DC interface voltage normally comes to pass as far as possible and consequently harms the capacitor or power electronic switches under framework deficiencies. Consequently, a chopper is utilized in corresponding with

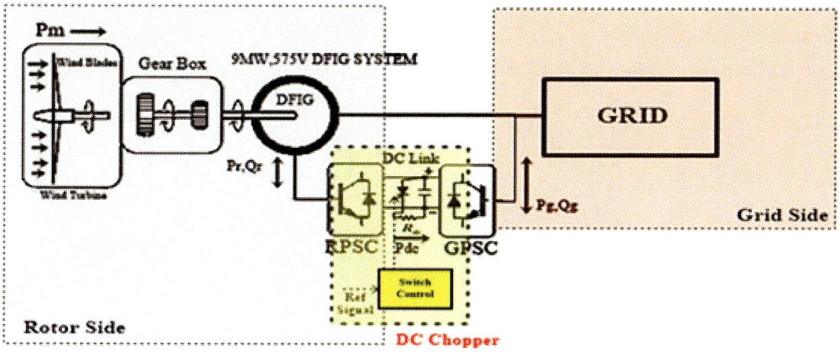

Figure 9.3 Schematic representation of DC voltage suppression unit FTM scheme.

the capacitor as displayed in Figure 9.3. During fault conditions, excess active power that cannot be transferred to the grid is dissipated through the resistor. This process helps maintain the DC-link voltage within safe limits, thereby protecting the system from overvoltage stress and ensuring operational stability.

9.3.1.3 FTM scheme based on series fault current limiter (SFCL)

Series fault current limiter (SFCL) is set in series with the shaft with a power electronic switch across it, as displayed in Figure 9.4 at the point when a plunge is detected SFCL to be enacted by the power electronic switches. Variety in shaft comparable opposition during the dunks helps in restricting the current overshoot, and mitigates DC interface voltage shoot and force motions.

9.3.1.4 FTM scheme based on voltage control

Figure 9.5 represents the voltage optimization strategy under which the voltage control FTM conspire, the inverter and the rectifier method of activity are supplanted with the voltage regulators where the voltage is infused involving the sources to keep up with the shaft side and framework side voltage and current. This plan helps in diminishing the motions in genuine power and shaft current.

9.3.2 Proposed integrated FTM schemes

The customary FTM plans examined in Section 9.3.1 perform well for controlling a few boundaries in the DFIG framework during flaws. To get the best

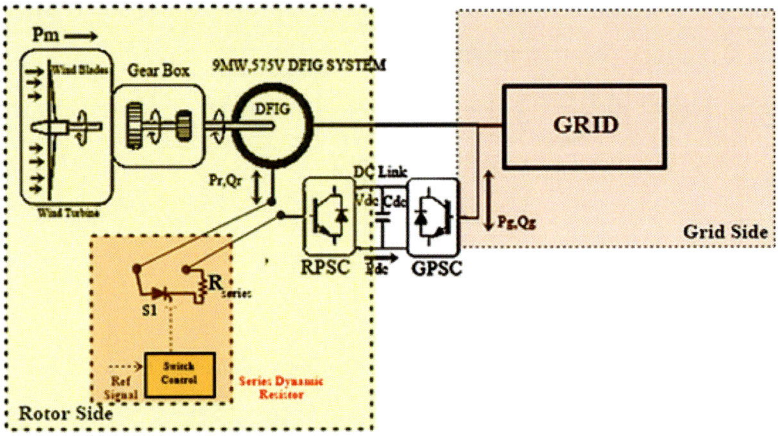

Figure 9.4 Schematic representation of dc voltage suppression unit FTM scheme.

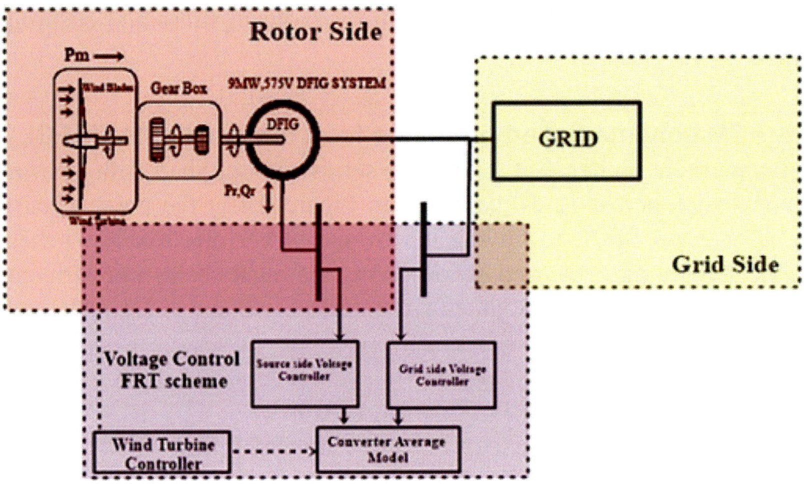

Figure 9.5 Schematic representation of voltage control FTM scheme.

attributes of each FTM plan, the regular FTM techniques are consolidated in a legitimate manner for better LVRT capacity. A portion of the half-breed FTM plans completed in this work are transient suppression resistor opposition with RL circuit, DC-interface chopper with transient suppression resistor obstruction, DC voltage suppression unit with series fault current limiter (SFCL), and voltage control conspire with transient suppression resistor obstruction.

9.3.2.1 Integrated transient suppression resistor and RL circuit

Integrated transient suppression resistor security circuit, designed with thyristor-based step-down resistors as shown in Figure 9.6, works in conjunction with the RL circuit to reduce shaft current and suppress DC interlink voltage. When the terminal voltage falls below the threshold value of the comparator, the FTM (fault-tolerant mode) design is triggered, deactivating the shaft controller. Conversely, when the terminal voltage exceeds the threshold value, the transient suppression resistor is turned off, and the shaft controller is reactivated.

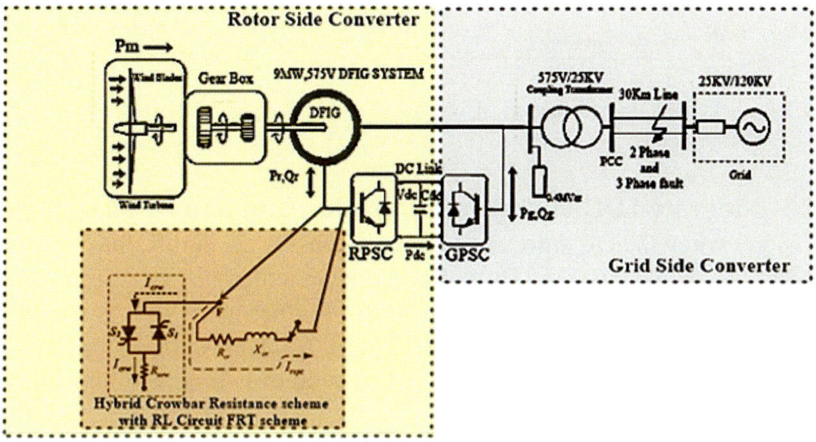

Figure 9.6 Schematic representation of integrated transient suppression resistor with RL circuit FTM scheme.

9.3.2.2 Integrated DC voltage suppression unit and transient suppression resistance

Transient suppression resistor opposition joined with DC voltage suppression unit associated in corresponding with capacitor tuned by the power electronic switch is displayed in Figure 9.7. This half-breed regulator safeguards the DC interface capacitor and changes from overvoltage during lattice shortcoming condition. The mix of DC voltage suppression unit and transient suppression resistor opposition helps in lessening DC connect voltage to the passable worth and the transient suppression resistor obstruction helps in keeping up with the shaft flows. This blend helps in decreasing DC interface voltage, shaft flows, and dynamic power dispersal.

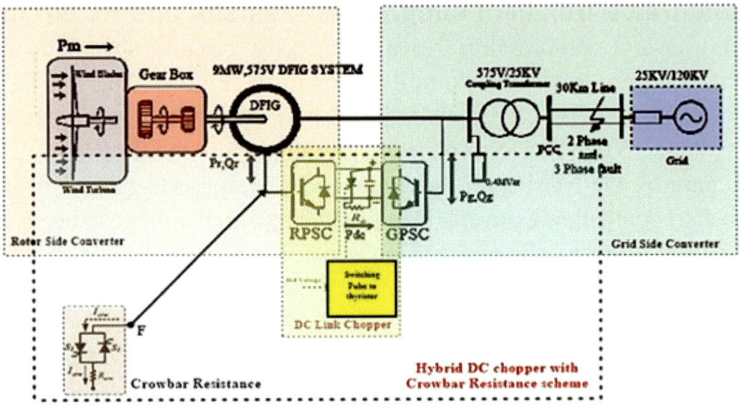

Figure 9.7 Schematic representation of integrated DC voltage suppression unit with RL circuit FTM scheme.

9.3.2.3 Integrated DC voltage suppression unit and SFCL

At the point when the DC-interface voltage surpasses the cutoff, the chopper will be turned on and when the DC voltage reduces under the limit esteem, the DC voltage suppression Unit will be crippled. Furthermore, with this cross-breed DC voltage suppression unit and SFCL, as displayed in Figure 9.8, SFCL can confine the shaft overvoltage and moreover limit the high shaft

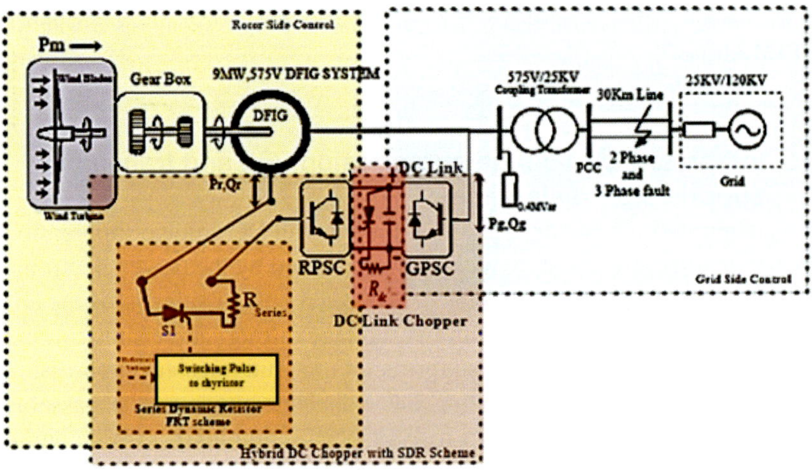

Figure 9.8 Schematic representation of integrated DC voltage suppression unit with SFCL FTM scheme.

current. The shaft current can diminish the charging current to the DC interface capacitor, and subsequently, lessen the DC-connect overvoltage.

9.3.2.4 Integrated voltage optimization strategy and transient suppression resistance

The combination of the voltage optimization strategy with the transient suppression resistor as shown in Figure 9.9 helps in reducing the oscillation of the real power and shaft current. The transient suppression resistor is adequate in damping the oscillation in the shaft current.

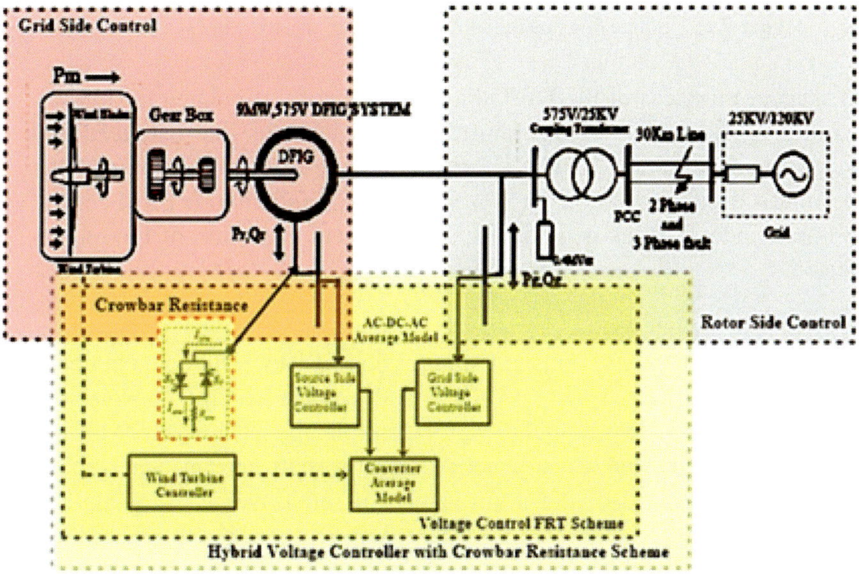

Figure 9.9 Schematic representation of integrated voltage optimization strategy with transient suppression resistor.

9.4 Results and Discussion

The presentation of the customary and mixture regulators are tried in a matrix associated 9 MW, DFIG test framework in MATLAB adaptation 2019a. Furthermore, the presentation of the half-breed regulator, DC voltage suppression unit with SFCL which gives a preferred reaction over the other FTM regulators is carried out in OPALRT climate and the test results are given. The test boundaries of the lattice and the breeze framework are given in Appendix A. The induction generator functions optimally at a typical wind speed of 12 m/s.

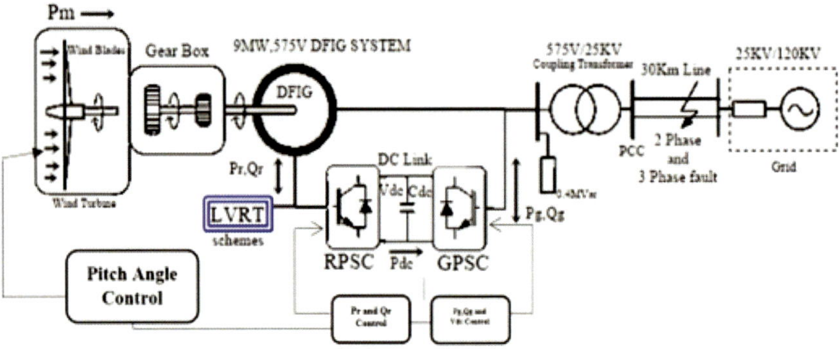

Figure 9.10 Single line diagram of 9 MW grid-connected wind system.

The transient practices of the DFIG-based framework associated framework are contemplated under various shortcoming conditions like balanced and unsymmetrical deficiencies. The shortcoming ride-through capacity of the framework is tried with various proposed traditional and mixture FTM control procedures under various deficiencies and transient execution of the plan is inspected.

(a) *Simulation results:*

Under regular FTM plans, transient suppression resistor, DC-connect chopper, series fault current limiter (SFCL), and voltage control conspire are carried out and the various boundaries, for example, the DC interface voltage, genuine and receptive power, magnetic core and shaft current, and voltage at PIP are explored. Also, the improvement of crossover FTM plans like (a) combined transient suppression resistor with RL circuit, (b) DC voltage suppression unit with transient suppression resistor, (c) DC voltage suppression unit interlinked with SFCL, and (d) voltage optimization strategy with transient suppression resistor under even and unsymmetrical issue conditions are executed and analyzed with various framework boundaries. Figure 9.10 presents the single-line diagram of the 9 MW grid-connected wind energy system, illustrating the main components, electrical interconnections, and power flow from the doubly-fed induction generator (DFIG) to the grid.

Scenario 1: study of proposed FTM schemes under symmetrical fault conditions

(a) Performance of conventional FTM schemes during symmetrical fault

Under this scenario, a symmetrical fault is simulated which produces low-voltage fault behaviors in the system. The DFIG is operating at the wind speed of 12 m/s at time $t = 4$ seconds. The following system parameters are analyzed under symmetrical fault condition.

(i) *DC interlink voltage*

Figure 9.11 shows the distinction of DC-connect voltage which shoots up promptly to 2110 V as a result of the inrush flows infused from the shaft side converter into the DC-interface capacitor. The voltage is smothered to 1700 V when the transient suppression resistor opposition insurance plot is presented. Whenever the DC-chopper is empowered in

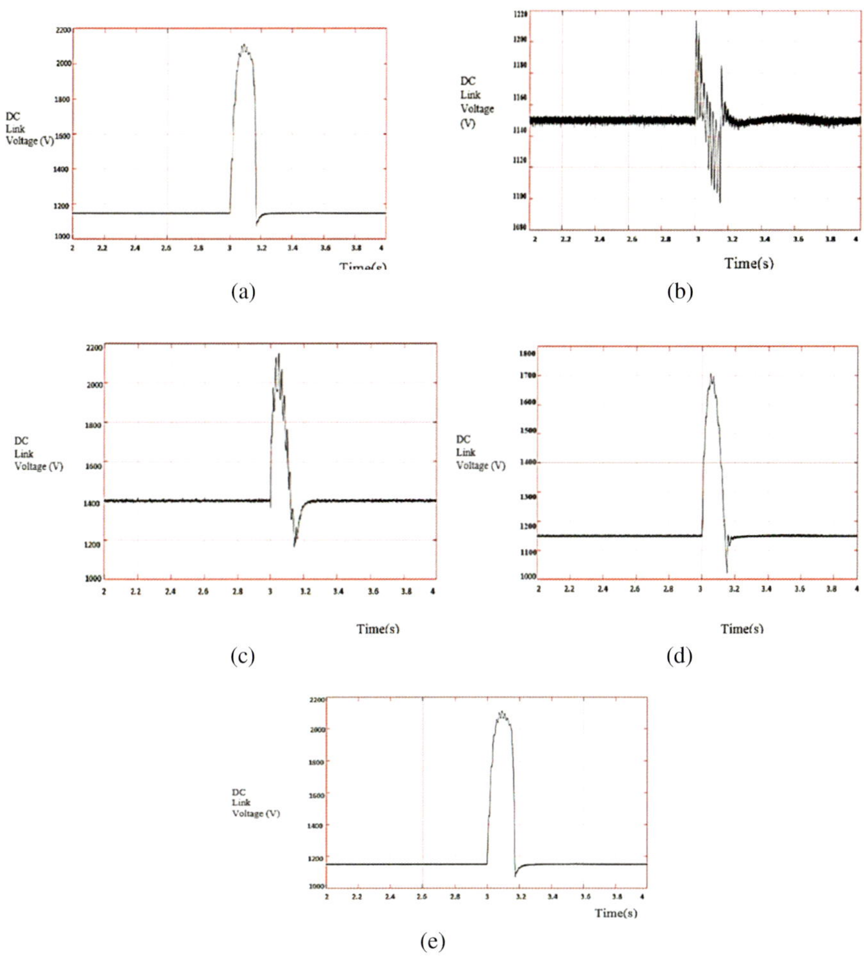

Figure 9.11 DC interlink voltage of the DFIG-WECS under the symmetrical fault scenario: (a) No FTM scheme; (b) DC voltage suppression unit; (c) SFCL; (d) transient suppression resistance; and (e) voltage optimization strategy.

the framework, DC interlink voltage is diminished to 1210 V, which is better compared to other FTM plans.

(ii) *Voltage at PIP point*

Figure 9.12 shows the voltage plunges, which happen at the power interface point (PIP) because of the event of the even issue in the framework. The shaft current increment prompts a weighty dunk of voltage in the PIP point. The plunge voltage ranges to zero when there is no FTM plot in the framework. Whenever the transient suppression resistor conspire is acquainted with the framework, the voltage plunges to 0.1 p.u and it recuperates back at 3.45 seconds. The DC voltage suppression unit is relatively better compared to other regular FTM plans which helps in decreasing the recuperation time thus the voltage is diminished to 0.2 p.u rather than 0 p.u during shortcomings and it is recuperated back to ostensible worth in 3.2 seconds.

(iii) *Shaft current*

Figure 9.13 shows the shaft side current which arrives at the pinnacle worth of 4.0 p.u under no FTM plot, When the FTM conspire is presented in the framework shaft current, the motions in the shaft current are limited. Whenever the series fault current limiter (SFCL) is presented in the framework, the shaft flow is decreased to 2.4 p.u, which shows that this plan delivers better outcomes as far as shaft current. The exhibition of the leftover FTM plans is likewise examined in the test framework. The motions in the shaft current are limited well when the DC voltage suppression unit conspire is presented.

(iv) *Real power*

Figure 9.14 shows the genuine power consumed by the framework when the framework is exposed to balanced issue condition. Weighty motions are knowledgeable about the genuine power under no FTM plot condition. At the point when the voltage control plot is presented in the framework, the wavering gets smothered and reaches 0.8 p.u during the issue clearing time. Consequently, the voltage control FTM plot assumes a significant part in lessening the genuine power motions of the framework during the shortcoming clearing time.

(v) *Reactive power*

Figure 9.15 shows the receptive power infusion by the framework when it is exposed to even blame condition. Under no FTM plot, weighty swaying happens in the responsive power during the postshortcoming condition and the infusion of the receptive power is exceptionally less

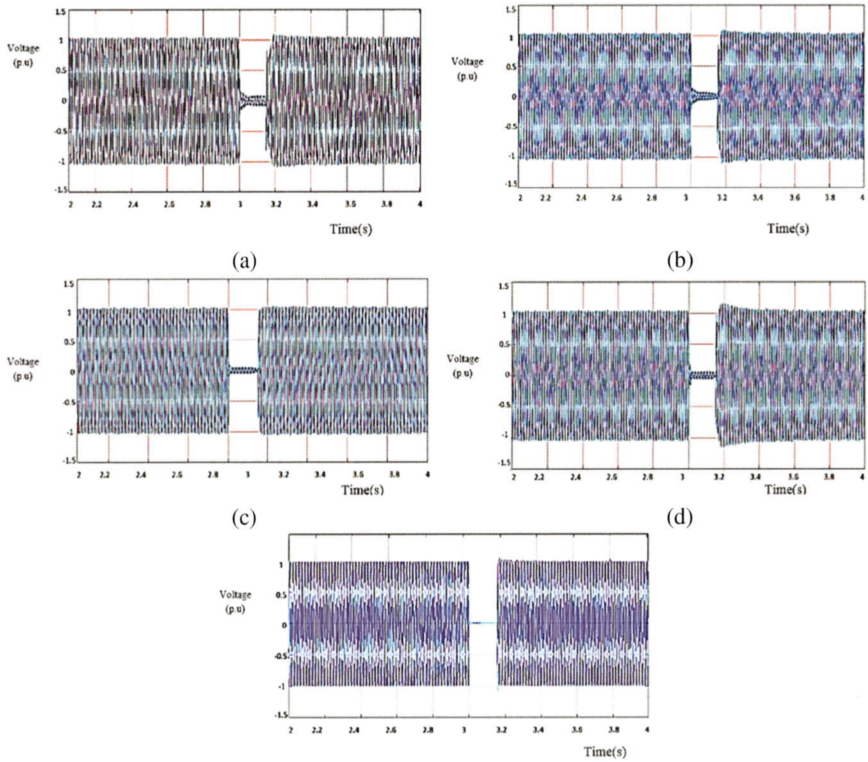

(e)

Figure 9.12 Voltage at PIP of the DFIG-WECS under the symmetrical fault scenario: (a) no FTM scheme; (b) DC voltage suppression unit; (c) SFCL; (d) transient suppression resistance; and (e) voltage optimization strategy.

during the issue time frame. DC voltage suppression unit FTM conspire is better than other FTM schemes, and thus, the receptive power infusion levels are high as 0.1 p.u.

The correlation of various FTM plans if there should be an occurrence of various framework boundaries is summed up in Table 9.2. SFCL conspire is better in restricting the shaft current and postissue transient current. Transient suppression resistor opposition conspire acts in well-diminishing postissue transient current. The voltage control plot is great in smothering the motions of genuine power during issue-clearing time. This extensive examination shows that the DC voltage suppression unit creates great transient conduct regarding numerous framework boundaries like DC interface voltage, voltage at PIP, responsive power, and motions in the shaft circuit.

Table 9.2 Performance of FTM schemes under symmetrical faults in DFIG-WECS.

FTM scheme	Shaft current	Oscillations in shaft current	Voltage @PIP	DC interlink voltage	Real power	Reactive power injection	Post fault transient current
Transient suppression resistor	Peak value of 3.75 p.u	Less oscillations when compared to no FTM scheme	Dips to 0.1 p.u and recovered at 3.45 seconds	Peak value 1700 V	Oscillations reached 1.3 p.u during fault clearing	0.05 p.u	Reaches to 1.2 p.u
DC voltage suppression unit	Peak value of 3.8 p.u	Less oscillations when compared to other FTM schemes	Dips to 0.2 p.u and recovered at 3.2 seconds	Peak value 1210 V	Oscillations reached 1.45 p.u during fault clearing	0.1 p.u	Reaches to 1.25 p.u
SFCL	Peak value of 2.4 p.u	Less oscillations when compared to no FTM scheme	Dips to 0.2 p.u and recovered at 3.25 seconds	Peak value 1325 V	Oscillations reached 0.6 p.u during fault clearing	0.05 p.u	Reaches to 1.2 p.u
Voltage optimization strategy	Peak value of 4 p.u	Heavy oscillations	Dips to 0.1 p.u and recovered at 3.35 seconds	Peak value 1800 V	Oscillations are very less reaches 0.8 p.u during fault clearing	0.01 p.u	Reaches to 2.4 p.u

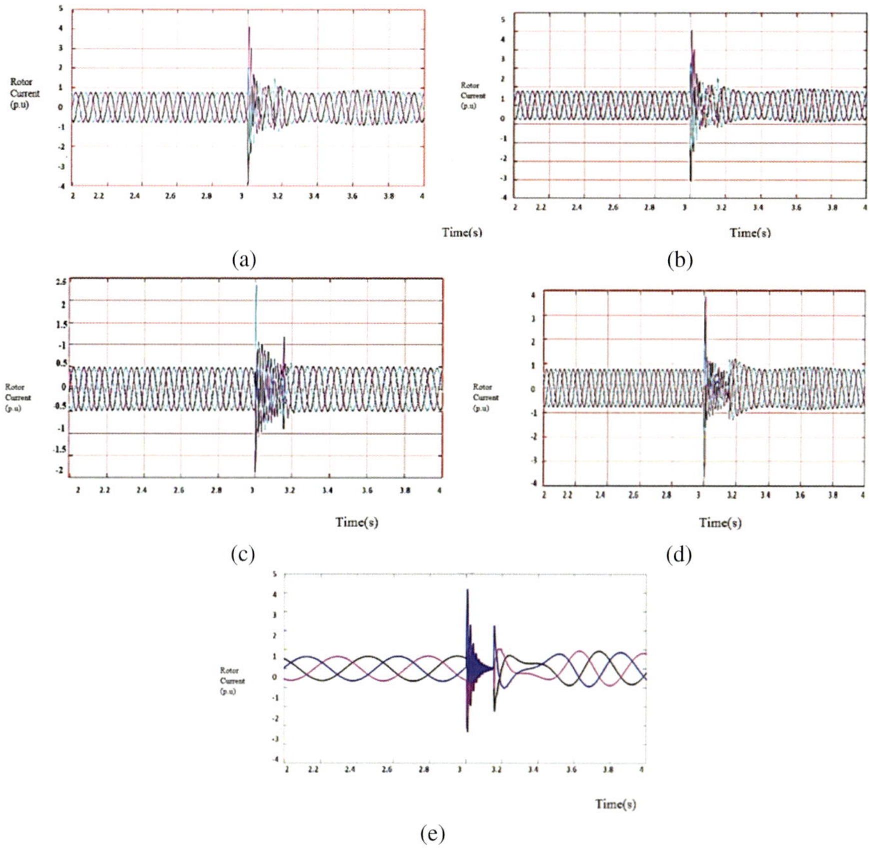

Figure 9.13 Shaft current of DFIG-WECS under the symmetrical fault scenario: (a) no FTM scheme; (b) DC voltage suppression unit; (c) SFCL; (d) transient suppression resistance; and (e) voltage optimization strategy.

(b) *Performance of integrated FTM schemes under symmetrical fault conditions*

In order to compare the good behavior of different FTM schemes during symmetrical faults, integrated controllers are implemented in the same test system.

(i) *DC interlink voltage*

Figure 9.16 shows the DC-interface voltage which shoots up in light of the fact that inrush current from the shaft converter. The integrated FTM controllers are more compelling in stifling the DC interlink voltage than a single FTM plot. Different half-breed plans are acquainted with

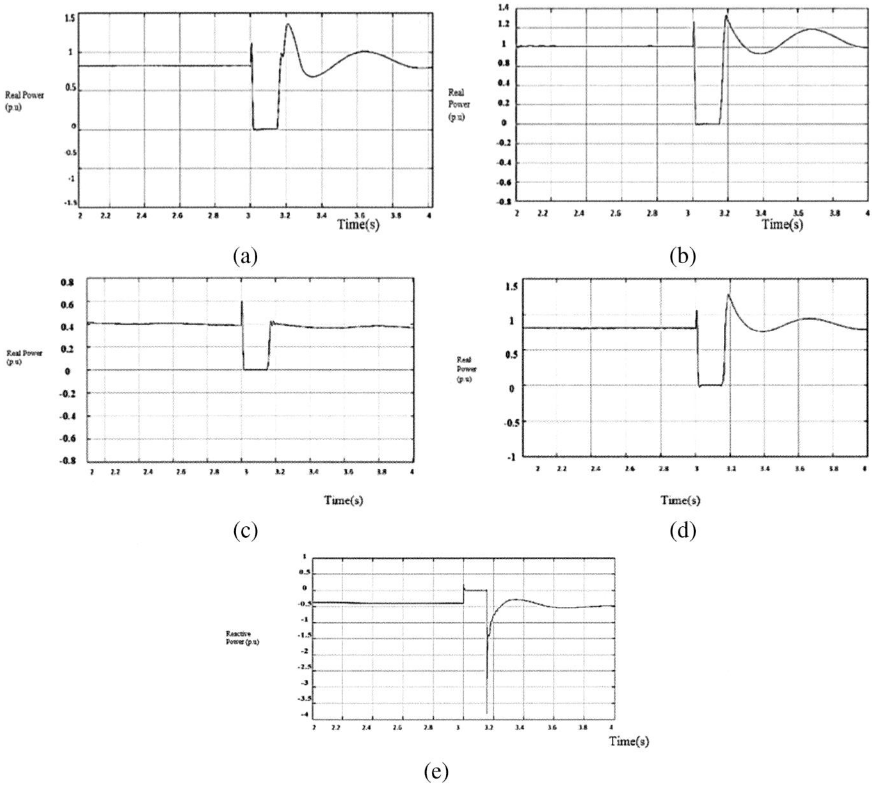

Figure 9.14 Real power of DFIG-WECS under the symmetrical Fault scenario: (a) no FTM scheme; (b) DC voltage suppression unit; (c) SFCL; (d) transient suppression resistance; and (e) voltage optimization strategy.

perceiving which mixture FTM plot is more powerful in diminishing the DC interlink voltage. The DC connect is stifled to 1155 V by utilizing the DC voltage suppression unit with series fault current limiter (SFCL), which is extremely less when contrasted with different mixes. DC voltage suppression unit with transient suppression resistor obstruction is the following viable strategy in keeping up with the DC interface voltage inside as far as possible.

(ii) *Voltage at PIP*

Figure 9.17 illustrates the dip in voltage at the power interface point during the short circuit fault. By introducing the integrated FTM scheme, the dip in the voltage is reduced better than conventional FTM scheme and the voltage recovery time is very less, and selecting the DC voltage

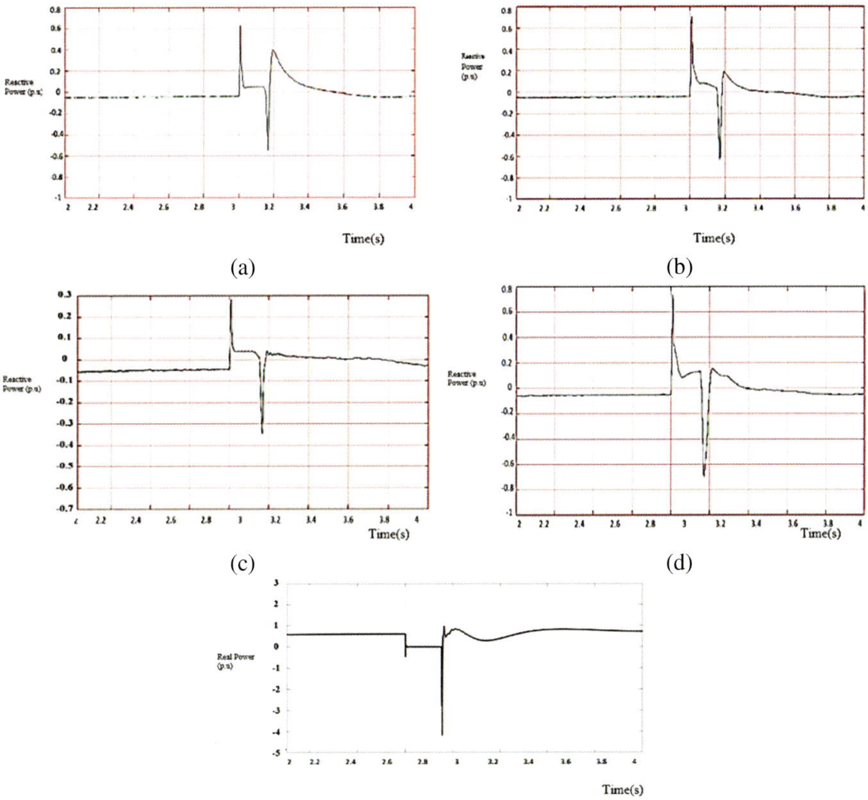

Figure 9.15 Reactive power of DFIG-WECS under the symmetrical fault scenario: (a) No FTM scheme; (b) DC voltage suppression unit; (c) SFCL; (d) transient suppression resistor; and (e) voltage optimization strategy.

suppression unit with the SFCL scheme helps in reducing the voltage dip to 0.2 p.u and recovery time of 3.17 seconds, and hence, the transient behavior of the system is improved.

(iii) *Shaft current*

Figure 9.18 outlines the shaft side current, when the half-breed FTM plans are presented in the framework, the shaft current is limited and the motions in the shaft current are decreased far superior to customary FTM plans. By utilizing a DC voltage suppression unit with a series fault current limiter (SFCL), the shaft current is abundantly diminished and arrives at a pinnacle worth of 1.7 p.u with less wavering. This joined

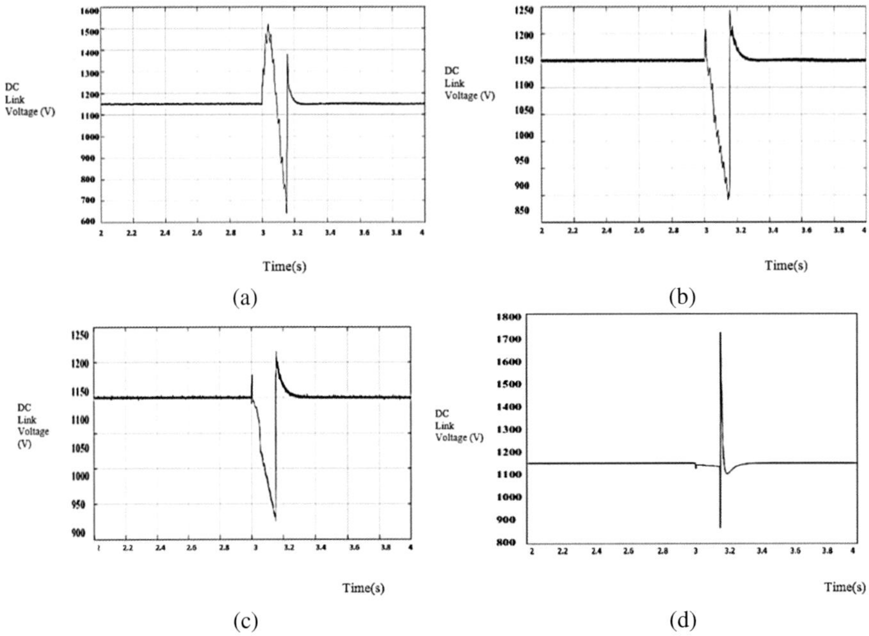

Figure 9.16 DC interlink voltage of the DFIG-WCES under symmetrical fault: (a) Transient suppression resistor with RL, (b) DC voltage suppression unit with CBR, (c) DC voltage suppression unit with SFCL, and (d) voltage optimization strategy with CBR.

plan is obviously superior to other regular plans, which diminishes the shaft current just to 2.4 p.u.

(iv) *Real power*

Figure 9.19 represents the genuine force of the framework when the framework is exposed to even blame condition. Whenever the voltage control conspire is presented in the framework, the wavering is decreased and comes to 0.8 p.u during the shortcoming clearing time. Whenever the cross-breed voltage control plot with transient suppression resistor obstruction is presented in the framework, the genuine power spans to 0.5 p.u and swaying is less when contrasted with the other FTM plans.

(v) *Reactive power*

Figure 9.20 shows the receptive power infused into the framework when it is exposed to balanced issue condition. At the point when the customary DC voltage suppression unit FTM plans are acquainted with the framework, the receptive power infusion levels arrive at the worth of

Table 9.3 Performance comparison of integrated FTM controllers under symmetrical fault condition.

FTM schemes	Shaft current	Oscillations in shaft current	Voltage @PIP	DC interlink voltage	Real power	Reactive power injection	Postfault transient current
Transient suppression resistor with RL circuit	Peak value of 3 p.u	More oscillations	Dips to 0.1 p.u and recovered at 3.25 seconds	Peak value 1500 V	Oscillations reached 1.2 p.u during fault clearing	0.1 p.u	**Reaches to 1.2 p.u**
DC voltage suppression unit with transient suppression resistor	Peak value of 3.3 p.u	Less oscillations when compared to transient suppression resistor with RL scheme	Dips to 0.1 and recovered at 3.25 seconds	Peak value 1200 V	Oscillations reached to 1 p.u during fault clearing	0.05 p.u	**Reaches to 1.25 p.u**
DC voltage suppression unit with SFCL	Peak value of 1.7 p.u	Very Less oscillations when compared to other FTM schemes	Dips to 0.2 and recovered at 3.17 seconds	Peak value 1155 V	Oscillations reached 0.8 p.u during fault clearing	0.5 p.u	**Reaches to 1.2 p.u**
Voltage optimization strategy with transient suppression resistor	Peak value of 2 p.u	Less oscillations during fault event but heavy oscillation at postfault period	Dips to 0.1 and recovered at 3.18 seconds	Peak value 1700 V	Less oscillations reaches 0.5 p.u during fault clearing	0 p.u	**Reaches to 2.4 p.u**

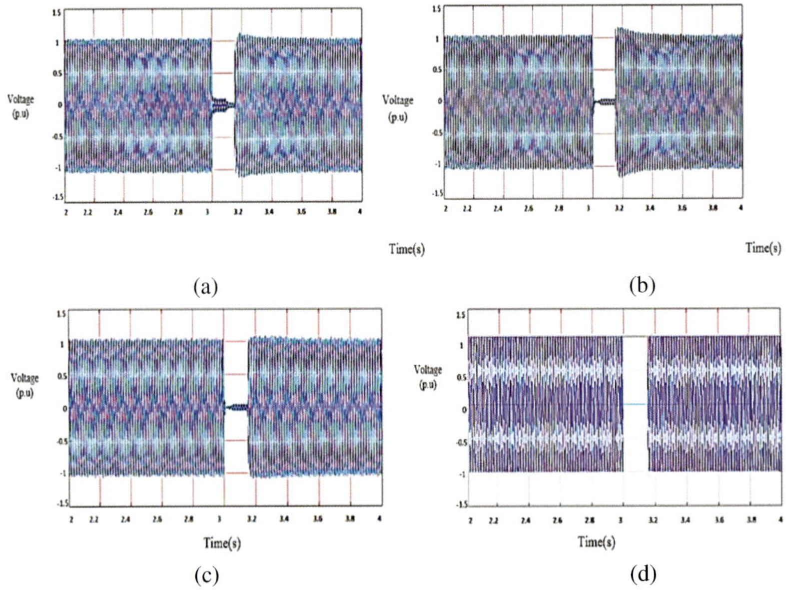

Figure 9.17 The voltage at PIP of the DFIG-WCES under symmetrical fault: (a) Transient suppression resistor with RL, (b) DC voltage suppression unit with CBR, (c) DC voltage suppression unit with SFCL, and (d) voltage optimization strategy with CBR.

0.1 p.u and when cross breed FTM conspire like DC voltage suppression unit joined with the SFCL is presented under a similar condition, it very well may be seen that the responsive power infusion is expanded to 0.5 p.u. This mixture mix is checked as the potential integrated FTM plot when contrasted with the other FTM plans if there should arise an occurrence of receptive power support.

The presentation correlation table of various crossover FTM plans produced for the test framework is displayed under even blames in Table 9.3. Transient suppression resistor opposition with RL plan and DC voltage suppression unit with SFCL plot is great in diminishing the postissue transient current. Voltage control plot with transient suppression resistor opposition is better in decreasing the motions in shaft and genuine power. The thorough survey shows that a DC voltage suppression unit with SFCL conspire is the best plan for lessening the overshoot shaft current. This plot likewise helps in limiting the voltage overshoot PIP during the issue period, and furthermore, the overshoot in the DC connect voltage.

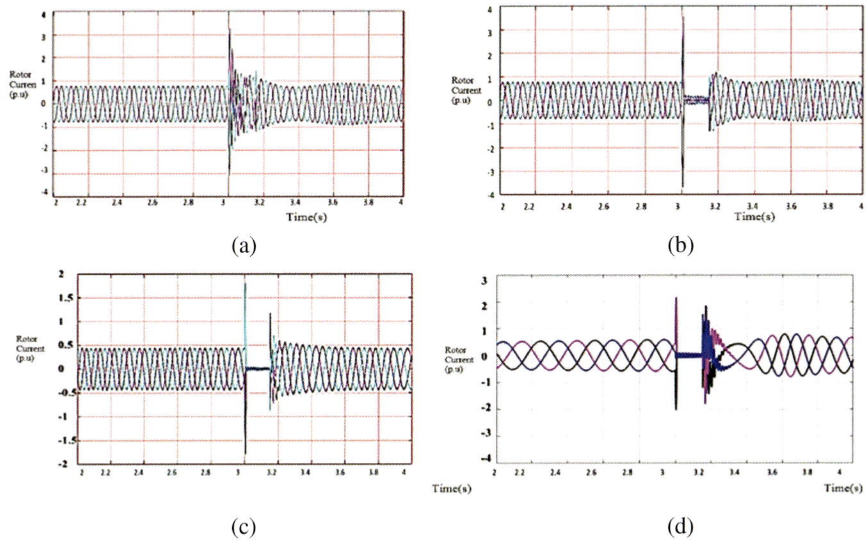

Figure 9.18 Shaft current of the DFIG-WCES under symmetrical fault: (a) Transient suppression resistor with RL, (b) DC voltage suppression unit with CBR, (c) DC voltage suppression unit with SFCL, and (d) voltage optimization strategy with CBR.

Scenario 2: study of proposed FTM schemes under unsymmetrical fault

(a) Performance of conventional FTM schemes under unsymmetrical faults

A Single-stage to ground shortcoming which is the most often happening issue is applied at time t =c 4 seconds and the transient practices of DFIG-based WECS are examined. The DFIG is to work at a wind speed of 12 m/s.

(i) *DC interlink voltage*
Figure 9.21 shows the DC interface voltage under various FTM plans. The series fault current limiter (SFCL) is presented which offers an enormous degree of DC connect voltage concealment and consequently, the worth decreases to 1150 V, which is supposed to be more successful than different plans in lessening DC interface voltage.

(ii) *Voltage at the PIP point*
Figure 9.22 outlines the plunge in voltage, which happens at the point of normal coupling during the unsymmetrical issue. The shaft current is expanded which prompts a weighty dunk in the PIP point voltage by presenting a regular FTM plot, the voltage level is kept up, and it additionally assists in lessening the recuperation with timing. By choosing the DC voltage suppression unit conspire the voltage plunge is

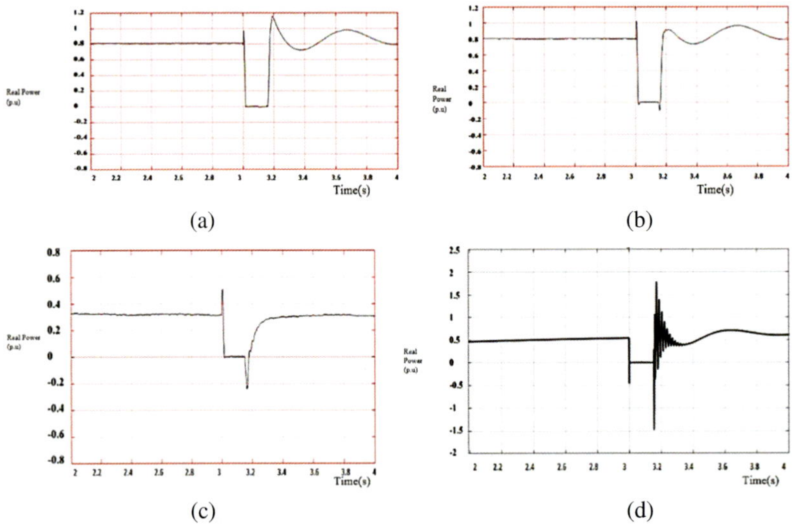

Figure 9.19 The real power of the DFIG-WCES under symmetrical fault: (a) Transient suppression resistor with RL, (b) DC voltage suppression unit with CBR, (c) DC voltage suppression unit with SFCL, and (d) voltage optimization strategy with CBR.

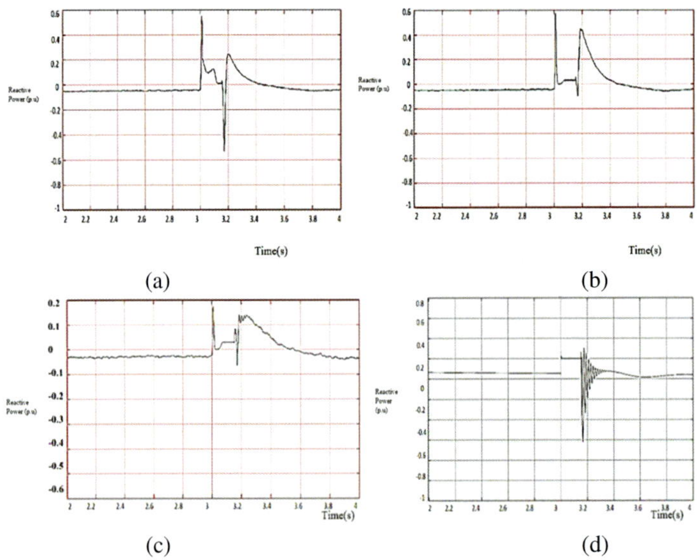

Figure 9.20 Reactive power of the DFIG-WCES under symmetrical fault: (a) Transient suppression resistor with RL, (b) DC voltage suppression unit with CBR, (c) DC voltage suppression unit with SFCL, and (d) voltage optimization strategy with CBR.

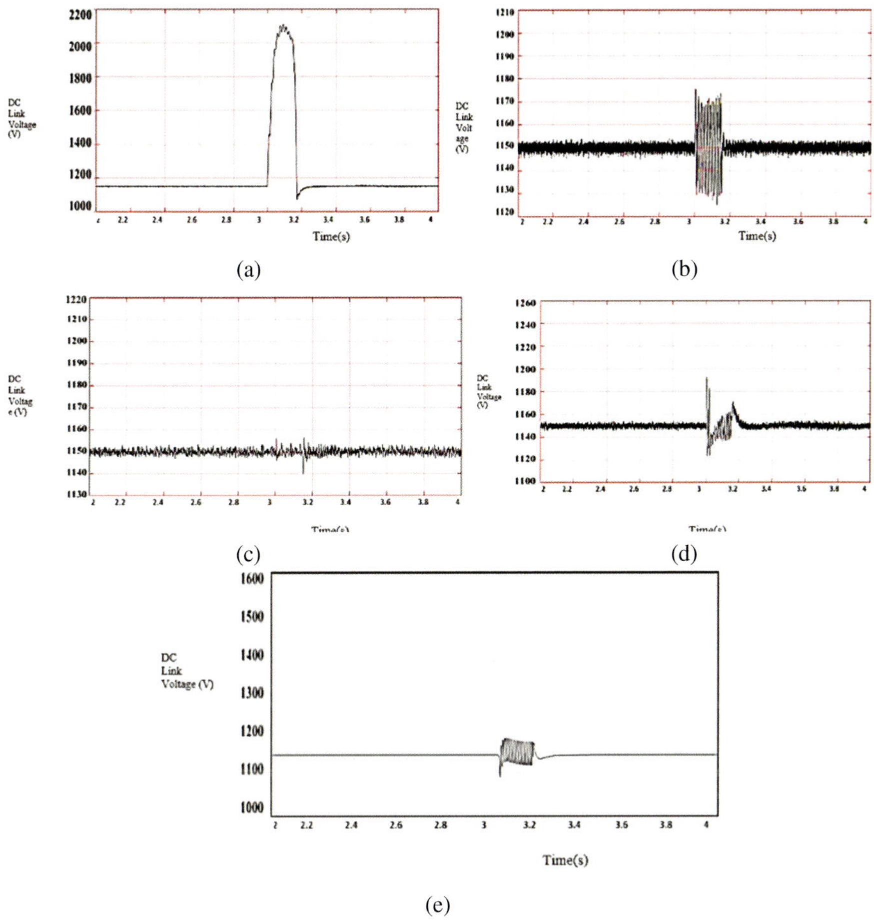

Figure 9.21 DC interlink voltage of the DFIG-WECS under the unsymmetrical fault scenario: (a) No FTM scheme; (b) DC voltage suppression unit; (c) SFCL; (d) transient suppression resistor; and (e) voltage optimization strategy.

decreased to 0.9 p.u and the recuperation time is 3.15 seconds whereas in the voltage controls plot, the voltage plunge is kept up at 0.9 p.u and the recuperation time is 3.1 seconds. Henceforth, the voltage control plot gives compromising outcomes connected with voltage at the PIP point than other FTM plans.

(iii) *Shaft current*

Figure 9.23 shows the shaft side current that arrives at the pinnacle worth of 0.8 p.u by utilizing the series dynamic obstruction conspire that is a lot

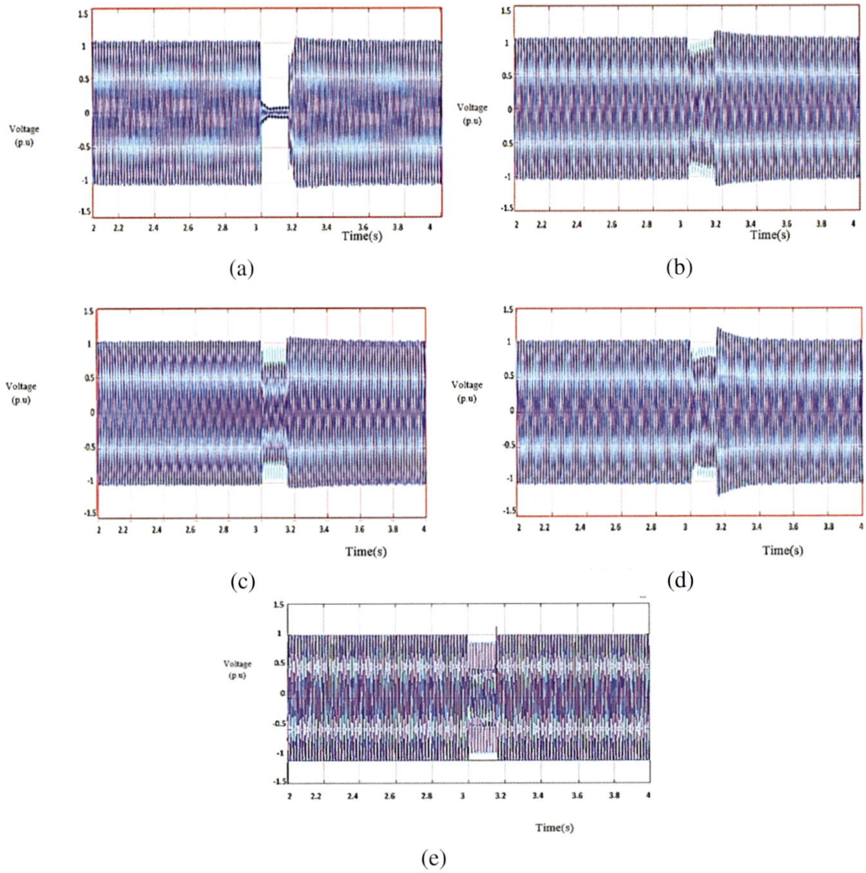

Figure 9.22 Voltage at PIP of the DFIG-WECS under the unsymmetrical fault scenario: (a) No FTM scheme; (b) DC voltage suppression unit; (c) SFCL; (d) transient suppression resistor; and (e) voltage optimization strategy.

lesser than 0.8 p.u without FTM plans, the motions in the shaft current is likewise decreased far superior to different plans.

(iv) *Real power*

Figure 9.24 shows the genuine force of the framework when the framework is exposed to unsymmetrical shortcoming condition. Weighty motions are knowledgeable about the genuine power under no FTM plot condition. Whenever the voltage control plot is presented in the framework the motions are diminished and reach to 0.6 p.u during the issue clearing time. The voltage control FTM plot does a significant job in lessening the genuine power motions than the leftover FTM plans.

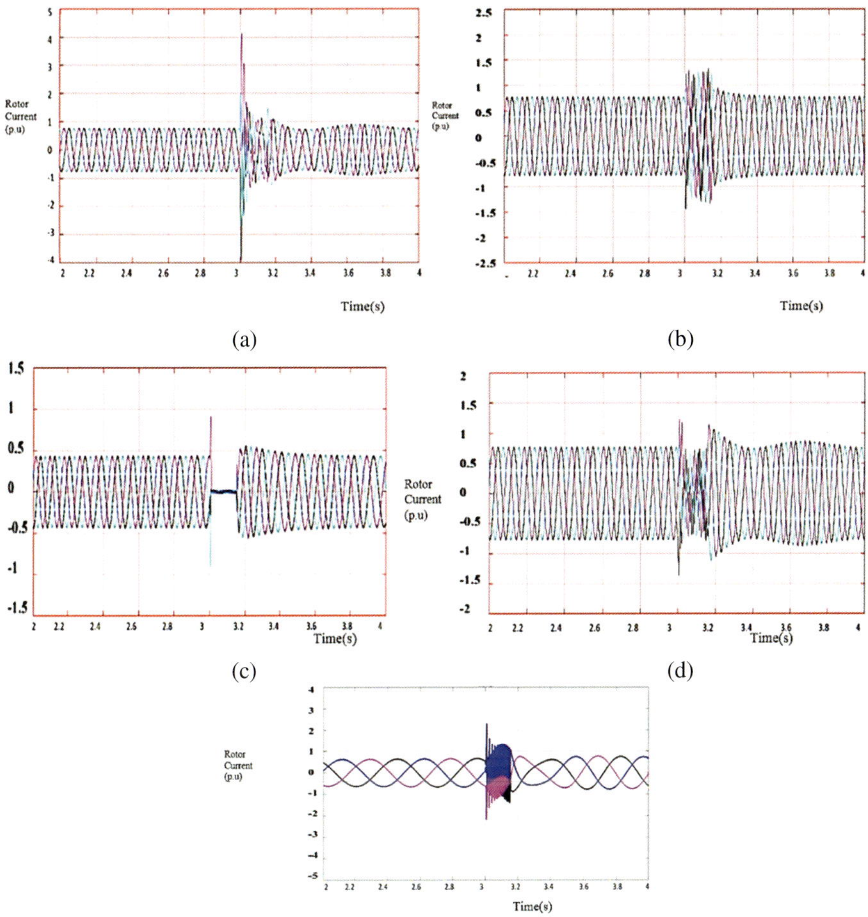

Figure 9.23 Shaft current of the DFIG-WECS under the unsymmetrical fault scenario: (a) No FTM scheme; (b) DC voltage suppression unit; (c) SFCL; (d) transient suppression resistor; and (e) voltage optimization strategy.

(v) Reactive power

Figure 9.25 shows the receptive power infusion by the framework when it is exposed to unsymmetrical issue condition. Under no FTM plot condition, weighty swaying happens during the postissue condition and the infusion of the receptive power is extremely less. At the point when the regular voltage control plans are acquainted with the framework, the receptive power infusion levels are high which comes to up to the worth of 0.4 p.u.

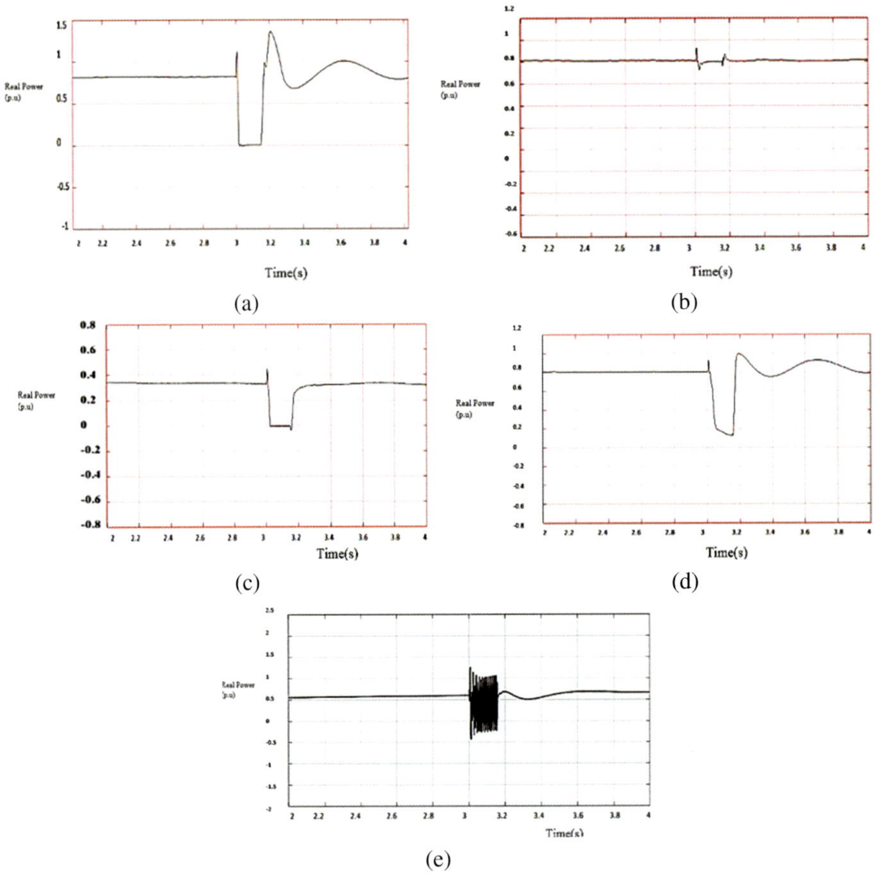

Figure 9.24 Real power of the DFIG-WECS under the unsymmetrical fault scenario: (a) No FTM scheme; (b) DC voltage suppression unit; (c) SFCL; (d) transient suppression resistor; and (e) voltage optimization strategy.

The presentation correlation table of various customary FTM plans under unsymmetrical shortcomings is displayed in Table 9.4. From this table, plainly DC voltage suppression unit is a decent plan for diminishing the motions in the shaft current. SFCL lessens the pinnacle worth of shaft current and the DC connect voltage is decreased to 1150 V, which is not exactly a different plan. The voltage control conspire diminishes the plunge in voltage at PIP highlight 0.9 p.u and the voltage is recuperated back to ostensible worth at 3.1 seconds. This plot likewise expands the receptive power infusion to 0.4 p.u during the issue period.

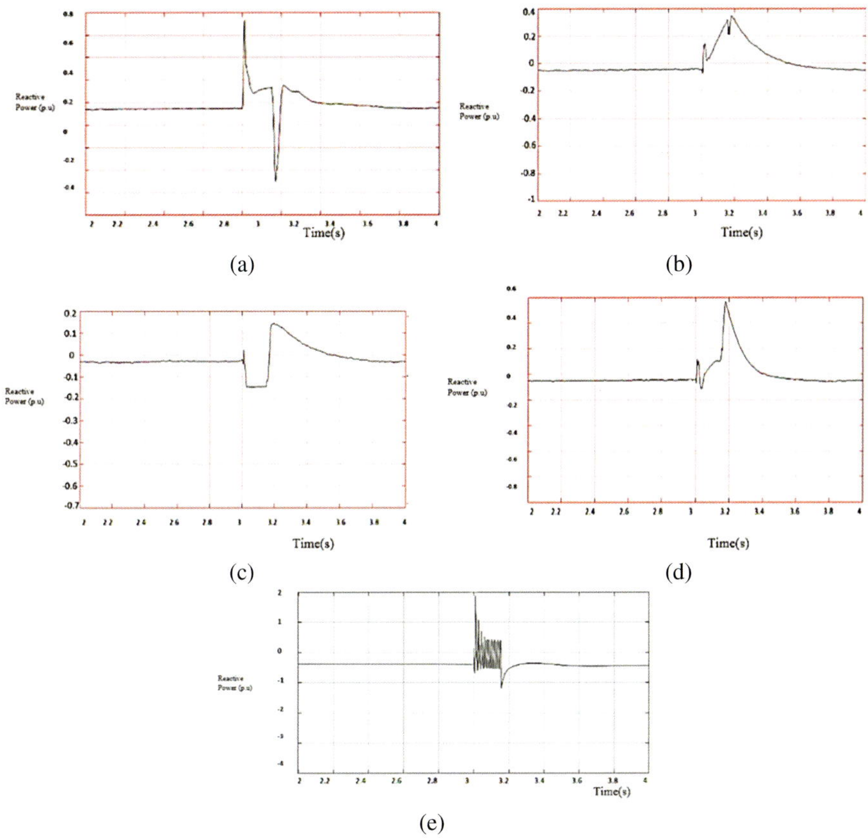

Figure 9.25 Reactive power of the DFIG-WECS under the unsymmetrical fault scenario: (a) No FTM scheme; (b) DC voltage suppression unit; (c) SFCL; (d) transient suppression resistor; and (e) voltage optimization strategy.

(b) *Performance of integrated FTM schemes under unsymmetrical fault*

The performance of DFIG-WECS is studied with the proposed Integrated FTM schemes under single phase to ground fault in the test system.

(i) *DC interlink voltage*

Figure 9.26 addresses the DC connect voltage shoot when contrasted with different regulators half and half regulators are more viable in stifling the DC interlink voltage than traditional plans. Different cross-breed plans are acquainted with perceiving which integrated FTM conspire is more compelling in stifling the DC interlink voltage.

Table 9.4 Performance comparison of conventional FTM controllers under unsymmetrical fault condition.

FTM schemes	Shaft current	Oscillations in shaft current	Voltage @PIP	DC interlink voltage	Real power	Reactive power injection	Postfault transient current
Transient suppres-sion resistor	Peak value of 1.25 p.u	Heavy Oscillations	Dips to 0.8p.u and recovered at 3.45 secs	Peak value 1190 V	Oscillations reached 0.9 p.u during fault clearing	0.1p.u	Reaches to 1.2 p.u
DC voltage suppres-sion unit	Peak value of 1.21p.u	Less when compared to Other FTM schemes	Dips to 0.9 p.u and recovered at 3.3secs	Peak value 1175 V	Oscillations reached 0.85 p.u during fault clearing	0.3p.u	Reaches to 1.25 p.u
SFCL	Peak value of 0.8p.u	Less oscillations when compared to No FTM scheme	Dips to 0.9 p.u and recovered at 3.15 secs	Peak Value 1150 V	Oscillations reached to 0.5 p.u during fault clearing	0.05 p.u	Reaches to 0.5 p.u
Voltage optimiza-tion strategy	Peak value of 2.1 p.u	Heavy oscillations	Dips to 0.9 p.u and recovered at 3.1 seconds	Peak value 1190 V	Oscillations are very less reaches 0.6 p.u during fault clearing	0.4 p.u	Reaches to 0.8 p.u

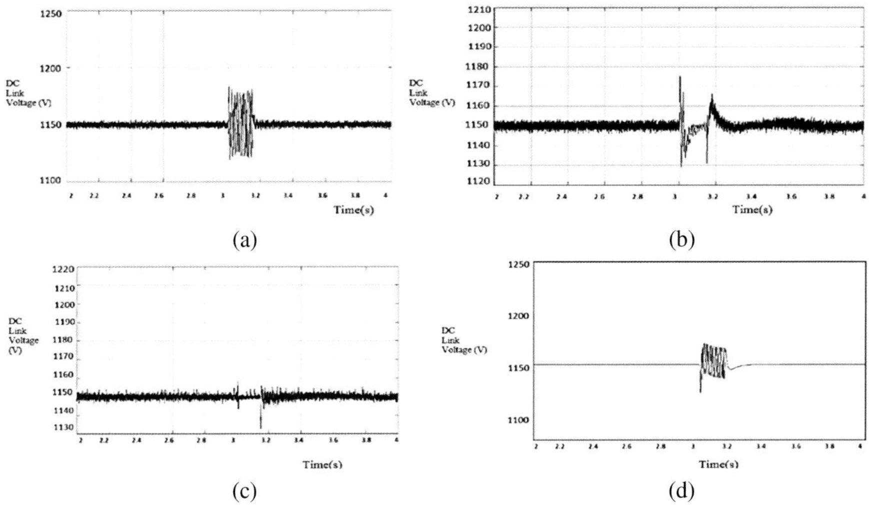

Figure 9.26 DC interlink voltage of the DFIG-WCES under unsymmetrical fault: (a) Transient suppression resistor with RL, (b) DC voltage suppression unit with CBR, (c) DC voltage suppression unit with SFCL, and (d) voltage optimization strategy with CBR.

The DC connect is stifled to 1150 V by utilizing a DC voltage suppression unit joined with a series fault current limiter (SFCL), and consequently, chosen as the reasonable cross-breed FTM conspire in smothering the DC interlink voltage. Transient suppression resistor opposition with the RL circuit is the next viable strategy in keeping up with the DC connect voltage of 1160 V.

(ii) *Voltage at the point of common coupling*
Figure 9.27 shows the voltage dip at the point of common coupling (PCC) during an unsymmetrical fault. Applying the hybrid FTM scheme reduces the voltage dip and improves the recovery time, thereby enhancing system stability. The shaft current increments which prompts a weighty dunk in the PIP point. By presenting the half-breed FTM plot, the voltage level is kept up and helps in diminishing the recuperation time. By choosing the DC voltage suppression unit with an SFCL plot, the voltage level is diminished distinctly to 0.7 p.u and recuperation with 3.16 seconds. This half-breed conspire is the best one when contrasted with other FTM plans by diminishing the voltage droop at the PIP point.

(iii) *Shaft current*
Figure 9.28 delineates the shaft side current which arrives at the pinnacle worth of 2.1 p.u by utilizing the regular FTM plot. Whenever the FTM

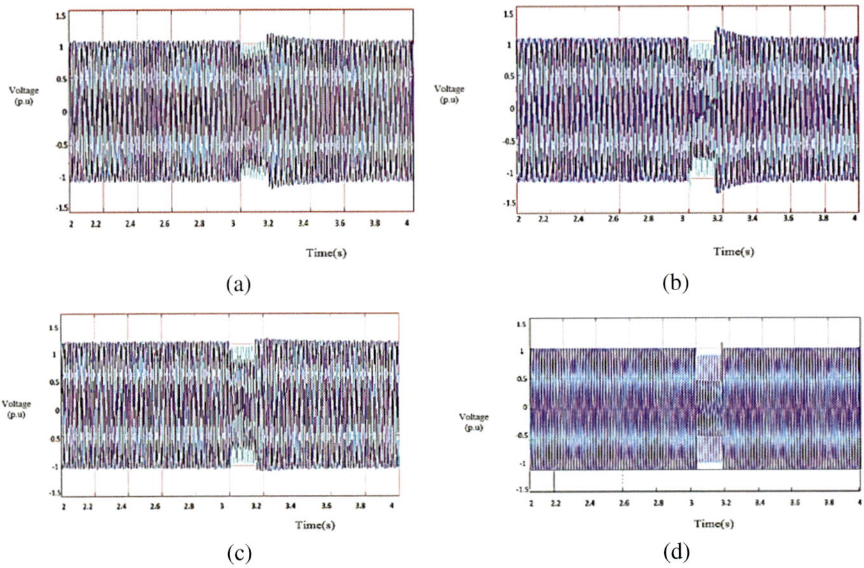

Figure 9.27 The voltage at PIP of the DFIG-WCES under unsymmetrical fault: (a) Transient suppression resistor with RL, (b) DC voltage suppression unit with CBR, (c) DC voltage suppression unit with SFCL, and (d) voltage optimization strategy with CBR.

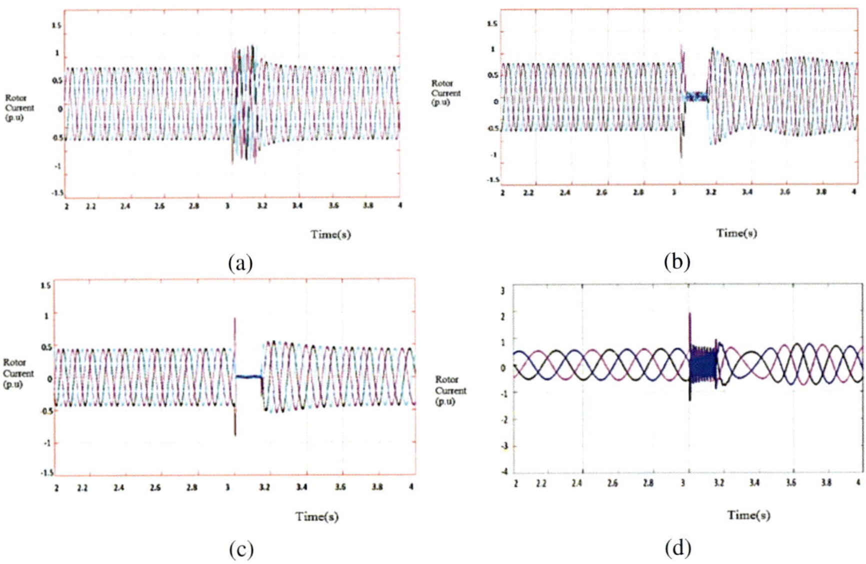

Figure 9.28 Shaft current of the DFIG-WCES under unsymmetrical fault: (a) Transient suppression resistor with RL, (b) DC voltage suppression unit with CBR, (c) DC voltage suppression unit with SFCL, and (d) voltage optimization strategy with CBR.

plot is presented in the framework, the shaft current is limited and the motions in the shaft current are likewise diminished. By utilizing the integrated FTM DC voltage suppression unit joined with the series fault current limiter (SFCL) plot, the shaft current is stifled and arrives at the pinnacle worth 0.9 p.u with less motions. Thus, this FTM is investigated to be preferable over other half-breed plans as far as shaft current.

(iv) *Real power*

Figure 9.29 delineates the genuine power consumed by the framework when it is exposed to unsymmetrical issue condition. Whenever the voltage control conspire is presented in the framework, the motions are decreased and reaches to 0.8 p.u during the issue clearing time. Voltage control FTM conspire acts a significant job in decreasing the genuine power motions. Whenever the DC voltage suppression unit with SFCL is presented in the framework at that condition, the genuine power spans to 0.5 p.u, and wavering is less when contrasted with the traditional FTM conspire.

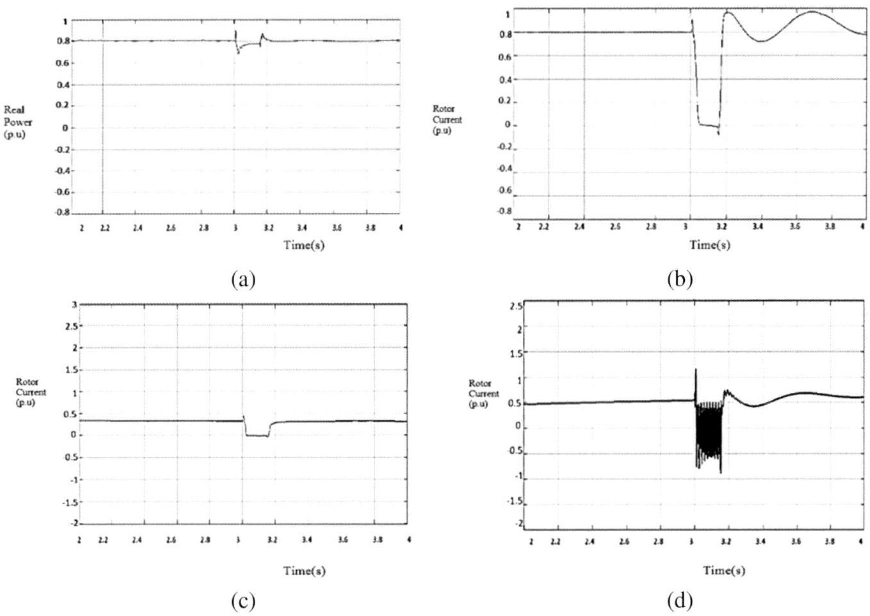

Figure 9.29 The real power of the DFIG-WCES under unsymmetrical fault: (a) Transient suppression resistor with RL, (b) DC voltage suppression unit with CBR, (c) DC voltage suppression unit with SFCL, and (d) voltage optimization strategy with CBR.

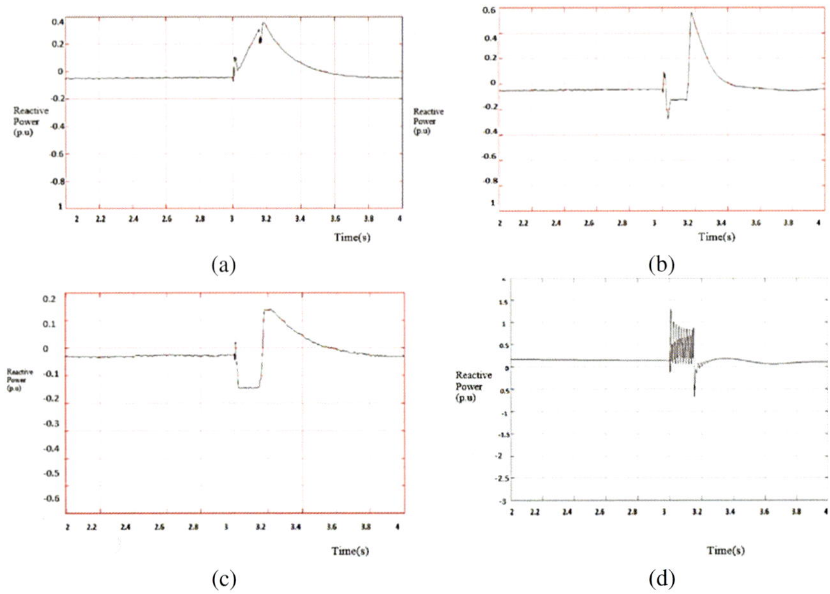

Figure 9.30 Reactive power of the DFIG-WCES under unsymmetrical fault: (a) Transient suppression resistor with RL, (b) DC voltage suppression unit with CBR, (c) DC voltage suppression unit with SFCL, and (d) voltage optimization strategy with CBR.

(v) *Reactive power*

Figure 9.30 shows the responsive power infused when exposed to unsymmetrical shortcoming condition. At the point when the DC voltage suppression unit FTM conspire is acquainted with the framework, the responsive power infusion levels are high and arrive at the worth of 0.4 p.u. The voltage control plot with transient suppression resistor obstruction is presented under a similar unsymmetrical issue situation and the receptive power infusion is continuously expanded to 0.5 p.u. This half-and-half mix is confirmed as a potential plan when contrasted with the other FTM plans in terms of responsive power infusion.

(b) *Experimental setup*

The proposed crossover regulator, DC voltage suppression unit with SFCL which performs better in FTM capacity in the mimicked climate is tried in the equipment arrangement. An equipment test system was planned and based on OPAL-RT RCP/HIL [Version-OP4510] interconnected with the RT lab test system programming introduced PC and dual stockpiling oscilloscope. Table 9.5 compares the performance of integrated FTM controllers under

Table 9.5 Performance comparison of integrated FTM controllers under unsymmetrical fault condition.

FTM schemes	Shaft current	Oscillations in shaft current	Voltage @PIP	DC interlink voltage	Real power	Reactive power injection	Postfault transient current
Transient suppression resistor with RL circuit	Peak value of 1.25 p.u	More oscillations	Dips to 0.9 p.u and recovered at 3.3 seconds	Peak value 1160 V	Oscillations reached 0.9 p.u during fault clearing	0.3 p.u	Reaches to 0.8 p.u
DC voltage suppression unit with transient suppression Resistor	Peak value of 1.1 p.u	Less oscillations when compared to transient suppression resistor with RL scheme	Dips to 0.7 p.u and recovered at 3.3 seconds	Peak value 1165 V	Oscillations reached to 0.8 p.u during fault clearing	0.1 p.u	Reaches to 1 p.u
DC voltage suppression unit with SFCL	Peak value of 0.9 p.u	Very less oscillations when compared to other FTM schemes	Dips to 0.7 p.u and recovered at 3.16 seconds	Peak Value 1150 V	Oscillations reached 0.5 p.u during fault clearing	0.05 p.u	Reaches to 0.5 p.u
Voltage optimization strategy with transient suppression resistor	Peak value of 1.9 p.u	Less oscillations during fault event but heavy oscillation at postfault period	Dips to 0.8 p.u and recovered at 3.16 seconds	Peak Value 1180 V	Oscillations reaches 0.7 p.u during fault clearing	0.5 p.u	Reaches to 0.9 p.u

unsymmetrical fault conditions. Key parameters such as shaft current, voltage at PCC, DC-link voltage, and power variations are evaluated. The results show the effectiveness of each scheme in reducing transients and improving system stability. The presentation of the proposed FTM regulator under constant climate is assessed under the even shortcoming condition. Figure 9.31 illustrates the hardware setup of the fault-tolerant mechanism

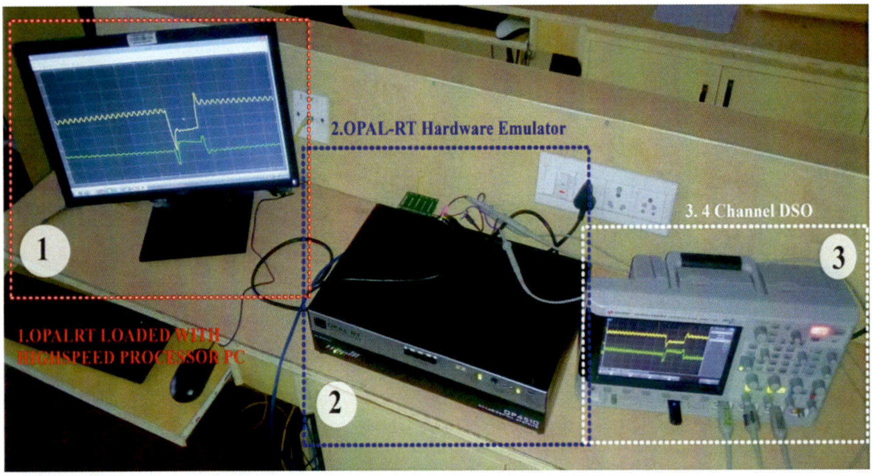

Figure 9.31 Hardware setup for FTM controller.

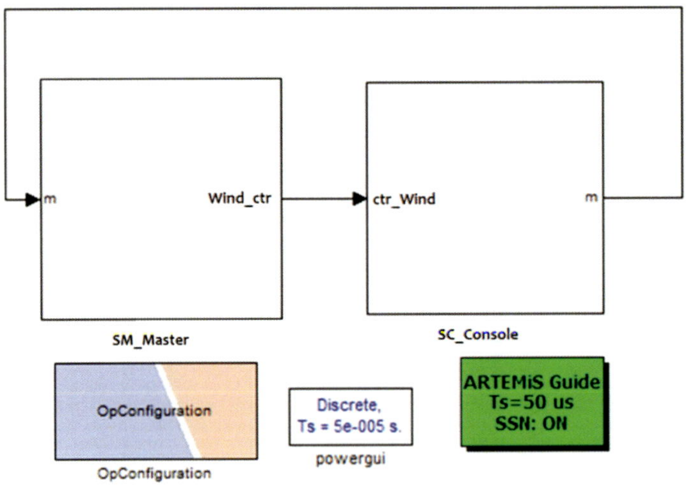

Figure 9.32 Hardware configuration.

(FTM) controller, highlighting the arrangement of key components, interconnections, and measurement systems used for experimental validation. Figure 9.32 shows the hardware configuration in which the OPAL-RT platform is employed with the SimPower Systems model, where the wind framework operates as the master subsystem (SM) and the FTM controller functions as the slave subsystem (SS).

(i) *DC interlink voltage*

Figure 9.33 shows the continuous DC connect voltage. It has been investigated in reenactment that the crossover regulators are more successful in smothering the DC interface voltage than customary plans. DC voltage suppression unit with SFCL mixture FTM conspire is more successful in stifling the DC interface voltage. The DC connect voltage is smothered to 1150 V in the double stockpiling oscilloscope by utilizing a DC voltage suppression unit joined with a series fault current limiter (SFCL), and subsequently chosen as the appropriate cross-breed FTM plot in stifling the DC interlink voltage.

(ii) *Real power*

Figure 9.34 shows the genuine power consumed by the framework when it is exposed to unsymmetrical issue condition. At the point when the DC voltage suppression unit with SFCL is presented in the framework at that condition, the genuine power ranges to 0.5 p.u, and swaying is less in the oscilloscope when contrasted with the traditional FTM conspire.

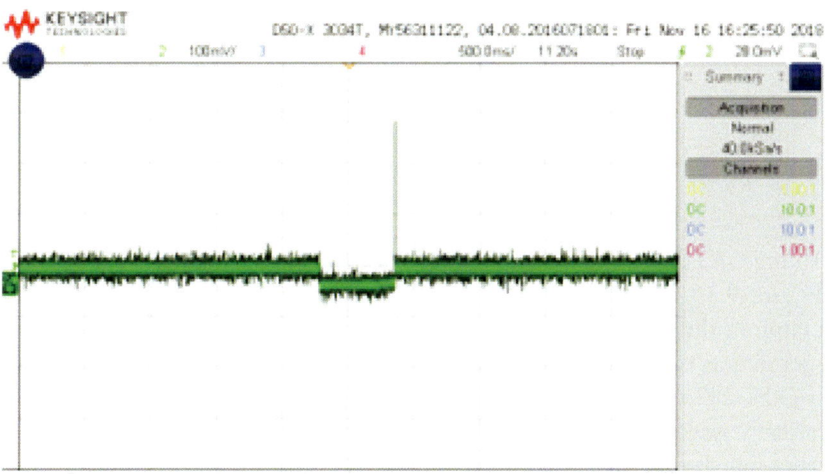

Figure 9.33 DC interlink voltage.

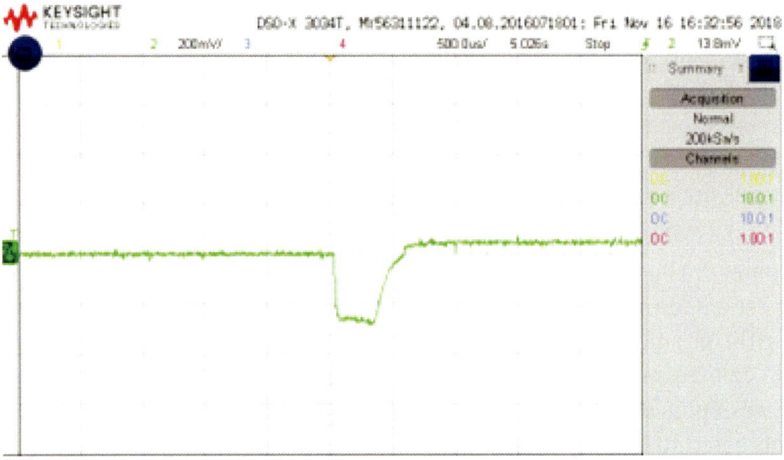

Figure 9.34 Real power.

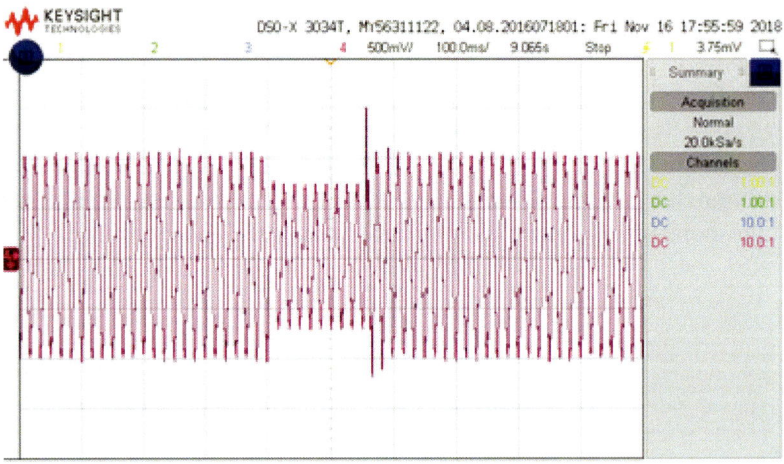

Figure 9.35 Voltage at PIP.

(iii) *Voltage at PIP*

Figure 9.35 represents the voltage plunge in the purpose-like manner coupling during the unsymmetrical issue condition. The shaft current increments which prompts a weighty plunge in the PIP point. By choosing the DC voltage suppression unit with SFCL plot, the voltage level is decreased distinctly to 0.7 p.u and recuperation inside 3.16 seconds. This half-breed plot is the best one when contrasted with other FTM plans by diminishing the voltage droop at the PIP point which is approved in the equipment emulator climate.

Table 9.6 Experimental results of integrated FTM controllers under unsymmetrical fault condition.

Voltage at PIP	DC interlink voltage	Real power
Dips to 0.7 p.u and recovered at 3.16 seconds	Peak value of 1150 V	Oscillations reached 0.5 p.u during fault clearing

Table 9.6 presents the experimental results of integrated FTM controllers under unsymmetrical fault conditions, highlighting PCC voltage, DC-link voltage, and real power performance. The findings validate the proposed schemes ability to suppress transients and enhance system stability.

Performance of FTM control scheme under unbalanced (LLG) fault with 60% dip in grid voltage

The exhibition of the proposed issue ride through control system in DFIG design is assessed under a twofold line to ground (LLG) issue with a 60% voltage plunge for a length of 0.15 s in the matrix side. The voltage at PIP, DC interface voltage, genuine power, responsive power, magnetic core current, shaft current, and shaft speed waveforms under the LLG issue are displayed in Figure 9.8. The dynamic power taken care of to the network during shortcoming is very nearly zero, and henceforth, there is high weight on the mechanical framework which prompts speed up to 1.26 p.u. The shaft current increments which prompts a weighty dunk in voltage in the PIP point. The voltage plunge ranges to 0.4 p.u when there is no GA-based FTM conspire in the framework. The proposed savvy control conspire, when contrasted and regular control techniques as far as different boundaries, is displayed in Table 9.7.

Table 9.7 Performance comparison of FTM controllers under LLG fault.

Controllers	Voltage at PIP (p.u)	Real power (p.u)	Reactive power (p.u)	DC interlink voltage (V)	Magnetic core current (p.u)	Shaft current (p.u)
FTM control scheme	*0.45*	*0.3*	*0*	*1256*	*1.6*	*3*
Transient suppression resistor resistance	*0.5*	*0.9*	*0.1*	*1270*	*1.5*	*0.85*
DC voltage suppression unit	*0.7*	*0.7*	*0.2*	*1265*	*1.25*	*1.21*
Series dynamic braking resistor	*0.5*	*0.5*	*0.05*	*1268*	*0.5*	*0.8*
Conventional PQ-FTM scheme	*0.5*	*0.6*	*0.35*	*1255*	*1.5*	*1.8*
Proposed FTM controller	*1*	*0.85*	*0.5*	*1254*	*0.9*	*1.0*

At the point when the GA based conspire is acquainted with the framework, the PIP voltage ranges to 1 p.u during the shortcoming time frame which shows 100 percent recuperation in PIP voltage. The proposed GA-based PQ FTM regulator keeps up with the genuine and responsive power by 0.85 and 0.5 p.u, which is better compared to the ordinary PQ-FTM conspire. The pace of ascending in shaft speed is restricted during lattice deficiencies utilizing the proposed plot which guarantees framework dependability.

Likewise, the proposed conspire consistently rides the shortcoming with less drifters. The proposed plot accomplishes 0.5 p.u of receptive ability to help lattice strength during flaws. The satisfactory scopes of magnetic core and shaft current are accomplished utilizing the control plot, and henceforth, FTM necessity is enormously fulfilled. The shaft wavering is enormously diminished to 1.23 p.u utilizing the proposed plot. Figure 9.36 shows the system's response to a double line-to-ground (LLG) fault with a 60% grid voltage dip, illustrating variations in real power, DC-link voltage, reactive power, magnetic core current, shaft speed, shaft current, and PIP voltage. The results demonstrate the proposed FTM's effectiveness in mitigating fault-induced transients and improving.

The framework boundaries improvement utilizing the proposed control procedure is approved and contrasted and a regular control plot and without regulators in exploratory arrangement. The genuine and responsive power infused into the matrix during lattice shortcomings is delineated in Figure 9.37. The responsive power infused to the framework by the proposed conspire during shortcoming is 0.8 p.u which is 40% higher than the ordinary PQ-FTM plot. What is more, genuine power motions are incredibly diminished during the shortcoming time frame and clearing time utilizing the proposed plot.

GA tuned control parameters under LLG fault.

K_p	K_i
493	14

Figure 9.38 shows the DC connect voltage which rises momentarily to 1600 V due to the shaft inrush flows infused from the shaft side converter into the DC interface capacitor. The proposed PQ-FTM conspire smothers the DC interface voltage to 1400 V, which is 60% more successful than the PI-based PQ-FTM plot.

The dip in the voltage happens at the power interface point because of the event of the LLG issue in the framework. The network voltage lessens

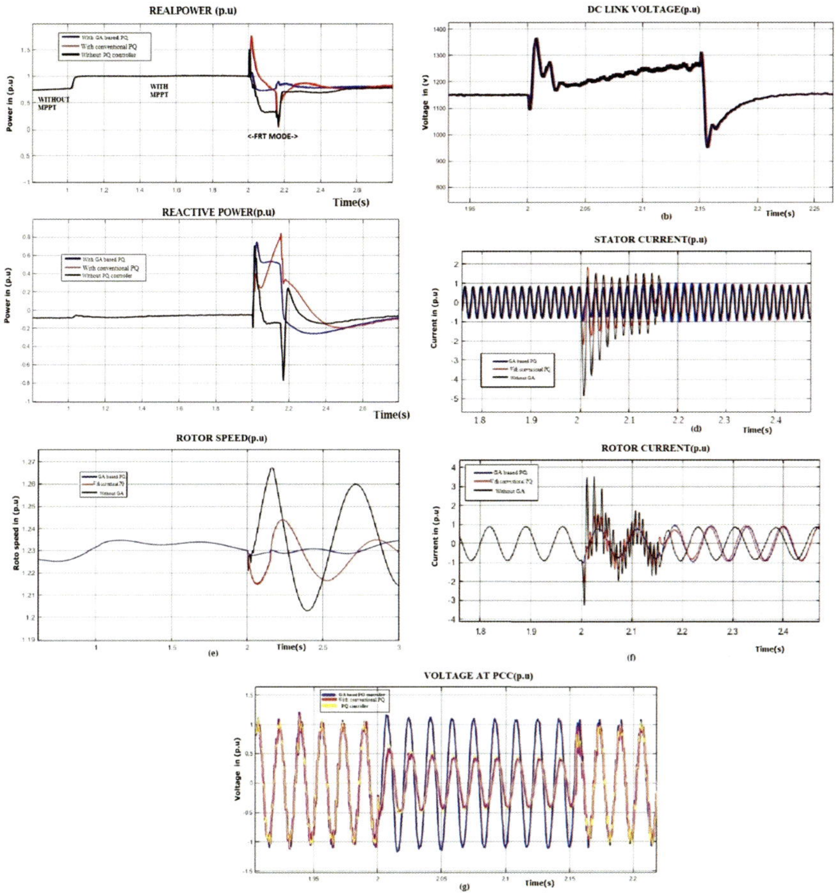

Figure 9.36 Response of the system under LLG fault with a 60% of dip in grid voltage. (a) Real power (p.u), (b) DC Interlink voltage (V), (c) reactive power (p.u), (d) magnetic core current (p.u), (e) shaft speed (ms), (f) shaft current (p.u), and (g) voltage at PIP (p.u).

to 0.1 p.u during shortcoming condition which is improved to 0.3 p.u by the ordinary FTM conspire as portrayed in Figure 9.39. Whenever the proposed control plot is presented in the framework, the voltage improves to 0.9 p.u and the recuperation season of 2.15 seconds is accomplished in the framework when the shortcoming is cleared. Consequently, an improvement of 75% in the network voltage is accomplished utilizing the proposed control procedure when contrasted with the traditional PQ-FTM conspire.

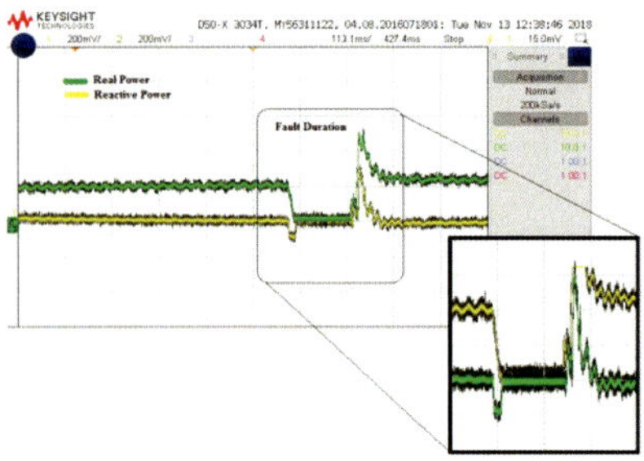

(a) Without FTM control scheme

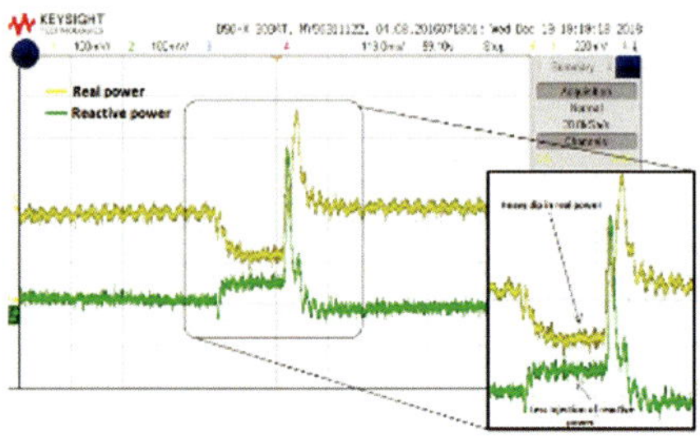

(b) Conventional PQ-FTM scheme

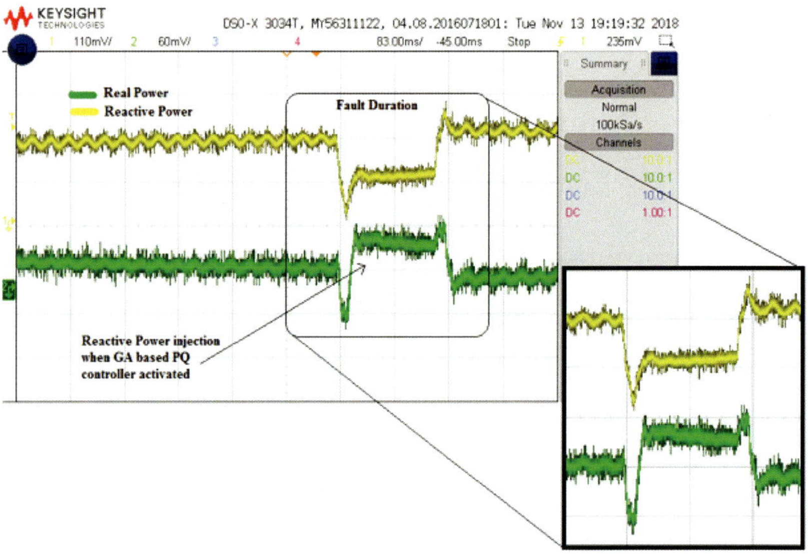

(c) Proposed GA-based PQ-FTM control scheme

Figure 9.37 Real power and reactive power during LLG fault.

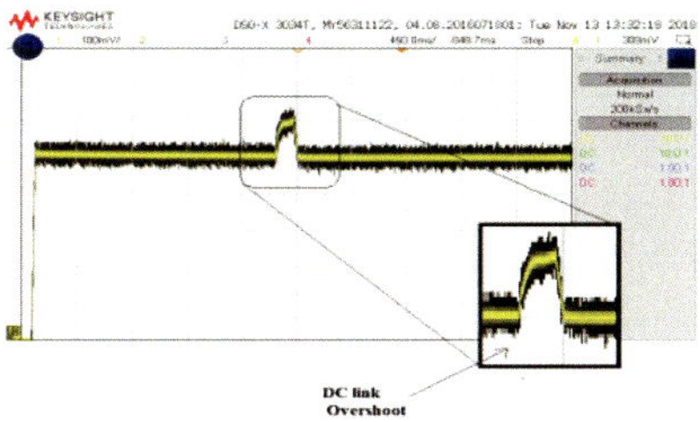

(a) Without FTM control scheme

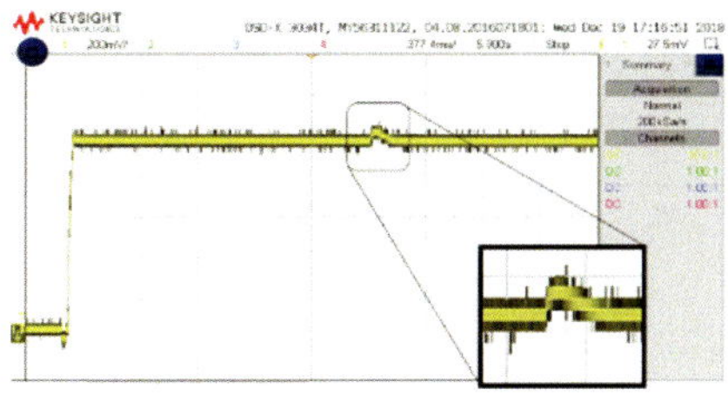

(b) Conventional PQ-FTM scheme

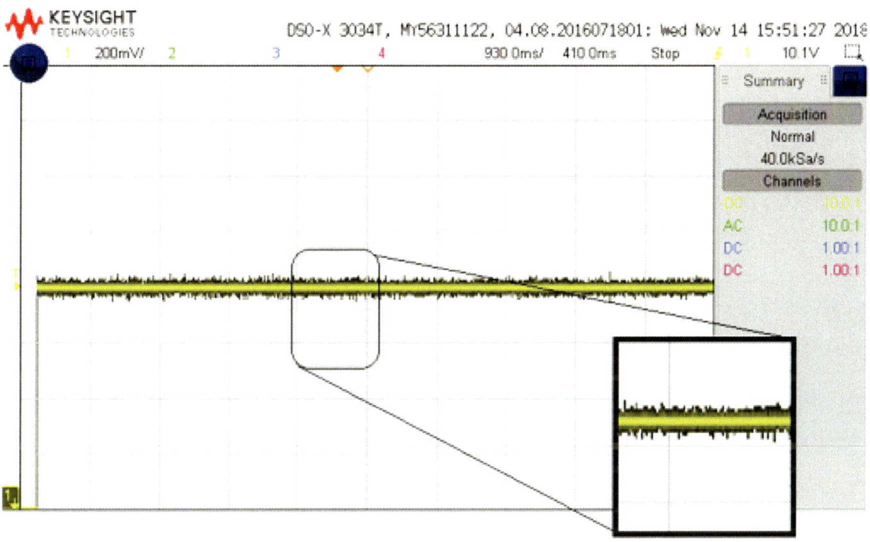

(c) Proposed GA-based PQ-FTM control scheme

Figure 9.38 DC interlink voltage during LLG fault.

(a) Without FTM control scheme

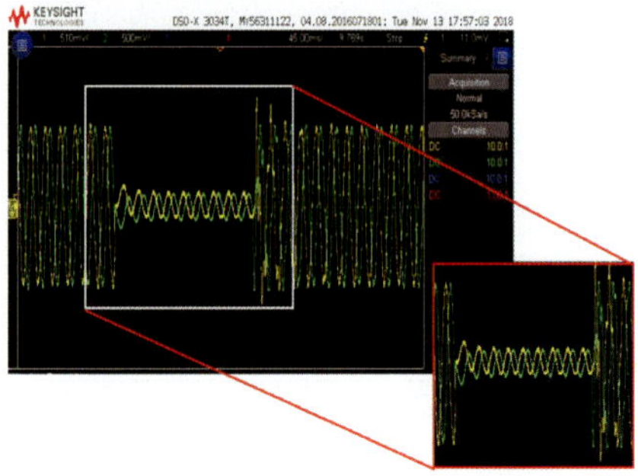

(b) Conventional PQ-FTM scheme

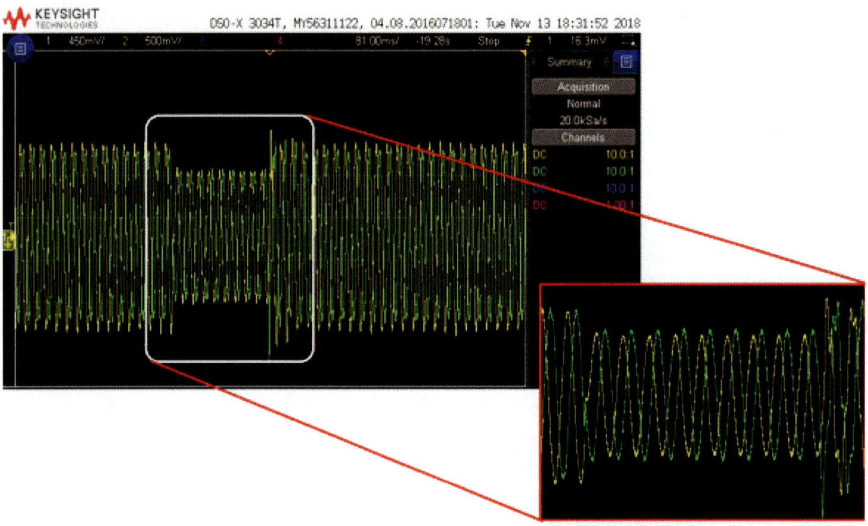

(c) Proposed GA-based PQ-FTM control scheme

Figure 9.39 Voltage at PIP during LLG fault.

9.5 Conclusion

To understand the FTM strategies for DFIG-based wind energy conversion systems (WECS), they are analyzed under various fault conditions. Variations in current levels, voltage levels, and PIP voltage are compensated for by both traditional and hybrid designs, which are thoroughly examined in this study. These FTM strategies are straightforward and easy to implement in practical settings, taking into account their operation, size, and structure.

The proposed FTM strategies are evaluated under various system parameters, such as changes in active and reactive power, shaft and magnetic core currents, DC-link voltage, and voltage at the point of common coupling. Additionally, several combinations of hybrid FTM strategies are designed and analyzed based on the system parameters of wind energy conversion systems, which enhance the stability of both the wind farm and the power grid. From a detailed analysis of simulation results and hardware testing in the OPALRT environment, it is evident that hybrid strategies perform effectively in improving system conditions and supporting the grid under both balanced and unbalanced faults. This section introduces a novel genetic algorithm-based fault ride-through control technique, which is effective in managing the system parameters of grid-connected DFIG-based WECS. The system parameters are significantly optimized by the incorporation of this control technique during grid faults. The reactive power is regulated, controlling the grid voltage and ensuring the protection of the WECS without the need for additional components. The simulation and experimental results in OPALRT confirm that the proposed fault ride-through control strategy improves key system parameters, including real power, reactive power, voltage at the PIP, magnetic core current, shaft current, and DC-link voltage in a grid-connected DFIG-based wind system. The required limits for magnetic core and shaft current are achieved using the proposed control scheme, satisfying the FTM requirements. The theoretical validation of the control scheme is tested using a root locus plot. The proposed control scheme is compared with conventional control strategies, demonstrating superior performance in both balanced and unbalanced grid fault conditions. Smooth transitions during switch-over actions are successfully carried out. Furthermore, the risk of damaging power converters is mitigated using the proposed control approach.

Declaration of conflicting interests

The authors declared no potential conflicts of interest with respect to the research, authorship, and/or publication of this article.

Funding

The authors received no financial support for the research, authorship, and/or publication of this article.

References

[1] Zeng Z, Yang H, Zhao R, Cheng C, Topologies and control strategies of multi- functional grid-connected inverters for power quality enhancement: a comprehensive review, Renewable and Sustainable Energy Reviews, 24(5), 2013, pp 223–270.

[2] Lund H, Munster E, Integrated energy systems and local energy markets, Energy Policy.34 (10), 2006, pp 1152–1160.

[3] M. Tazil,V. Kumar , R. C. Bansal , S. Kong, Z. Y. Dong , W. Freitas , H. D. Mathur Three-phase doubly-fed induction generators: an overview. IET Electrical Power Applications.4 (2), 2010, pp 75–89.

[4] Asuhaini AB, Zin BM, Pesaran MHA, Khairuddin AB, An overview on doubly-fed induction generators control and contributions to wind based electricity generation. Renewable and Sustainable Energy Reviews. 27, 2013, 692–708.

[5] Mwasilu F, Justo JJ, Ro KS, Jung JW, Improvement of dynamic performance of Doubly-fed Induction Generator-based wind turbine power system under an unbalanced grid voltage condition. IET Renewable Power Generation. 6(6), 2012, 424–434.

[6] Eriksen PB, Ackermann T, Abildgaard H, Smith P, Winter W, Rodriguez Garcia, System operation with high wind penetration. IEEE power & energy magazine. 3(6),2005, 65–74.

[7] Pannel G, Atkinson DJ, Zahawi B, Minimum-threshold crowbar for a fault-ride- through grid-code-compliant DFIG wind turbine. IEEE Transactions on Energy Conservation. 25(3), 2010,750–759.

[8] Noureldeem O. Behavior of DFIG wind turbines with crowbar protection under short circuit. International Journal of Electrical and Computer Science. 12(3), 2012, 32–37.

[9] Jadhav HT, Roy R, A comprehensive review on the grid interaction of doubly-fed induction generator. International Journal Electrical Power Energy System. 49,2013, 8–18.

[10] Johan morren, and sjoerd w. h. de haan, ride through of wind turbines with doubly-fed induction generator during a voltage dip, *IEEE transactions on energy conversion*, 20. (2), 2005, 435-441

[11] Lopez J, Gubia E, Olea E, Ruiz J, Morroyo L, Ride through of wind turbines with doubly-fed induction generator under symmetrical voltage dips. *IEEE Transactions on Industrial Electronics*.56(10), 2009, 4246–4254.

[12] Patric da Costa J, Pinheiro H, Degner T, Arnold G, Robust controller for DFIGs of grid-connected wind turbines. IEEE Transactions on Industrial Electronics. 58(9), 2011, 4023–4038.
[13] Justo JJ, Mwasilu F, Lee J, Jung JW. AC-micro grid versus DC-microgrid with distributed energy resources: a review, Renewable Sustain Energy Reviews. 24, ,2013, 387–405.
[14] Morren J, de Haan SWH, Ride through of wind turbines with doubly-fed induction generator during a voltage dip. *IEEE Trans Energy Conversation*. 20(2), 2005,435–441.
[15] Kasem AH, El-Saadany EF, El-Tamaly HH, Wahab MAA,An improved fault ride- through strategy for doubly-fed induction generator-based wind turbines. IET Renewable Power Generation.2 (4), 2008, 201–214.
[16] Rodriguez M, Abad G, Sarasola I, Gilabert A,Crowbar control algorithms for doubly-fed induction generator during voltage dips. In: Proceedings 11th European. *Conference Power Electronics Application*.90-75815-09-3. 2005.
[17] Hachicha F, Krichen L, Rotor power control in doubly-fed induction generator wind turbine under grid faults, Energy.44 (1), 2012, 853–861.
[18] Jin C, Wang P., Enhancement of low voltage ride-through capability for wind turbine driven DFIG with active crowbar and battery energy storage system. In: Proceedings of the Power and Energy Society General Meeting. 978-1-4244-6551-4,2010.
[19] Pannell G, Atkinson DJ, Zahawi B.,2010. Minimum-threshold crowbar for a fault- ride- through grid -code- compliant DFIG wind turbine. *IEEE Transactions on Energy Conversion*. 25,750–759
[20] Jesús López, Eugenio Gubía, Eneko Olea, Josu Ruiz, and Luis Marroyo.,2009. Ride Through of Wind Turbines with Doubly-Fed Induction Generator Under Symmetrical Voltage Dips, *IEEE Transactions on Industrial Electronics*,56(10), 4246-4254.
[21] Okedu KE, Muyeen SM, Takahashi R, Tamura J, Protection schemes for DFIG considering rotor current and DC-link voltage. In: Proceedings of *the International Conference on Electrical Machine System*,2011, 2–7.
[22] Okedu KE, Muyeen SM, Takahashi R, Tamura J., Improvement of fault ride through capability of wind farms using DFIG considering SDBR. In: Proceedings of the 14th Europe Conference on Power Electronic Applications. 2011. 1–10.
[23] A.Causebrook, D. J. Atkinson, and A. G. Jack. Fault ride-through of large wind farms using series dynamic braking resistors, *IEEE Transactions on Power Systems* .22(3), 2007, 966-975.

[24] Wessels C, Gebhardt F, Fuchs FW, Fault ride-through of a DFIG wind turbine using a dynamic voltage restorer during symmetrical and asymmetrical grid faults. IEEE Transactions on Power Electronics. 26(3), 2011, 807–815.

[25] Ibrahim AO, Nguyen TH, Lee DC, Kim SC,A fault ride-through technique of DFIG wind turbine systems using dynamic voltage restorers. IEEE Transactions on Energy Conversion. 26(3), 2011, 871–882.

[26] Zeng Z, Yang H, Zhao R, Cheng C., 2013. Topologies and control strategies of multi- functional grid-connected inverters for power quality enhancement: a comprehensive review. Renewable and Sustainable Energy. 24,223–270.

[27] Wind turbine transient response and fault ride through improvements with optimal control.Shanoob ML, Iqbal K, Al,Maliki A ,*International Transactions on Electrical Energy System*. 2017;27:e2412

[28] Impact of fault ride-through requirements on fixed-speed wind turbine structural loads Anca D. Hansen, Nicolaos A. Cutululis, Helen Markou and Poul E. Sorensen,*Wind Energy*. 2011; 14:e1–11

[29] Awad H, Svensson J, Bollen M., Mitigation of unbalanced voltage dips using static series compensator. *IEEE Transactions on Power Electronics*. 19(3), 2004, 837–846.

[30] Hossain MJ, Saha TK, Mithulananthan N, Pota HR., Control strategies for augmenting LVRT capability of DFIGs in interconnected power systems. IEEE Transaction on Power Electronics. 60(6),2013,2510–2522.

[31] Hu S, Lin X, Kang Y, Zou X, An improved low-voltage ride through control strategy of doubly-fed induction generator during grid faults. IEEE Transaction on Power Electronics. 26(12), 2011, 3653–3665.

[32] Okedu KE, Muyeen SM, Takahashi R, Tamura J., Protection schemes for DFIG considering rotor current and DC-link voltage. In: Proceedings of the International Conference Electrical Machine System.19 (3), 2011, 2–7.

[33] Justo JJ, Ro KS., Control strategies of doubly-fed induction generator-based wind turbine system with new rotor current protection topology. *Renewable and Sustainable Energy*. 2012, 4(4), 043123.

[34] Mahyar Gholizadeh, Ashknaz Oraee, Sajjad Tohidi, Hashem Oraee, Richard A. McMahon. An analytical study for low voltage ride through of the brushless doubly-fed induction generator during asymmetrical voltage dips, *Renewable Energy*,115,2018,64-75.

[35] Md. Arifujjaman, M.T. Iqbal and J.E. Quacioe, "Development of an isolated small wind turbine emulator," *Open Renewable Energy Journal*, 4, 2011, pp. 3-12.

[36] Dipesh Kumar & Kalyan Chatterjee "Design and analysis of artificial bee-colony-based MPPT algorithm for DFIG-based wind energy conversion systems" International Journal of Green Energy, 4(14), 2017, pp 416-429.

[37] A. Kunwar & R. C. Bansal "Effect of high-voltage direct current link on small signal stability of a power system with different penetration level of doubly-fed induction generator and direct drive synchronous generator based wind farms", International Journal of Green Energy,4 (13),2016,pp-335-343.

Raghavendran C R received his B.E in Electrical and Electronics Engineering from Anna University, M.E (Power System) from Anna University, and Ph.D from SRM Institution of Science and Technology. His areas of interest include machine modeling, power system stability, and grid integration issues of renewable energy. He is currently working as an assistant professor at Easwari Engineering College.

J. Preetha Roselyn received her B.E in Electrical and Electronics Engineering from Madras University, MS (By Research) from Anna University, and Ph.D from SRM Institution of Science and Technology. Her areas of interest include voltage stability, soft computing applications in power systems, grid integration issues of renewable energy, and smart metering infrastructure. She has published 20 papers in international journals, 15 papers in international conferences, and 1 book. She is a faculty mentor of the SRM MTS student chapter, United Kingdom.

D. Devaraj completed his B.E and M.E in Electrical and Electronics Engineering and Power System Engineering in the years 1992 and 1994, respectively, from Thiagarajar College of Engineering, Madurai. He has organized 2 international conferences, 9 national conferences, 6 seminars, and conducted 16 workshops. He has authored two textbooks and coauthored three textbooks. He has published 126 papers in journals and presented 250 papers at conferences. He is the project director of the Power System Automation Group attached to TIFAC-CORE in Network Engineering which is sponsored by DST, Government of India. Currently, he is the principal investigator for the DST-funded research project titled "Energy Saving through IoT-based

Building Automation and Nonlinear Predictive Controller." His research interests include power system optimization, power system security, voltage stability, renewable energy, smart grid, evolutionary algorithms, neural networks, IoT, and data mining. Presently, he is the director-academic and dean-SEET at Kalasalingam Academy of Research and Academy (KARE), Krishnankoil. He is a senior member of IEEE and actively involved in the student branch activities of KARE, since 2006.

Appendix A

A.1 Parameters of DFIG.

DFIG rating	
Rated power (MW)	**9**
Rated voltage (V)	**575**
Rated frequency (Hz)	**50**
Stator resistance (p.u)	**0.004843**
Rotor resistance (p.u)	**0.004377**
Stator leakage inductance (p.u)	**0.1248**
Rotor leakage inductance (p.u)	**0.1791**
Mutual inductance (p.u)	**6.77**
Transmission line rating	
Positive sequence resistance (ohm/km)	**0.1153**
Zero sequence resistance (ohm/km)	**0.413**
Positive sequence inductance (henries/km)	**0.00105**
Zero sequence inductance (henries/km)	**0.00332**
Positive sequence capacitance (farads/km)	**11.33e-9**
Zero sequence capacitance (farads/km)	**5.01e-9**
Transformer rating	
Rated power (MVA)	**12**
Turns ratio	**575V/25KV**
Impedance (p.u)	**0.0017+j0.05**
Transformer(T2) parameter	
Rated power (MVA)	**47**
Turns ratio	**25KV/120KV**
Impedance (p.u)	**0.00534+j0.16**
Grid impedance value	
Impedance (p.u)	**0.0004+j0.004**

Appendix B

A.2 List of symbols

Subscripts	Expansion
s	DFIG machine stator circuit quantity
r	DFIG machine rotor circuit quantity
e	Excitation reference frame
$*$	Reference quantity of the controller
v ,i	Instantaneous voltage and current quantities
E_g,v_g	Terminal voltages at the GSC converter and the network connecting point
s	Slip
ωm	Rotor speed in rad/s (electrical)
ωs, ωr	Synchronous frequency and the slip frequency in rad/s (electrical)
f	Synchronous frequency in Hz
$\theta\ r$	Angle of the rotor frame with respect to the stationary frame
Pm	Mechanical power
$T\ a$, $T\ m$	Aerodynamic torque and mechanical torque
τ	Time constant of system response
ϕm	Phase margin
$\theta\ s$	Twisting angle between high speed shaft and low speed shaft
$T\ e$	Electromagnetic torque
Pr, Qr,	Rotor active and reactive power
$R\ r$, $R\ s$	Rotor and stator resistance

10

End Use Application of Intelligent Techniques for Sustainable Development

Amarsingh B. Kanase-Patil[1], Avinash P. Kaldate[1], and Shashikant D. Lokhande[2]

[1]Department of Mechanical Engineering, Sinhgad College of Engineering, Savitribai Phule Pune University, India
[2]Department of Electronics and Telecommunication Engineering, Sinhgad College of Engineering, Savitribai Phule Pune University, India
E-mail: amarbkanase@yahoo.co.in; er.avinashkaldate@gmail.com; sdlokhande.scoe@sinhgad.edu

Abstract

Sustainable development is a development that meets current needs without compromising the ability of future generations to meet their own needs. The number of intelligence technology applications is increasing, including in the water management, health and hygiene, education, social sector agriculture, economic, and energy sectors. It is important to review the long-term development effects of the advent of intelligence and its growing impact on many areas of society. The purpose of this chapter is to examine the existing evidence of the positive and negative effects of intelligence approaches on achieving sustainable development goals. Intelligence techniques are helping to achieve the goals of sustainable development. Intelligence techniques enable new technologies that drive efficiency and output. Proper legislation and regulation are needed to facilitate the rapid growth of intelligence systems. Technology based on intelligence methods will have gaps in transparency, accountability, safety, and ethical standards, which is detrimental to the development and long-term use of intelligence techniques. This chapter also examines the impact of intelligence techniques on sustainable development due to the growing impact on the energy sector. Intelligence

DOI: 10.1201/9788743809852-10

techniques increase the efficiency and reliability of power generation in a smart city environment. Furthermore, an intelligence technique helps to integrate renewable energy into smart networks by effectively controlling risk and bridging the demand-supply gap. This chapter explores how the use of intelligence techniques is beneficial for sustainable development. Also, this chapter has been analyzing the role of intelligence techniques for sustainable development in the different sectors through sustainable energy supply.

Keywords: Artificial intelligence, sustainable development, intelligent techniques, smart grid, renewable energy, water management, health, education, energy efficiency.

10.1 Sustainable Development

Eternity is accomplished in the near future, but in the long run, it is not always possible. At present, there is an attempt to use the concept of all areas of development at the same time. The problem is that it is difficult for us to define sustainable development. The report of the World Environment and Development Commission, established by the United Nations General Assembly, provides a preliminary discussion on sustainable development. Sustainable development is a "development that meets current needs without jeopardizing the ability of future generations to meet their own needs." Sustainable development has four dimensions: society, environment, culture, and economics, all intertwined without being separated [1].

Sustainability is a way of thinking about the future that balances environmental, sociological, and economic concerns in the pursuit of a higher standard of living. While sustainability is sometimes envisioned as a long-term goal, sustainable development refers to the many processes and ways used to achieve it (e.g., sustainable agriculture and forestry, sustainable production and consumption, good governance, research and technology transfer, and education and training). Education for sustainable development has been incorporated into a number of global frameworks and norms relating to key aspects of sustainable development.

Intelligence practices and sustainability are two new technologies and organizational innovations that are influenced or influenced by increasing productivity and ensuring long-term productivity. Global competitiveness, market fluctuations and demand, more personalization through communication, information and intelligence, and declining creativity are some of the issues that intelligence technology is trying to address. Intelligence technology has the potential to make a significant contribution to organizational and social stability, as well as to its limitations. Reduced set-up time, less time, less labor and material costs, better product flexibility, higher productivity, and enhanced customization are all possible from an economic standpoint. Intelligence technology reduces energy and resource consumption by finding and analyzing data from industrial and supply chain activities, which is good for the environment [2].

Through data-driven and verifiable carbon footprint assessments, it helps to reduce waste and CO_2 emissions. Intelligence technology provides civility with many obstacles and limitations. Low employment, information security concerns, data complexity, electronic waste, and poor quality are just a few examples. Some research has explored the relationship between intelligence technology and environmental sustainability. According to proponents of the principle of legitimacy, businesses are embracing sustainability to address stakeholder concerns and expectations. The United Nations Sustainable Development Goals (SDGs) provide a common framework and goals for enterprises, regions, and governments to pursue sustainable development, despite their shortcomings [3].

The goals are poverty alleviation, job creation, and economic growth, as well as industry, innovation and infrastructure, and reduced inequality. Eliminating hunger, promoting better health and well-being, providing quality education, promoting gender equality, promoting peace and justice, and building strong institutions are all features of social sustainability. Environmental impact qualities are closely related to clean water and sanitation, cheap and clean energy, sustainable cities and communities, responsible use and production, climate action, underwater life, and land life. This pattern serves as the basis for our investigation of the link between intelligence technology and the long-term viability of society. Low emission levels – a meteorological action objective – benefit the manufacturing company's urban development efforts, and shorter life than the water objective – able to help manufacturing businesses avoid damage to freshwater ponds [4].

10.1.1 Operational criteria for sustainable development

As shown in Figure 10.1, there appear to be three operational requirements for sustainable development based on different definitions and concepts. With the following three precautions, these criteria should be used to evaluate any objective.

Economic objectives should not be pursued at the expense of environmental and social concerns. Environmental benefits should not be increased

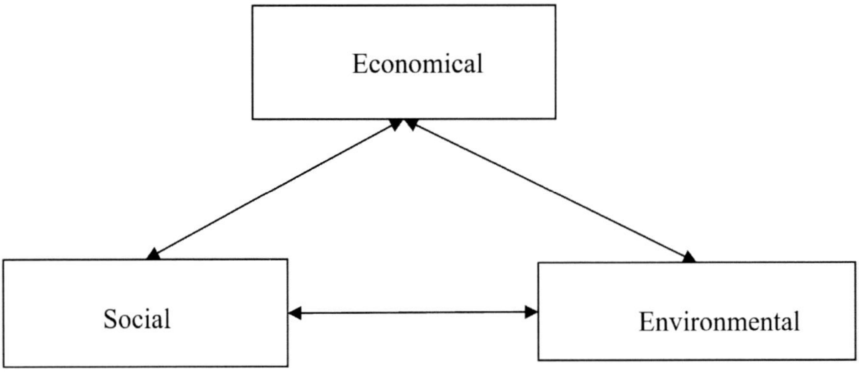

Figure 10.1 Criteria for sustainable development.

if economic and social constraints are not met. Social benefits should not be increased if economic and environmental constraints are not met.

10.1.2 Influencing factors for sustainable development

Consumption, production, and distribution are the three components of sustainable development. Naturally, the issue in terms of consumption is the exploitation of resources beyond the natural boundaries given by reproduction. Total inefficiency and mismanagement in the use of water, energy, and minerals define a sustainable product [5]. When discussing consumption, it is important to consider not only the number of resources used but also the patterns used. There are five compelling reasons to examine these patterns.

1. According to the current usage pattern, financial efficiency alone is not enough to satisfy the hunger for natural resources. Many people focus on eco-efficiency, assuming that achieving high environmental efficiency alone will fulfil the current generation's needs. However, this is not sufficient; consumption itself must be reduced through stronger environmental regulations and revised policies. No matter how environmentally capable, the current generation's need for natural resources is never met by environmental efficiency alone. Consumption must be reduced. This is achieved by adopting strong rules of environmental regulation and by modifying the policy framework to reflect the extent of environmental damage to the world caused by current practices.
2. Use is the key to understanding policy issues while focusing on the demand side. Valuable resources such as water and energy are being

wasted as a result of increasing demand in response to falling prices. This creates a vicious circle of waste and pollution.

3. Consumption patterns reflect what is consumed and if it meets people's fundamental requirements. It is critical to consider what is being consumed, whether it is essential commodities or high-end ones since it will reveal if the consumption pattern fits people's basic requirements and whether development will be sustainable.
4. This trend shows that the poor not only use less but also reduce pollution and are directly affected by environmental degradation. Polluted water has a direct effect on the health of the poor. Poor people in urban slums, where 40% of the population of developing countries lives, are exposed to polluted air 24 hours a day, seven days a week, endangering their health. Similarly, agricultural chemicals and solid waste will damage water and soil, endangering the health of farmers.
5. Consumption patterns reveal much about the intricate link between economic progress, meeting basic needs, and human ambition. GDP growth in a country may be slow, yet large numbers of people are living in poverty. GDP growth may be similar in other countries, but some people live in poverty. Initially, production and consumption habits may be more focused on luxury goods and increasing economic rewards at the expense of environmental and social benefits. In the second case, GDP growth is more likely to be used to meet the basic needs of the people.

10.2 Intelligence Techniques

Intelligence tactics are affecting different areas. The impact of artificial intelligence on global productivity, equality and inclusion, environmental impact, and many other areas is expected to be immediate and long-term. The potential impact of intelligence on sustainable development has been shown to have both positive and negative effects. This is a critical research gap, as it is found that intelligence techniques affect our ability to meet all sustainable development goals. The rationale is how intelligence either enables or prevents the fulfillment of all 17 objectives and 169 goals identified in the 2030 Sustainable Development Agenda.

To meet the United Nations Sustainable Development Goals (SDGs) by 2030, development practitioners are quickly finding ways to rediscover how they make choices and address challenges. Intelligence developed through imaginary applications of new technology is allowing such progress. Intelligent techniques are the most promising intelligence. Intelligent technique typically seeks to automate tasks related to human thought, including learning, decision-making, and natural language processing, using algorithms that mimic human learning and comprehension to solve strictly defined tasks [6].

Due to the increasing availability of large data sets created in various fields such as agriculture, education, human health, commerce, and transportation, as well as the continuous advancement in the field of computing, progress in the field of intelligent technique cannot be stopped. Problems in various sustainable developments are being solved by developing advanced machine learning algorithms and techniques.

Development is helping practitioners and beneficiaries if an intelligent technique is used responsibly. For example, machine learning algorithms are used in child-growth monitors to consistently detect malnutrition and increase their efficiency by collecting more data. In poorer nations and situations where internet connections are restricted, processing data and algorithms locally, rather than in the cloud, on smartphones will be beneficial and increase privacy protections. As a result, intelligent techniques are helping to predict risks and opportunities by bridging information gaps, and identifying patterns and relationships. It has the potential to improve knowledge and situational awareness. New types of intelligent techniques impact assessment and prediction are also on the rise, which will help maximize impact and identify the best solutions. Intelligent technique paves the way for new types of smart service delivery, humanitarian and environmental solutions, and revenue streams [7].

10.3 End Use Application

In addition, intelligent technique helps individuals and communities improve their skills and abilities by facilitating peer-to-peer information exchange and distance learning.

Finally, new intelligent technique applications will improve the agility and efficiency of interventions by automating them, freeing human operators to focus on more difficult tasks (such as analyzing large amounts of data). Machine learning will evaluate a large amount of data in a fraction of the time it takes a person to do so and then propose actions based on the results. Poor road conditions are a safety hazard for vehicles and disrupt traffic, both of which hinder economic progress.

Responsible intelligent technique use is helping to achieve these goals, which are important for development initiatives, but development practitioners need to be careful. They should keep an eye on new negative externalities and barriers and examine whether and how much they will be reduced. The risks associated with the employment of developing technologies are being reduced. Given that intelligent techniques rely on high-quality training data, it is necessary to purchase data collection technology or collaborate with data-holding or data-generating partners to obtain data. Lack of high-quality training data in terms of diversity and representation will lead to bias, discrimination, and erroneous conclusions, which are ineffective in reaching development goals.

Furthermore, failure to incorporate ethical criteria into machine learning algorithms will result in potential adverse development projects. While intelligent technique approaches will help production organizations become more efficient, they will also lead to a reduction in low-skilled jobs. Development practitioners are playing an important role in ensuring that the shift to more use of intelligent technique approaches is inclusive. Furthermore, since data is the lifeblood of intelligent technique, data privacy issues play an important and often limited role in its development. Creating an intelligent technique system designed to remain private by default usually comes at a higher cost. It is worth noting that measures that allow intelligent technique models to learn from data sets without compromising their privacy are becoming more important. By building trust and mitigating risks, responsible intelligent technique use, along with the consistent implementation of data security standards, will contribute to the success of intelligent technique applications in development.

The following cases are discussed for the use of intelligence techniques in sustainable development.

10.3.1 Water management

Water is essential to survival. A constant supply of clean, safe water has been controlling human habitat and progress for thousands of years. Unstable supply systems have driven people to prosper and migrate. The SDG resource triangle revolves around water and sanitation. It has impacts on issues such as women's empowerment and gender equality, food and agriculture, energy and climate change, and infrastructure and technology. Water is widely used in agriculture, industry, and even at home. Cities and towns around the world operate on millions of tons of raw water, treated water, and wastewater every day to meet the needs of human civilization.

As shown in Figure 10.2, intelligence techniques for water resource management have different aspects such as water distribution, efficient water

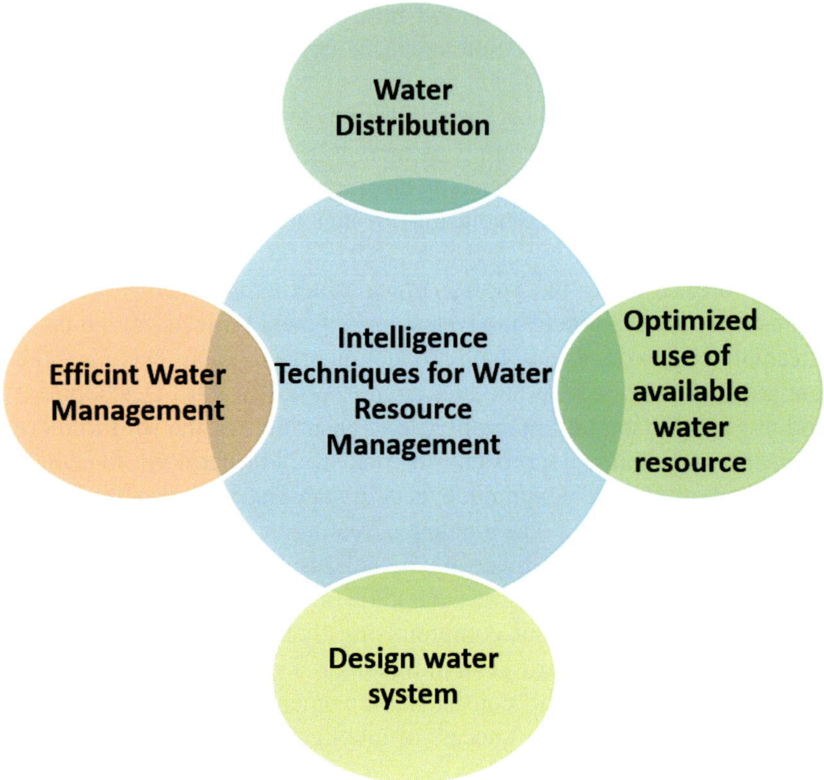

Figure 10.2 Intelligence techniques for water resource management.

management, optimized use of available water resources, and designed water systems. The decisions that villages, towns, and nations make on water resource management have far-reaching consequences for our future well-being. By compromising and overexploiting freshwater sources, overflowing groundwater, contaminated natural water resources such as river banks and beaches, and damaged ecosystems that nourish the ocean and food chain, some have lost their future security or have compromised stability. Humanity must harness technological advancements and intelligent technique innovations to meet short-term economic demands while ensuring the long-term environmental sustainability of natural ecosystems and biological regions.

Although water sources and supplies have remained relatively constant over time, the management techniques available on the farm have changed. Water utilities will benefit from smart water management powered by intelligent techniques. Advances in intelligent techniques could make water treatment plants cheaper and easier to monitor, as well as help protect public health. When using expert systems or rule-based algorithms to select native intelligent technique-based software outputs or evaluate options in a field of options, new intelligent technique tools use applications over time to learn how people learn.

During the learning phase, the input data is tied to the known output, allowing the algorithm to learn over time. The algorithm then enters the "operational phase," when it begins to recognize the pattern when new data is supplied. Intelligent technique is a perfect tool for managing water resources in a constantly changing environment and water business, allowing water utility managers to optimize current revenues and successfully plan for years to come, due to its ability to continuously convert and process data.

10.3.2 Health and hygiene

In a world where billions of people are infected with waterborne diseases, access to clean water is a major problem for many. This is a never-ending problem that is difficult to solve as it requires financial support and time to learn how to make the most of the new pure water system. As shown in Figure 10.3, intelligence techniques in the health sector are able to solve the problems of disease diagnosis, treatment, health records, and treatment and clinical records. The clean water system employs artificial intelligence to identify the size of molecules under a microscope. Aquatic microorganisms are identified by distinct forms, colors, densities, and edges in each evolutionary neural network. In its conceptual evidence, the researchers focused on identifying the microorganisms that cause Escherichia coli and cholera. However, because various species of bacteria have distinct forms and features, identification of other bacteria is able to be done. The intelligent technique testing system will detect hazardous germs and contaminants on a map in real time, and it is simple to use, requiring little training. First, it creates a method for towns to employ to install monitoring devices in water sources, and then it creates particular regional variants that require installing devices in water pipelines. Innovation has far-reaching implications for enhancing the safety and effectiveness of water and sanitation systems and practices all around the world. Artificial intelligence-based clean water has far-reaching implications for towns and cities all over the world, particularly for those who are willing to make modifications and early expenditures to take advantage of this low-cost breakthrough.

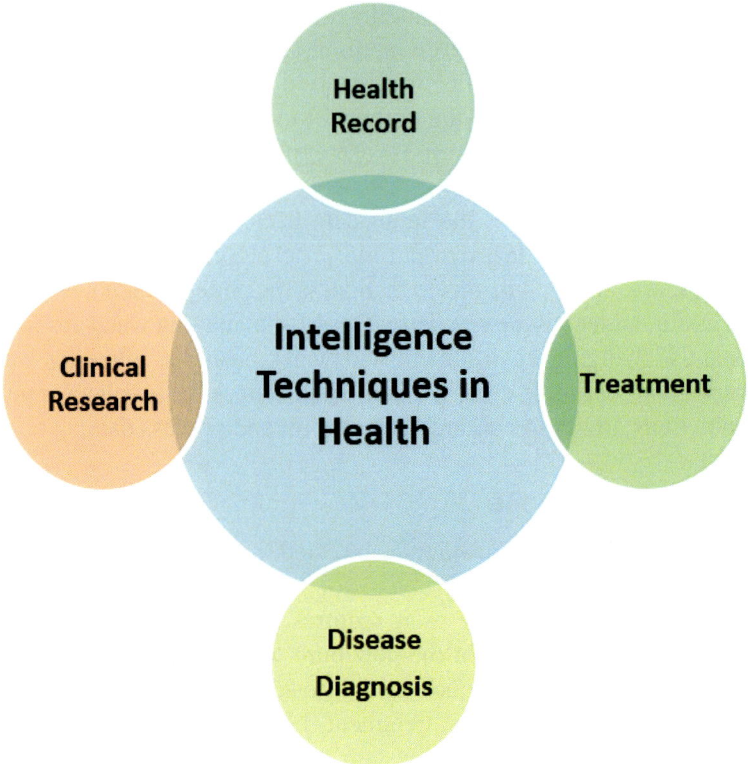

Figure 10.3 Intelligence techniques in health care.

10.3.3 Education

Students and future corporate executives and strategists should learn about the costs and advantages of specific, micro-level applications, as well as the larger macro-level difficulties in high-tech intelligent technique infrastructure. In many regions of the world, science, technology, engineering, and mathematics education will become more crucial; otherwise, a significant and expanding digital gap between nations will emerge. Introducing intelligent technique applications has several advantages. The decision-making capacity of researchers and students in higher education is transforming the way they study, execute, and ultimately assess solutions to problems in this new environment based on intelligent techniques.

Humans are far slower at learning, storing, comparing, and recommending than networks. Students either learn to use this information for the greater

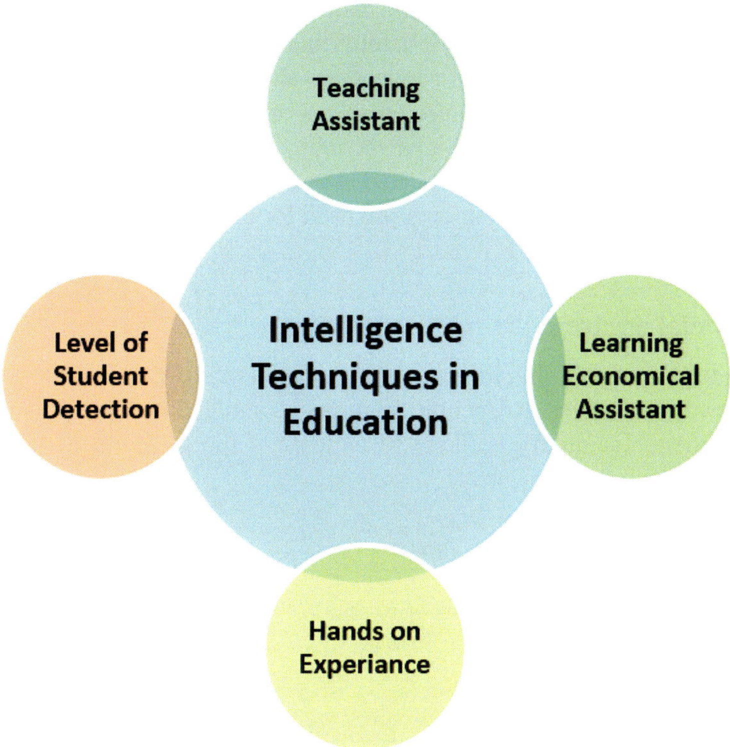

Figure 10.4 Intelligence techniques in education.

good of mankind and tackle some of the problems highlighted in the SDGs, or they will try to stop this new wave of knowledge from emerging, although innovation is nearly hard to stop. This innovative remote data-gathering strategy enables smartphone-level diagnostics at the level of a single farmer in a remote part of the world where he or she would not otherwise have access to such help.

Knowledge is exchanged in a totally different way than it was in the past when higher education was divided into silos. Education should prepare students to collaborate with artificial intelligence and new technologies accessible in diverse businesses, as highlighted in the instance of smart water management. Intelligent technique aids smart water management. Educators must educate students on how to work with intelligent techniques and comprehend the role intelligent technique plays in making management choices today and in the future, so that they are able to use what they have learned in the classroom when they enter the workforce. Professors should also instruct students on how to appropriately enter data into these intelligent technique decision-making neural networks, as the output is only as accurate as the data input. Many of the future prospects open to students will require in-depth information input in intelligent techniques for management decision-making, thus students should be educated about in-depth learning. As shown in Figure 10.4, intelligence techniques are used for different problem-solving in the education sector.

10.3.4 Social sector

Unfortunately, complex intelligent technique systems are still expensive, and low-income families and people from disadvantaged backgrounds cannot afford them. Inequality is likely to increase if the government fails to

Figure 10.5 Intelligence techniques in the social sector.

regulate how intelligent technique profits are distributed to different parties. For example, large producers may benefit, while small farmers may lag behind because they cannot afford expensive intelligent technique systems to increase production. Furthermore, if there is not enough transparency and diversity, the current inequality between hate crimes will increase.

Figure 10.5 explains the use of intelligent techniques in the field of social sectors. Some studies have shown that most of our intelligent technique systems, especially in the NLP and computer vision domains, are still showing signs of systematic racism and bias. This is because the data used for training is already tainted by existing social biases. Transparency and diversity are key to keeping models as objective as possible. One method would be a decentralization process where intelligent technology could be implemented across teams with different cultural, ethnic, and gender backgrounds.

10.3.5 Economic outcomes

Figure 10.6 shows the use of intelligent techniques in different economical sectors. Intelligent technique adds value to one's work, making them more productive and increasing their productivity. From smart agriculture technology to smart security algorithms that keep an eye on criminals and potential invaders, they inform farmers when, where, and how to grow. However, if current trends continue and future markets rely heavily on data-driven economies, the income gap could widen dramatically, especially in low- and middle-income nations where resources for human capital development are restricted.

Some may argue that this split will occur organically as a result of the creative destruction process, but the government is able to surely take action to reduce the repercussions. Retraining and reskilling will be critical in closing the inequality gap through government-led efforts.

10.4 Energy Sector

The number of intelligent technique applications in the environmental sector is increasing, including smart grids, agriculture, and monitoring. Recent advances in IoT hardware and intelligent technique algorithms, vision, and sensor fusion have made it feasible. For example, the introduction and possible adoption of electronic vehicles (EVs) and smart grids in a smart city environment is able to improve the efficiency and reliability of power generation. Furthermore, intelligent techniques are possible to incorporate

Figure 10.6 Intelligence techniques in the economic sector.

renewable energy into a smart grid, accurately control its risk, and efficiently fill gaps between supply and demand. In countries where coal-based power is still generated, increased efficiency will be detrimental.

Figure 10.7 explains the use of intelligence techniques in the energy sector. By 2030, it is estimated that information and communication technology will consume up to 20% of the world's total electricity demand. The number of intelligent technique applications in the field of environment including smart grid, agriculture, and monitoring is increasing. Recent advances in IoT technology and intelligent algorithms, as well as the merger of vision and sensors, have made this possible. For example, the introduction and potential adoption of electronic vehicles (EVs) and smart gadgets in a smart city environment will improve the efficiency and reliability of power generation.

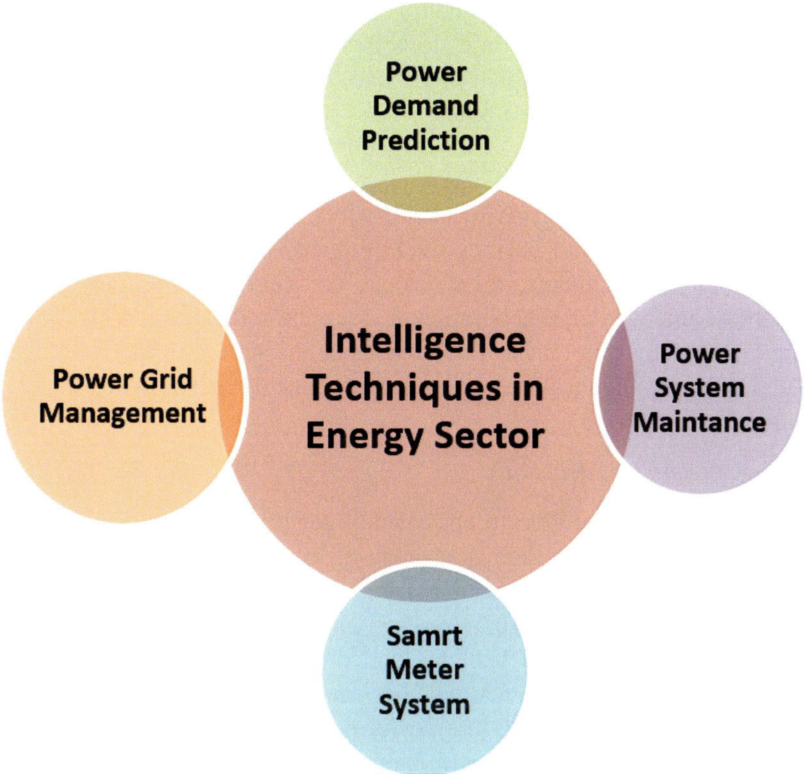

Figure 10.7 Intelligence techniques in the energy sector.

In addition, intelligent techniques integrate renewable energy into the smart grid, precisely control its risk, and effectively fill supply-demand gaps.

10.4.1 Applications of intelligence technique in the energy sector

Prediction of faults

Intelligence techniques are used extensively in the energy sector, including fault prediction, real-time maintenance, and the development of optimal maintenance plans. In the energy sector, equipment failure is prevalent, with potentially serious repercussions. With the correct sensors incorporated, intelligent techniques are used to monitor and detect equipment failure before it occurs, saving resources, money, time, and lives. It is being considered as a possible source of base load energy (the electricity required to supply the power system at any given time) to help support the spread of less reliable renewable. Rich data-enabled predictive diagnostics are used to detect problems that could shut down a plant.

Image processing makes maintenance easier

Drones are being used by grids to check wires that carry electricity from power plants to homes and businesses. Equipped with high-resolution stationary and infrared cameras, these drones are very useful for detecting faults as they can cover large geographical areas and rough terrain. Overhead lines are covered by this material. The state of the power property is then monitored using intelligence techniques to determine if it needs to be replaced or repaired.

Making energy-efficient decisions

Automatic meters will be able to apply intelligence techniques to optimize energy consumption and storage as a result of the digital transformation of residential energy management and consumer equipment. When power is expensive or cheap, or when solar rooftop energy is abundant, equipment that stores electricity through vehicles and other batteries may be turned down. Grid data, smart meter data, weather data, and information on energy use – to study and improve building performance – will play a key role in artificial intelligence using data, optimizing resource utilization and comfort for residents, as population growth and urbanization in emerging markets and cities expand. In view of demand, it can be used to inform consumers of the degree of renewable energy demand. Intelligence techniques aid in improving power demand and production estimates, as well as decision-making in terms of productivity.

This is especially crucial in transitions to renewable energy, which is frequently inconclusive owing to their reliance on weather, wind, and water flow, as well as fossil fuels for backup. Using intelligence algorithm-based forecasts in conjunction with energy storage infrastructure can help to lessen the requirement for backup systems. Finally, when distributed products become more common, customers will be able to contribute to power generation, effectively acting as producers. As customers become more essential actors in the system, rather than being taken out of it, it will be easier to determine the best moment for distributed generation to contribute to the intelligent grid. Traditional manufacturers and system operators can also benefit from intelligent techniques, which assist them in balancing new demand-side trends such as greater intermittent renewable energy, distributed distribution to customers, and the rise of electric vehicles (EVs).

10.5 Conclusion

This chapter deals with the medium and long-term effects of the intelligence approach. Therefore, the importance of strengthening the global discussion about the use of intelligence methods and providing the necessary regulatory knowledge and control for technology based on intelligence techniques has been discussed. These new technologies allow for ever-increasing levels of productivity. It has the potential to have a significant impact on social and environmental sustainability. Institutions should consider the contribution of technology to sustainability. There is not enough guidance in scientific and scholarly literature in this regard. This business seeks to understand the hybrid multisituation with the decision process combined with the intelligence approach. It is advisable to use technology to maximize the impact of sustainability, but each technology must be carefully analyzed as each technology will have a different impact on industry and sustainability aspects. The priority and basis of such technology should be considered while investing in it. Increased consumption, efficiency, changing supply and demand patterns, and the lack of analysis required for successful management are all issues facing the energy sector around the world. These problems are exacerbated in emerging market countries. Expansion of the informal connection to the power grid means that huge amounts of electricity are not measured or even paid, resulting in losses and high CO_2 emissions and little incentive for users to use nonexistent energy wisely. Intelligent technologies that allow communication between smart grids, smart meters, and Internet of Things devices are already being used in the energy sector in developed countries. These technologies will help improve energy management, efficiency, and transparency, as well as increase the use of renewable energy sources.

References

[1] M. A. Goralski and T. K. Tan, "Artificial intelligence and sustainable development," Int. J. Manag. Educ., vol. 18, no. 1, 2020.

[2] I. Worighi, A. Maach, and J. Van Mierlo, Advanced Intelligent Systems for Sustainable Development (AI2SD'2018), vol. 912. 2019.

[3] A. Chakraborty, D. Goswami, and A. E. Hassanien, Studies in Computational Intelligence 912 Artificial Intelligence for Sustainable

Development: Theory , Practice and Future Applications, vol. 10, no. 17. 2017.

[4] A. Nazarov, D. Kovtun, and S. Talu, "Using artificial intelligence technologies for sustainable development," E3S Web Conf., vol. 291, p. 04010, 2021.

[5] M. E. Mondejar et al., "Digitalization to achieve sustainable development goals: Steps towards a Smart Green Planet," Sci. Total Environ., vol. 794, no. June, 2021.

[6] C. Bai, P. Dallasega, G. Orzes, and J. Sarkis, "Industry 4.0 technologies assessment: A sustainability perspective," Int. J. Prod. Econ., vol. 229, p. 107776, 2020.

[7] B. A. Jnr, M. A. Majid, and A. Romli, "Application of intelligent agents and case based reasoning techniques for green software development," Ttem, vol. 12, no. 1, pp. 30–43, 2017.

Index

About the Editors

Anuradha Tomar is an accomplished academic and researcher currently serving as an Assistant Professor in the Department of Instrumentation and Control Engineering at Netaji Subhas University of Technology (NSUT), New Delhi. With over 12 years of experience in teaching and research, she is widely recognized for her contributions to the fields of power systems, renewable energy integration, and smart grid technologies.

Dr. Tomar earned her B.E. in Electronics Instrumentation & Control from the University of Rajasthan, followed by an M.Tech in Power Systems from NIT Hamirpur. She completed her Ph.D. in Electrical Engineering from the prestigious Indian Institute of Technology Delhi. She also carried out post-doctoral research at the Eindhoven University of Technology, Netherlands, where she contributed to EU Horizon 2020 projects such as UNITED-GRID and UNICORN. Her research interests lie in microgrid operation, solar photovoltaic systems, energy-efficient industrial automation, and AI applications in power systems. She has published over 65 research papers in leading international journals including IEEE Transactions and holds several patents. She has also co-edited multiple technical books with reputed international publishers such as Springer and Elsevier. Dr. Tomar is a Senior Member of IEEE and a recipient of several academic honors, including the Premier Researcher Award and Commendable Patent Award from NSUT. Her name features among Stanford University's global list of top 2% scientists, underscoring her impact in the research community. Beyond academia, she actively mentors undergraduate, postgraduate, and Ph.D. students and serves as a faculty advisor for various student-led technical initiatives. With a strong focus on sustainable and socially relevant research, Dr. Tomar continues to lead in developing energy solutions for a cleaner, smarter, and more inclusive future.

Yog Raj Sood is affiliated to Department of Electrical Engineering, National Institute of Technology, Hamirpur, where he is currently working as Professor (HAG) and has also served as Director of NIT Puducherry and

Vice-Chancellor JIIT Noida. Dr. Yog Raj Sood has authored and co-authored more than 370, national and international publications and also works as an Editor/Editor-in-Chief/reviewer for reputed professional journals. Dr. Yog Raj Sood has an active association with different societies and academies around the world and he has made his mark in the scientific community with his contributions and wide recognition worldwide. His one patent was granted and three patents were published by the Indian Patent Office. Dr. Yog Raj Sood has received several awards for his contributions to the scientific community. He has completed three MHRD, Research projects and one DST, TIFAC-CORE Project of 5.30 Cr. Twelve research scholars have already completed Ph.D. under his guidance. Two more research scholars are pursuing their Ph.D. under his guidance. Recently one of his research articles, certified by Elsevier Publishers, was linked to the United Nations Sustainable Development Goals (UN SDGs), helping to tackle some of the world's greatest challenges. He has major research interest in the field of AI application to power systems, deregulated power sector, congestion management, microgrid and renewable energy technologies.

Ritu Kandari is a distinguished researcher and academic in the field of renewable energy integration, power electronics, and smart grid technologies and is presently working as an assistant professor in KIET group of Institutions, India. She earned her Ph.D. from Indira Gandhi Delhi Technical University for Women (IGDTUW), New Delhi, in 2023, where her research centered on control strategies for microgrids with large-scale renewable energy penetration. Dr. Kandari has held notable positions in both academia and research, including her role as a Research Support Officer at the University of Malta, where she contributed to international projects on sustainable energy systems, and as an Assistant Professor at HMR Institute of Technology and Management, New Delhi, where she guided students in cutting-edge areas of electronics and renewable energy applications. Her contributions to the scientific community include publications in SCIE and Scopus-indexed journals, several book chapters, and her role as co-editor of multiple Elsevier volumes on energy technologies. She has also organized international conferences, workshops, and faculty development programs, reflecting her commitment to knowledge dissemination and academic leadership. A recognized innovator, Dr. Kandari holds a patent on an intelligent hybrid AC/DC microgrid, showcasing her ability to translate research into practical solutions. Her expertise spans renewable energy systems, IoT-enabled smart grids, and power conversion technologies, with a strong focus on advancing sustainable and resilient energy infrastructures. With a blend of academic rigor, industry collaboration, and innovative research, Dr. Kandari continues to contribute toward shaping the future of clean energy and intelligent power systems.

For Product Safety Concerns and Information please contact our
EU representative GPSR@taylorandfrancis.com Taylor & Francis
Verlag GmbH, Kaufingerstraße 24, 80331 München, Germany